*Hominid Evolution and Community Ecology*

This is a volume in

*Studies in Archaeology*

*A complete list of titles in this series appears at the end of this volume.*

# *Hominid Evolution and Community Ecology*

*prehistoric human adaptation in biological perspective*

*edited by*

*Robert Foley*

*Department of Anthropology*
*University of Durham*

1984

Academic Press

*(Harcourt Brace Jovanovich, Publishers)*

London · Orlando · San Diego · San Francisco · New York
Toronto · Montreal · Sydney · Tokyo · São Paulo

ACADEMIC PRESS INC. (LONDON) LTD.
24/28 Oval Road
London NW1

*United States Edition published by*
ACADEMIC PRESS INC.
(Harcourt Brace Jovanovich, Inc)
Orlando, Florida 32887



British Library Cataloguing in Publication Data

Hominid evolution and community ecology.
1. Man, Prehistoric
I. Foley, Robert
930.1 GN740

ISBN 0–12–261920–X

LCCCN 83–72771

Phototypeset by Deltatype Limited, Ellesmere Port
Printed in Great Britain by
St Edmundsbury Press, Bury St Edmunds, Suffolk

# *Contributors*

*Robert Foley* Department of Anthropology, University of Durham, 43 Old Elvet, Durham DH1 3HN, UK

*Clive Gamble* Department of Archaeology, University of Southampton, Southampton SO9 5NH, UK

*Andrew Garrard* British Institute at Amman for Archaeology and History, PO Box 634, Amman, Jordan

*John Gowlett* Research Laboratory for Archaelogy, University of Oxford, 6 Keeble Road, Oxford, UK

*Andrew Hill* Department of Anthropology, Peabody Museum, Harvard University, Cambridge, Massachusetts, 02138, USA

*Richard Potts* Department of Anthropology, Yale University, Newhaven, Connecticut 06520, USA

*Neil Roberts* Department of Geography, Loughborough University, Loughborough, UK

*Katharine Scott* Department of Archaeology, University of Cambridge, Downing Street, Cambridge, UK

*Chris Stringer* Department of Palaeontology, British Museum (Natural History), Cromwell Road, London SW7 5BD, UK

*Alan Turner* Department of Palaeontology, Transvaal Museum, Paul Kruger Street, PO Box 413, Pretoria, South Africa

# Preface

Palaeoanthropology is a peculiar discipline. Among the biological sciences it is uniquely concerned with a single family, rather than broad inter-specific comparisons, and the principal species is somewhat different from the run of the mill organisms that clutter the fossil beds, zoos and wildlife parks of the world. Perhaps its oddest characteristic, though, is that interest has centred not on what early man was, but what he became. In palaeoanthropology hindsight has proved to be the most widely used analytical framework. By noting the direction of evolutionary change to modern man, the characteristics of fossil man were ticked off in a simple linear manner. While the trilobites and dinosaurs and mammoths were worthy of study for what they were, the australopithecines were principally of interest for what they became.

Fortunately another characteristic of palaeoanthropology is that it changes rapidly. Admittedly much of this is imposed reluctantly by the appearance of inconvenient fossils and dates, but shifts in the explanatory framework have also taken place recently. Fossil hominids are being interpreted in the ecological and evolutionary context in which they lived. It has become more interesting to see how Ramapithecus relates, taxonomically and ecologically, to contemporary Miocene hominoids, than to later hominids. The "Neanderthal problem" is now less the problem we have solving it, than the problems that the Neanderthals faced in the Upper Pleistocene glacial environment. This collection of papers extends this approach by looking at early human prehistory in relation to the ecological communities to which man belonged.

The volume arises out of a symposium I organized at the Third Theoretical Archaeology Group Meeting held at the University of Reading in December 1981. The general theme of that symposium was the use of ecological models in interpreting the Pleistocene archaeological record and the behaviour of prehistoric hunter-gatherers. From the papers presented there a less diffuse theme emerged that has formed the basis for this book. That theme is that the first requirement in palaeoanthropology is to understand both the fossil and the archaeological record in the context of the environment and biological system of which hominids were a part. This requirement has certain important implications. First, that early hominids are best treated not as precursors of modern man, but as what they were at the time—large mammals with certain phylogenetic and ecological characteristics. Secondly, that as large mammals the appropriate analytical framework is evolutionary ecology—in other words, the direct application of the principles employed to account for variability in the biological sciences as a whole, not ones designed to apply specifically to hominids. Thirdly, that the most important element of the environment in which the early hominids lived was the other species on which they depended and with which they competed—hence community ecology is the most suitable level of analysis. This collection of papers explores various issues in palaeoanthropology and palaeolithic archaeology in the light of these implications.

The structure of the book proceeds from these considerations. The first three chapters review respectively the principles of community ecology, the patterns of climatic and environmental change, and the fossil evidence for human evolution. With the setting described the remaining chapters deal with hominids first in the context of tropical Africa during the Plio-Pleistocene, and then in more temperate latitudes as they dispersed into the rest of the world throughout the Pleistocene. Thus the topics covered by the papers include man as a large mammal, his relationship to the carnivores, behavioural and mental adaptive capacities and their recognition in the past, and human impact on the communities to which the hominids belonged.

Given this structure and intent it is clear that this volume is an interdisciplinary one. The various papers belong or relate to several traditional fields—archaeology, physical anthropology, palaeontology and Quaternary studies. However, they all share both a common concern with the continuities of human evolution that cannot be partitioned into separate datasets, and a theoretical basis derived from evolutionary and ecological principles. As such it is hoped that the book will be of use to all those interested in the living world of the last several million years, a living world of which the hominids were an integral and not necessarily extraordinary part.

Most of these papers are based on those presented at the Reading meeting. However, Nicholas Flemming of the Institute of Oceanographic Sciences was unable to contribute to the published version, and Neil Roberts, Chris Stringer, Richard Potts and John Gowlett were either unable to attend the conference or were asked to contribute subsequently. I should like to thank all the contributors for their help and cooperation in this compilation, as well as for their ingenuity in discovering nine different ways of putting together a bibliography. Thanks must also be offered to Robert Chapman for inviting us to take part in the Reading conference, and to David Harris who acted as a discussant. Many of the comments and suggestions that he made then are hopefully now incorporated or discussed. At Academic Press Ruth Gadsby provided much help and encouragement. Finally, I should like to thank my wife Jill, not only for drawing all the figures in this volume, but also for her support and ability to think more rationally than I about the timetable for this book.

*April 1984*

*Robert Foley*
*Durham*

# *Contents*

# 1 *Putting People into Perspective: an Introduction to Community Evolution and Ecology*

*Robert Foley*

## 1. INTRODUCTION

"Descended from apes! My dear, let us hope that this is not so, but if it is, that it does not become known." So commented a cleric's wife on hearing of the theory of evolution. Were she to read much of the archaeological literature on early man some hundred years later she might be reassured to discover that at least the second part of her hopes have been fulfilled. While certain of the important implications of Darwinian theory have become cornerstones of archaeology and anthropology—the antiquity of man, sequences of fossils and stone tools, the notion of different species of hominid—the fundamental core of evolutionary theory has had little impact as an explanatory device in human prehistory. The notion of an evolutionary story, of a pattern of gradual change that can be traced in the archaeological and palaeontological record, is a key element of the discipline, but the use of the neo-Darwinian theory of natural selection analytically remains rare.

Natural selection is the differential survival and reproduction of biological entities (Futuyma, 1979, p 292). This differential survival produces the pattern of biological diversity and evolutionary change. An evolutionary analysis, therefore, is one that relates the differential survival of biological phenomena to the properties of the environment. This involves not just a description of the relationship between an organism and its ecological context, nor an examination of the patterns of the relationships, but a demonstration of how certain characteristics have a relative advantage over others in particular contexts, and the ways in which these are reproduced. This is what may be termed *analytical evolution*, rather than descriptive evolution, and it is

HOMINID EVOLUTION
*ISBN 0-12-261920-X*

analytical evolutionary studies that have been rare in the anthropological sciences.

The principle of natural selection then, is the central dogma of evolutionary theory. Historically though, for archaeologists, and indeed to some extent for biologists in general, the observable consequences of natural selection — evolutionary epiphenomena—were more of a research focus than natural selection itself. Thus an evolutionary analysis would frequently centre on the pattern of organization of life, on phylogenetic and classificatory relationships, and on temporal sequences, rather than the mechanism by which these phenomena were brought about. To some extent natural selection and its operation were simply assumed. This is now changing, though, and increasingly an evolutionary analysis is one that specifically attempts to test a hypothesis relating to natural selection. A flourishing of theory over the last decade or more has equipped biologists with the methods to investigate directly the processes as well as the pattern of evolution (Williams, 1966; Hamilton, 1964; Trivers, 1971, 1972; Wilson, 1975; MacArthur, 1962; Dawkins, 1975, 1981; Maynard Smith, 1974a, 1976a). The result has been a renaissance of evolutionary studies, and in fact studies more closely related to Darwin's original interests than was found in the intervening years. According to Ghiselin (1982, p 158)

> anyone equipped with a sound knowledge of late twentieth century biology can go back to Darwin's writings on such matters, even though we now use a rather different jargon. However, the fundamental principles of evolutionary ecology were long forgotten and their rediscovery has taken place in the last twenty years or so.

The current wealth of evolutionary theory is beginning to have its impact on archaeology and anthropology (see Dunnell, 1980), and the principle of natural selection is being explored as a means of explaining the many patterns of variation to be found in human morphology and behaviour. Several growing fields can be identified, achieving various degrees of success. Sociobiology has been one such, attempting to account for patterns of behaviour among living populations in terms of inclusive fitness (Wilson, 1975, 1978; Chagnon and Irons, 1979). Behavioural ecologists have employed a comparative, interspecific framework to place human characteristics within a general context of animal variability (e.g. Harcourt *et al.*, 1981; Martin and May, 1981), and the same principle has been applied to physiological characteristics (e.g. Martin, 1981). Equally attempts have been made to build on evolutionary theory to produce specific mechanisms for the evolution of culture (Wilson and Lumsden, 1981; Cavalli-Sforza, 1982). This book represents yet further evidence for the growing links between biology and anthropology, although in a somewhat different way from the examples cited above. All the papers in this volume are concerned with the analysis of human adaptation from a biological perspective. The patterns of adaptation in prehistory are the product of natural selection, and therefore explicable in these terms. Thus they all belong to the general field of evolutionary biology. A further, more specific theme is that such a biological perspective requires that hominids, and human prehistory, cannot be seen in isolation. Rather the problems of human evolution must be placed into the context of community evolution and ecology. Prehistoric human adaptation is a product of the ecological and evolutionary interactions between hominids and other species that constitute the biological community, and the papers in this volume represent attempts to understand the nature of those interactions.

Such a perspective is a valid and powerful one. It is not, though, sufficient on its own. While the study of human prehistory may be a branch of evolutionary biology, it is distinct from other branches in terms of the data base that it must employ. Blind application of theoretical principles developed for studying living populations stand little chance of being adequately tested in relation to fossil and archaeological data unless some effort is made to relate those theories to the way their predictions will be manifested in the archaeological record. Thus another equally important theme linking these papers is the problem of working directly with the evidence from the past. This is an important prerequisite for attempting to understand the relationship between humans and the rest of the biological world.

## 2. MAN'S PLACE IN THE BIOLOGICAL WORLD

The central themes of this book are that observable adaptations are the product of natural selection, and that these adaptations should be seen in the broader context of the biological community. It thus belongs to a long anthropological tradition, that of man–environment relationships. Prior to the publication of the *Origin of Species* by Charles Darwin in 1859 the nature of this relationship was not a particularly problematic one. Human kinship with the rest of the organic world was sufficiently distant, indeed was qualitatively different, not to require any subtle thinking. The establishment of the theory of evolution altered this, replaced apparent clarity with confusion, and thus opened up a major field of research.

Historically this field of research can be seen focusing on two central questions—what is the nature of the relationship between humans and their natural environment, and can the same principles be employed to explain this relationship as for non-human animals? The answers to these questions would provide the basis for understanding man's relationship to the biological world.

Much of the early work in this field was descriptive and empirical. It is best exemplified by the physical anthropologists who employed anthropometry to show how the environment influenced the shape of humans. Variation in body size and shape reflected variation in environmental conditions. Correlations could show that skin colour varied with solar intensity, lung capacity with altitude, and limb proportions with temperature (Harrison *et al.*, 1964). Several difficulties quickly arose in this work though. Historical factors such as migration were clearly an important consideration; deviations from the expected correlations could always be explained away in terms of the plethora of human movements known to have taken place over the last two thousand years. Furthermore, not only was fitness in relation to these supposedly adaptive features difficult to measure, but also some of the characteristics were shown to be "acquired" as a result of environmental factors, and thus strictly speaking not a product of natural selection. Perhaps most significant, though, was the obvious and often overriding importance of behavioural features in the survival of strategies of human populations. Despite some triumphs, such as the elucidation of the relationship between sickle cell anaemia and malaria (Allison, 1954; Motulsky, 1960), physical anthropology was severely constrained by the complexity of the issues with which it had to deal.

Studies in man–environment relationships in the more social branches of anthropology have proved equally problematic. These studies were all underlain by the principle that humans adapt to environmental conditions through behavioural means. In White's (1959), terms, culture was exosomatic adaptation and therefore the patterns of cultural variability ought to relate to the pattern of environmental problems faced by human populations. Pioneering attempts to document this were Daryll Forde's influential book *Habitat, Economy and Society* (1934), and the work of Julian Steward in North America (Steward, 1955, 1977), which was later developed into the theory of cultural ecology (Vayda, 1969; Rappaport, 1967). While their approach circumvented some of the difficulties faced by the physical anthropologists, others nonetheless arose. Principal among these was the incompleteness of most culture–environment correlations. All too frequently populations could be found in the same environment and yet living very differently. The environment could not be shown to determine behaviour. A second difficulty lay in the influence of European sociologists arguing that social form and behavioural patterns conformed to the logic of social, structural and largely internal factors, not the external environment (see Harris, 1968). Their models showed that humans manipulate the environment, both practically and symbolically, for social ends, and not the other way around (for a critique of these views, and a recent attempt to re-establish the importance of external factors in social anthropology, see Harris, 1979).

Man–environment studies have been more successful in the third branch of anthropology, archaeology. While relationships between physical form and environment, and between cultural behaviour and environment, would not hold, those between technology and environment did. Grahame Clark (1952) in Europe, and Desmond Clark in Africa (1970) demonstrated patterning in prehistoric technology in relation to environment and subsistence behaviour. The relationship between the archaeological record and environmental conditions became one of the strongest and liveliest branches of archaeology (see Butzer, 1972, 1982).

However, while archaeology has maintained a stronger research interest in man–environment relationships than social anthropology, it has by no means been problem free, and it is the nature of these problems and those that have constantly dogged man–environment studies that give rise to some of the contributions to this volume. First there is the simple dichotomy between man and environment. In most studies humans are the actors and the environment the stage props. This reduces the role of the environment to a static background, rather than an interactive agent. Secondly, the environment has usually been narrowly defined in terms of simple geological and biological properties. The narrowness of definition has led to a narrowness of approach. Current definitions are broader, including a continuum of physical features, other species, and conspecifics (Pianka, 1978). The environment in this view is anything that impinges on the organism, and the focus of interest is less what is impinging as how it impinges. Furthermore, this leads to a consideration of environmental conditions less in terms of general habitat classifications and more in terms of the ecological variables that are directly relevant. Thirdly, humans have usually been treated in isolation from, and independent to, the rest of the biological world. By adopting this approach the problems of integrating humans into the environmental continuum have been aggravated. While human independence may be a fact of twentieth century life, assuming that it is so will not be useful in explaining how such a situation should have come about in the past. Fourthly, many of the ecological principles employed by anthropologists to account for the patterns they observed have been divorced from the evolutionary mechanisms on which they are based.

Renewed emphasis on evolutionary mechanisms, seeing humans as one part of a biological community, rather than in isolation or dichotomized with their environment, and thus concern with the ecological principles controlling inter-and intra-specific interactions, are all important elements in the papers presented here, and represent attempts to avoid the shortcomings and pitfalls of earlier work on human interactions with the conditions of the environment.

## 3. THE EVOLUTION AND ECOLOGY OF COMMUNITIES

A biological community is a set of interacting biological organisms (Pianka,

1978, p 4). Community evolution and ecology is therefore the study of these interactions and the principles governing them. However, it is important to establish what these principles may and may not be. Most important in this respect is the idea that community evolution and ecology is not the same as treating biological communities as the unit of evolution and selection. That there is structure and pattern to communities, such as in succession, is not grounds for arguing that communities are in any way a reproductive, self-perpetuating unit, and thus subject to processes of selection (Colinvaux, 1973; Pianka, 1978). Rather the structure and pattern of the community is a product of orthodox Darwinian selection operating at the level of the individual. That such selection occurs in the context of many species—i.e. a biological community—means only that the consequences of selection, not the mechanism itself, occur at a community level. Community evolution is thus the product of ecological interactions at the level of individuals and breeding populations (Hoffman, 1979).

With this definition in mind it is possible to identify three basic types of problem in community evolution. First, problems concerned with the extent to which a population's adaptation is explicable in terms of ecological interactions within the community—in other words the biological community acting as a selective agent. Secondly, those concerned with the extent to which a population's adaptation affects the structure of the community as a whole. And thirdly, the overall patterns of community structure and function. From these three fields certain overlapping themes may be identified that represent the principal topics of the subject.

### 3.1. Coevolution

Ecology is concerned with interactions between populations. These interactions will often involve competition and the differential reproductive success of the organisms involved. It follows, therefore, that these ecological interactions will lead to evolutionary changes in the organisms involved in the interactions, and coevolution consequently may take place. Coevolution thus occurs where taxa that have a close ecological relationship impose reciprocal selective pressures (Ehrlich and Raven, 1964). Obviously the degree and nature of this inter-relationship varies, and hence so too does the degree of coevolution. At one extreme are forms of total mutual interdependence—pollinators and their flowers, ruminants and intestinal bacteria, humans and their domestic plants and animals—but less intense degrees of ecoevolution may also occur. Table 1.1 suggests the main types of coevolution.

Table 1.1. Types of coevolutionary relationship.

| Type of relationship | Nature of coevolution |
|---|---|
| Competitive | coevolution occurs through competition for a shared resource |
| Predatory A | a predator evolves in response to increasing elusiveness of prey |
| Predatory B | a prey species evolves in response to increasing fitness of predators |
| Parasitic A | a parasite evolves in response to increased resistance of host |
| Parasitic B | a host evolves in response to increasing virulance of parasite |
| Mutualistic A | coevolution of adaptations that are mutually beneficial and obligatory to both species |
| Mutualistic B | coevolution of adaptations that are mutually beneficial and obligatory to one species |
| Mutualistic C | coevolution of adaptations that are mutually beneficial but obligatory to neither species |
| Amensalistic | evolution in one species that adversely affects another, but does not benefit that species |
| Mimic | evolutionary change to mimic another species |

Given that coevolution is a common feature of biological systems, a broad distinction can be made in work on coevolution between the description and explanation of specific types of interaction and coevolution, and the development of general theories that account for the overall pattern of coevolutionary relationships. An example of the former, the empirical documentation of coevolutionary relationships, is Janzen's (1966) work on *Acacia*–ant interactions. He was able to show that certain *Acacia* species produce nectaries that attract ants, and that the presence of these ants will then deter foiliovores, thus enhancing the photosynthetic potential of the *Acacia*. Many other examples could be cited, going back to Darwin's work on orchids (1862), all illustrating the point that adaptive features of organisms can be explained when relationships with other organisms are considered.

Work on the general theory of coevolution has tackled the problem from a different perspective—that is, the construction of models to describe under what conditions types of coevolutionary relationships should develop—or, in other words, when species should impose reciprocal selective pressures (Maynard Smith, 1974b). Three areas of theory building can be recognized. First, theories that can explain which particular organisms should interact sufficiently to coevolve. Secondly, theories that can explain the degree of interdependence between species. And thirdly, theories that can account for the number of links involved, and the patterning of these links through time. Most of the models that have been produced are extremely complex, but some contributions that have direct relevance to problems in human evolution are worth mentioning.

One of these is van Valen's Red Queen hypothesis (1973). This is discussed more fully in Chapter 4, but briefly it suggests that organisms are in an evolutionary race, in which changes in a species adaptation represent a deterioration in the environment of others, and thus a shift in their selective pressures. Van Valen thus suggests that organisms should evolve rapidly to maintain the ecological *status quo*. Failure to do so will result in extinction. Indeed, van Valen put forward the hypothesis as a means of predicting the rates of extinction of taxa. While the Red Queen has been criticized in various ways (Maynard Smith, 1976b), it does give a good indication of the nature of evolutionary change and the way organisms coevolve. What it suggests is that because selective pressures are themselves evolving organisms, they are both continuous and continually changing. This is the broadest view of coevolution, and is important in showing the extent to which there may be an ecological steady state (Gould, 1977) and at the same time rapid evolutionary change. As Chapter 4 will attempt to demonstrate, aspects of human evolution are explicable in terms of the evolutionary changes among the community with which the early hominids were sympatric.

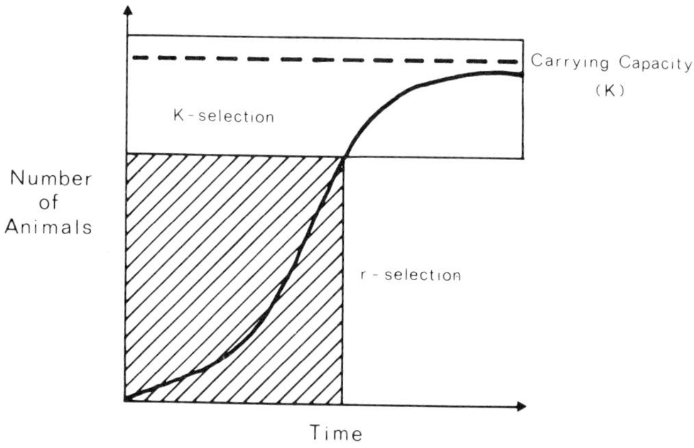


Fig. 1.1. *r* and *K* selection and the sigmoid curve of population growth. Populations in the shaded area of the curve are *r* selected, those in the unshaded part are *K* selected.

A second field of theory that may be used to look at coevolutionary patterns and that in fact has been applied very widely in ecology in recent years, is that of $r$ and $K$ selection (MacArthur and Wilson, 1967). These terms are derived from Verhulst's logistic equation that accounts for the pattern of animal population growth. From Fig. 1.1 it may be seen that when the population is small (shaded area of graph) population growth is entirely a product of the relationship between the intrinsic reproductive rate and the number of the population. Thus

$$R = rN \qquad (1)$$

where R = observed rate of increase at any one time, $r$ = the maximum intrinsic rate of increase, and N = the number of organisms. However, as the population increases, in the real world, resources become a limiting factor (unshaded part of graph), thus impinging on the rate of increase. As a population increases in size and competition for resources becomes more critical, the carrying capacity, or the number of individuals that may be supported by the environment, ($K$), becomes the primary control on population growth rather than the intrinsic reproductive rate. The equation therefore becomes:

$$R = rN\,(1 - \frac{N}{K}) \qquad (2)$$

It is from the terms of equation (2) that $r$ and $K$ selection models have been derived. An "$r$-selected" species is one that normally exists below "carrying capacity" and has the capacity to reproduce rapidly. A $K$-selected species normally exists close to carrying capacity. As a result there is little advantage in a high potential rate of increase, since being close to the resource ceiling it cannot increase its population size through reproduction alone. It is thus selected instead for its competitive efficency. In summary, $r$-selected species are opportunitistic ones in which evolutionary advantage is derived from a high reproductive rate, while a $K$-selected one gains advantage through increases in its competitive ability.

How, though, does $r$ and $K$ selection affect problems of coevolution? It is useful because it may predict the conditions under which coevolution will occur. For example, $K$-selected organisms are more likely to enter into intense inter-specific competition for resources. In this sense they are more likely to become involved in the sort of evolutionary race predicted by the Red Queen. As many large mammals tend to be $K$-selected, then the scope for coevolution is enhanced. Furthermore, the larger the number of $K$-selected species within a community, the greater the chances for close competitive interaction, and hence coevolution. Hominids are $K$-selected organisms, belonging to some diverse and complex communities, and the prediction may be made that their behaviour and adaptation may be partly explicable in terms of coevolution. Thus the *r* and *K* concepts do provide a set of expectations that may be formulated into theories of community evolution and ecology.

Another useful set of concepts in this context is that of *generalist* versus *specialist*. These terms have been employed rather loosely in biology, but here the terms are restricted to a measure of dietary breadth in particular, and the consequences for niche structure. Specialists occupy narrower niches than generalists. On the basis of this pattern of variation it is possible to make certain predictions about coevolution. For example, the more specialized an organism, the fewer other species with which it will interact. However, those few interactions are likely to be intensive. A generalist, on the other hand, will interact with a broad range of organisms, but only superficially. From this hypothesis it is possible to deduce that a specialist will be involved in a few, highly specific, coevolutionary relationships, whereas generalists may have many weaker coevolutionary interactions. Hypotheses such as these can then form the basis for model building and analysis of hominid evolution and adaptation.

These three examples show that coevolution is a concept of broader interest than the minutiae of orchids and moths. Rather, evolutionary change is integrally related to the ecology and evolution of the community as a whole. These relationships are patterned in various ways, and thus may be used to construct models of human evolution in the context of their biological community.

Table 1.2. Types of ecological stability (these are not necessarily exclusive). Based on discussion in Preston (1969), Lewontin (1969), Holling (1973), Orians (1975), Margalef (1975 May 1973), and Pianka (1978) (see also Fig. 1.2).

| Type | Characteristics |
|---|---|
| Constancy | extent to which value (species diversity, etc) changes through time |
| Persistance | length of survival or time to extinction of population, species, community, etc. |
| Periodicity | period of any cyclicity that may be evident |
| Amplitude | variation around equilibrium point involved in any cyclicity that may be evident |
| Elasticity | rate of return to equilibrium point after deviation |
| Resiliance | amount of fluctuation that can be tolerated and after which the system can return to equilibirum |

3.2. Stability

A second major focus of research in community ecology has been that of stability. Considerable effort, not to say mathematical sophistication, has gone towards first trying to understand the meaning of the term, and secondly towards discovering why and where it may occur in the biological world (Woodwell and Smith, 1969; May, 1973; van Dobben and Lowe McDonnell, 1975; Holling, 1973). Concern with this question derives from the problem of whether, and if so when, ecological systems are constant in various forms against a background of climatic, geophysical and evolutionary change, and thus ultimately whether ecological processes are dissipatory or self-sustaining.

Much of the difficulty associated with this problem comes from the elusive and varied quality of the term "stability". Intuitively a definition such as that of Orians (1975, p 141) that stability is the "tendency of a system to remain near an equilibrium point, or to return to it after a disturbance" is self-explanatory, but confidence in this dissolves when the varied properties and factors involved in this definition are identified (Table 1.2). It turns out that stability is not an elemental characteristic, but the term applied to a range of biological properties that may or may not be present in a system, either singly or in association. In tracking the behaviour of a population, species or community through time it is possible to identify and measure these properties (Fig. 1.2). The problem for the biologist, therefore, is to identify and account for the patterning of these properties through time.

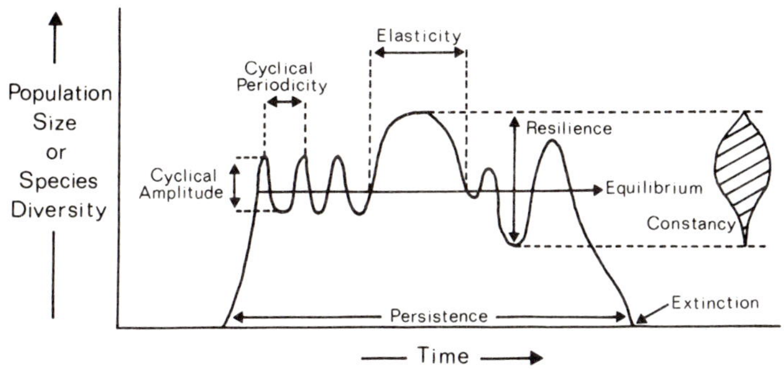


Fig. 1.2. Properties of ecological stability. See Table 1.2 for definitions.

Various hypotheses have been put forward to account for the presence and absence of ecological stability. Of these the most widely discussed is the complexity/diversity hypothesis. MacArthur (1955) suggested that the most diverse communities (i.e. communities with a large number of species) are the most stable. The principle behind this is that the more links in the system, the more control there will be on fluctuations in population size of any component species and hence the greater the stability. While this has been challenged by May (1973), it would broadly speaking reflect a consensus. However, the hypothesis can be regarded as true within a very limited view of stability—that is, diversity is related specifically to constancy (see Table 1.2 and Fig. 1.2), but not to resilience. If stability were to be measured in terms of resilience or persistance, then a different set of communities would be considered to be stable.

The diversity–stability hypothesis represents one approach to community evolution, an approach which minimizes historical events and specific biological characteristics, and emphasizes certain fixed forces that behave with the predictability of physical laws (Lewontin, 1969, p 22). Alternative approaches to the problem of stability—that is, the problem of the extent to which biological communities vary through time, and the period over which they can be maintained (Fig. 1.2)—would rather attempt to account for stability in terms of particular ecological and evolutionary properties of the system. Three such general explanations may be identified—those based on energetics and resource ecology; those based on evolutionary history; and those based on environmental history. All of these may incorporate the stability–diversity theories, but would see both stability and diversity as a product of other factors.

The energetics and resource ecology explanations suggest that stability is a function of the gross ecology of the community. The quantity of production, its distribution in time and space, and its predictability, would all be important factors contributing to the stability of the ecosystem. For example, in environments where either the growing season is short or the dry season prolonged the energy will not be available to sustain a diverse environment, and consequently stability will not be maintained. Increased production will increase the number of potential niches, and thus stability. It should then be possible to map out the resource ecology at both a local and a global scale, and relate this to the stability of the systems. When this is done, it is possible to see increased diversity and hence stability with increased rainfall in tropical environments, and with increased solar input in the mid and high latitudes (Fig. 1.3; Whittaker, 1970).

However, this explanation raises questions that lead directly towards the two other modes of explanation—evolutionary and palaeoenvironmental. Evolutionary explanations arise because the patterning of stability and diversity might indicate simply that richer environments have a greater "ecological space", and thus speciation is more likely to occur, and consequently diversity increased and stability enhanced. And palaeoenvironmental explanations suggest that the distribution of stable and diverse ecosystems

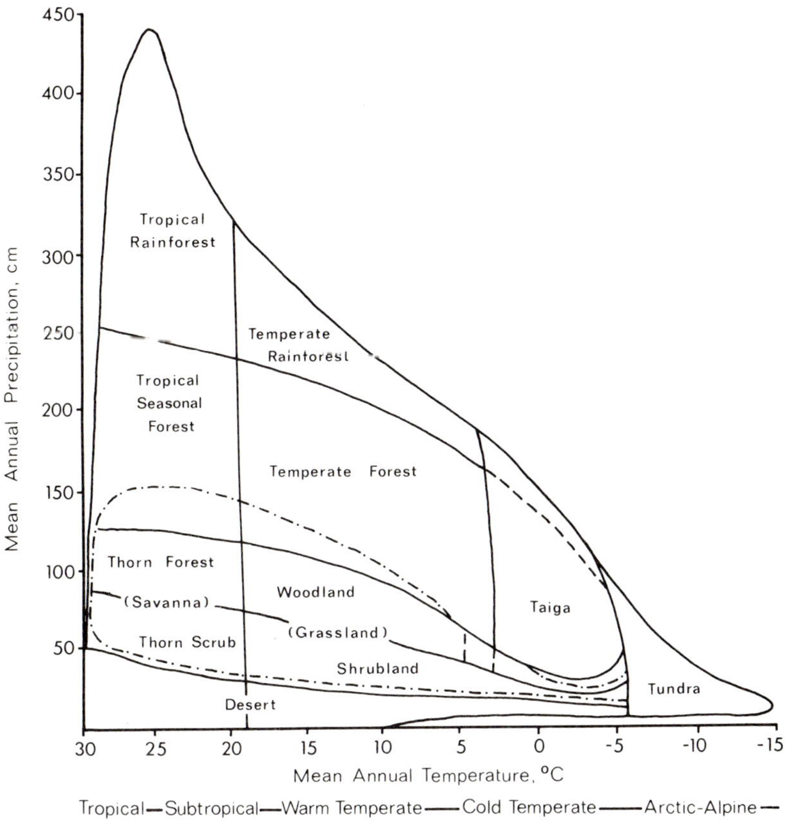


Fig. 1.3. Biotic communities in relation to climatic factors (redrawn with permission from Whittaker, 1970, Fig. 3.8).

can indicate the extent to which different regions have been subject to climatic change during geological time. The higher latitudes are less stable and diverse than the low latitudes, and it is these that were most affected by the reduced temperatures and fluctuating environments of the Pleistocene.

While no simple explanation of stability is likely to be forthcoming, the problem remains an important element of evolutionary ecology at the community level. In dealing with it, though, the central importance of palaeoecology becomes clear. Rigorous mathematical theory and certain relatively straightforward biological principles lead to expectations about the past. Studies of palaeoenvironments should both act as a critical test of these expectations and also provide further expectations about the way in which evolutionary events such as the emergence of the hominids may have been influenced by their ecological context. Palaeoenvironmental complexity such as that discussed in Chapter 2 certainly show that extensive climatic and environmental change has occurred over the last few million years, and thus the extent to which palaeoecology can contribute to solutions of the problem of ecological stability.

## 3.3. Diversity and Abundance

It has already been shown that diversity and stability are related. Diversity, though, may also be related to abundance. The problem may be stated fairly simply. Why do some areas have more species than others, and why, when species diversity is similar, are individuals of one species more abundant than another? Table 1.3 gives the main explanations that have been put forward to account for species diversity. Of greater consequence here is not why diversity may occur, but the consequences for adaptation.

Table 1.3. Main theories of species diversity (adapted from Pianka, 1978).

| Hypothesis | Explanation |
|---|---|
| Evolutionary time | species diversity increases through time because of speciation until saturation is reached |
| Ecological time | species diversity increases through time because of time for dispersal and colonization |
| Temporal heterogeneity | variability in climate through time at several scales—seasonal, annual, up to the long term geological—and the unpredictability of this variation will reduce species diversity |
| Spatial heterogeneity | in contrast, spatial complexity allows greater scope for patch specialization, and thus higher diversity |
| Productivity | high productivity will increase both the width and height of the trophic pyramid, and may thus enhance diversity |
| Predation | predation is often frequency dependent, and so will maintain diversity by preventing any one species from becoming dominant |
| Rarefaction | reduced abundance per species will allow greater species diversity within a community |
| Competition | as diversity increases there is a greater premium on competitive ability, resulting in the success of a large number of species with highly specialized competitive ability |

From the perspective of evolutionary change and adaptation high diversity has several implications. First, high diversity can itself promote further diversity—each new species added may be a potential source of food for other species. This is a factor affecting the height of a trophic pyramid (Colinvaux, 1973). Secondly, within diverse communities the niche's breadth is narrower resulting in a greater degree of specialization. The implications of this are extremely complex, and form one of the main concerns of optimal foraging theory (Pyke *et al.*, 1977). Furthermore, diversity may be thought to promote competition for particular resources; however, as the niches become more and more partitioned competition avoidance rather than competition itself become evolutionarily significant. This suggests that levels of diversity have a direct affect on the way evolution works. Thirdly, high diversity will enable a species with broad dietary requirements to subsist omnivorously from a wide range of resources. Low diversity will, in contrast, promote the inverse of these characteristics.

Diversity and abundance provide the structure and organization of the community, and consequently the problems and opportunities of any component species. The more diverse, the greater are these problems and opportunities, and it should not be ignored that the hominids evolved in a highly diverse community (see Chapter 4, this volume), and at a later time came to occupy a wide range of communities (see Chapters 8, 9, 10 and 11 this volume).

### 3.4. Extinction and Speciation

Coevolution, ecological stability, diversity and abundance all give structure to the community. Differential survival and reproduction within this structure will lead to evolutionary events, and, in particular, to speciation and extinction. Speciation refers to the appearance of new species, either through continuous transformation of existing species (anagenesis), or through branching (cladogenesis) (Rensch, 1947). Extinction usually refers to the local and total disappearance of species, although it should be recognized that a form of extinction occurs when a species evolves into a new form (chronospecies). Stanley (1979) has termed this "pseudoextinction". As extinction and speciation must relate, by definition, to co-evolution, stability and diversity, then rates and patterns of species formation and loss must be a field of community evolution and ecology.

Several factors may be said to affect speciation. Speciation may be said to occur when permanent reproductive isolation develops between two populations that formerly interbred. This process may be gradual or it may be sudden and indeed, there has been considerable debate recently as to which is the normal type of speciation (Eldredge and Gould, 1972; Gould and Eldredge, 1977; Williamson, 1981; Mayr, 1982). The actual factors that might lead to speciation are diverse. When populations become geographically isolated or distanced such that gene flow is interrupted or reduced, then genetic reproductive barriers may evolve (allopatric speciation). However, species may also separate when there is no geographical isolation (sympatric speciation). This may occur as a result of several selective factors. First, there may be disruptive selection (Mayr, 1970), such that within a population different individuals come under different selective pressures. For example, large and small individuals may be selected for different characteristics. Secondly, selection may favour competition avoidance, leading to selection for two ecological and morphological extremes within a population, and thus to speciation. Thirdly, environmental patchiness, through time and space, may result in speciation as a result of differences in selection for scheduling of behaviour and foraging. Fourthly, speciation may relate to the resource availability, which would clearly inter-connect with patterns of diversity and abundance as discussed above. A further model of speciation has been put forward by Vrba (1980), known as the "effect hypothesis", in which rates of speciation relate more closely to the ecological characteristics of the taxa. Specialized taxa will speciate more frequently than generalist taxa, as they can tolerate more easily the existence of closely related groups. Within a community, therefore, variable rates of speciation should occur.

These are the main processes of speciation. Although they differ in detail, all of them relate to the structure of the ecological community—its richness, diversity and stability—and thus show the ways in which ecological parameters directly affect evolutionary processes and events.

Obviously the first factor to be considered to affect the rate of extinction is that of speciation. Several authors (van Valen, 1973; Raup *et al.*, 1973; Stanley, 1977) have suggested that speciation and extinction are related. New species evolve in relation to characteristics and frequencies of species becoming extinct. Alternatively, species become extinct because novel species have evolved. Indeed, the close relationship between these two processes has been used as evidence for ecological stability. Apart from the effect of speciation, other factors affecting extinction are environmental change and habitat loss; inter-specific competition and predation (coevolution); disease (a form of predation); and behavioural and reproductive abnormalities.

By combining speciation with extinction a third property may be considered—that of species survivorship, or the length of time between speciation and extinction. This is the equivalent of an individual longevity. Some authors (Stanley, 1979; Eldredge and Cracraft, 1980) have argued that this is a "real" intrinsic character of species, and that a mean species duration is a valuable biological measurement. Alternatively, survivorship may merely be an effect of adaptive success. However, whether species longevity reflects external conditions or the life history of a species, it does provide information about the way a population fits in with its biological community as a whole.

Speciation, extinction and survivorship are important evolutionary processes that must be considered in relation to patterns of community biology. Their importance in consideration of human evolution during the Pleistocene is all too apparent. During the last three or more million years several hominid species have evolved, all but one of which have become extinct or pseudo-extinct (see Chapter 3, this volume). Many other species have also come into existence during this time, and the Pleistocene is notable for the high rate of extinction (see Chapter 11, this volume). And in more recent times, humans have been directly responsible for the production of new species and varieties through the process of domestication. It is necessary in considering human adaptation to relate all these events to the field of community evolution and ecology.

In summary, community evolution and ecology is concerned with the analysis of adaptation and evolution as products of interactions between members of a biological community. These interactions may be examined either in terms of the way the community influences individual adaptation, or in terms of the impact that specific adaptive strategies have on the community. The structure of a community is the product of these interactions and should thus provide information on how biological systems work—the occurrence of coevolution, the nature and types of stability, the patterns of diversity and abundance, and the means by which new species are formed and existing ones become extinct. In turn this promotes a series of questions we are able to ask about hominids, given that they have evolved, that they are biological organisms, and are members of a biological community. We can ask whether they have influenced the biological community, and if so, how? We can ask whether human evolution has been shaped by the structure of the community. We can ask whether hominids fit into expected patterns of stability, of diversity, and of abundance. And we can also ask how the formation of the human species occurred, as well as how several hominids became extinct. Community evolution and ecology is thus the appropriate context for the problems of human adaptation discussed in this volume.

## 4. COMMUNITY ECOLOGY AND PREHISTORIC HUMAN ADAPTATION

Two main conclusions to be drawn from this discussion of community evolution and ecology. One is that even if the focus of study is a particular population or taxon the characteristics of the biological context of that unit must be taken into account in an interactive rather than passive manner. And the other is that a community perspective on the evolution of a particular group does not imply that evolution occurs at the level of communities, nor that communities are the unit of selection. A logical consequence of this is that communities do not actually exist, and that like ecosystems, they are useful abstractions that help us to simplify the biological world for the purposes of analysis (Colinvaux, 1973). It therefore follows that the definition of a particular community is arbitrary, and must be justified in the context of the analytical objectives. In the papers presented in this volume two overlapping approaches may be identified. The first of these is geographically based—that is, that the community of the hominids can be defined in terms of spatial coexistence. Thus for the earlier periods of human evolution the tropical African community is the centre of interest (see Chapters 4, 5, 6 and 7, this volume), while for later periods more temperate and high latitudes are relevant (see Chapters 8, 9, 10 and 11, this volume). The second approach is to look more specifically at the types of organisms with which the hominids may usefully be compared, and in the context of which their adaptations may be assessed. In this case it is the large mammals, that group of animals that most closely share the ecological problems of prehistoric hominid populations (see Chapters 4, 5, 8, 9, 10 and 11, this volume).

These then are the communities of the early hominids. Having identified them, though, what problems can usefully be tackled from this perspective, and thus provide the rationale behind this volume? While groups that may be uniquely ancestral to later human populations may possibly be identified in the Miocene (see Chapter 3, this volume), it is really only in the Pliocene and Lower Pleistocene that sufficient evidence of distinctive hominid forms occur. It is this period that has formed the focus for palaeoanthropological research in recent years. It is also a period that coincides with the culmination of a gradual decrease in temperatures during the Late Cenozoic and the onset of the climatic oscillations associated with high latitude glaciation. In the discussion of the structure of communities above (see p 12) one of the factors in relation to environmental stability mentioned was that of palaeoclimatic conditions. If climate can affect the stability and diversity of communities, which in turn is a factor in human adaptation, then it is important to establish the nature of climatic change and its environmental manifestations. This is the problem that Roberts tackles in the following chapter.

If environmental conditions influence the biological community as a whole, including the hominids that were part of that community, then it should be reflected in the palaeontological record of fossil hominids, and in the morphology of the specimens recovered. In Chapter 3 Stringer reviews the evidence for Pliocene and Pleistocene hominid evolution and discusses the patterns that emerge in relation to rates of evolution, climatic context and body size. Each of these parameters has attracted the interest of evolutionary biologists in general, and their effect on hominid evolution is of both specific and general interest.

The pattern of human evolution is one of adaptive changes; these adaptive shifts are the product not only of morphological changes, but also of behavioural ones. In Chapter 4 Foley examines what might be the characteristics of early hominid behavioural evolution by relating them to the hunter–gatherer adaptive strategy. Adaptive strategies represent attempts by populations to solve the environmental problems they face. By identifying these problems it is possible to relate the characteristics of early hominids to the selective agencies operating on them. These adaptive problems derive partly from the phylogenetic and physiological characteristics of the hominids, and partly from the ecological and evolutionary characteristics of the tropical African community to which they belong.

Many of these problems, though, are not unique to hominids. This is the strength of a community based approach, for it enables comparison of adaptive problems and strategies. However, this lack of uniqueness can also lead to methodological difficulties. Because prehistoric adaptation must be investigated through the fossil record, then the processes of fossil formation must be understood. If similar problems may lead to similar behaviours, they may also lead to a similar representation in the fossil and archaeological record. In Chapter 5 Hill highlights the difficulties that this may lead to by examining the processes of archaeological formation in relation to the behaviour of hyaenas, and their potential for producing an "archaeological record" of their own. This in turn leads to a consideration of the criteria by which we refer archaeological patterns to behaviour.

Similarities between hominids and hyaenas arise from similarities of behaviour, amongst which are the consumption of meat. Hindsight informs us that at some time hominids did incorporate meat eating into their adaptation. This, however, is not sufficient grounds for assuming its occurrence at any time period. Thus in Chapter 6 Potts looks at what selective agents might be operating to increase meat consumption among hominids, and the means by which it might be possible to identify it in the archaeological record, and, more problematically still, whether it is possible to distinguish between hunting and other forms of meat acquisition.

Adaptation is the means by which organisms survive in the context of the biological community that acts as the selective agents on the organisms. While Chapters 3, 4, 5 and 6 all show ways in which human behaviour and morphology may have been adaptive, and the means by which adaptation hypotheses may be tested, it must be remembered that adaptation must be mediated through some genetic, neurological, biochemical or cultural mechanism. Environmental problems may be the stimulus, and adaptation the product, but between the two must lie the capacity to implement the adaptation. While for many species this may be provided by means of fixed genetic programming, in many mammals, including man, learned behaviour is critical. Certainly a substantial increase in brain size is a distinctive characteristic of human evolution. In Chapter 7 Gowlett examines the direct evidence for human mental abilities during the Lower Pleistocene, and their role in solving environmental problems.

All these papers relate directly to the tropical African community in which the current evidence suggests the very early hominids evolved. However, the Pleistocene saw the expansion of the hominid range into many parts of the world, and consequently new communities against which to measure hominid adaptation. As they moved out of their original environment the hominids were both influenced by novel adaptive situations, and also had an impact on those new communities. Turner, in Chapter 8, compares the pattern of hominid migration into temperate environments with those of other species of large mammal.

Among the most severe adaptive problems faced by the early hominids in these new environments was that of intense seasonal variability, and in particular the difficulties of surviving extreme glacial winters. By looking at one of the most extreme habitats ever occupied by hominids, that of Britain during the Late Pleistocene glaciation, Scott (Chapter 9) is in a position to relate the characteristics to the way other organisms adapt to cold conditions and thus act as competitors and resources for hominids.

Analysis of hominids under extreme condition is one means of understanding the selection operating on hominids. Another is to examine patterns of variability on a regional scale. This is the problem that Gamble tackles in Chapter 10, where he constructs a model of regional ecological variation in Pleistocene Europe, and then considers the impact this would have on the behaviour of hominids and the structure of the archaeological record.

Finally, the hominids expansion into diverse communities did not take place against a static ecological and evolutionary background. Rather the Pleistocene saw many changes in fauna and flora, including a high rate of extinction of mammals. The role of hominids in these extinctions has long been a subject of debate, and in Chapter 11 Garrard analyses the patterns of extinction in the Levant in the context of community ecology, and, in particular, inter-specific competition, in which the hominids are treated as just one element of the ecosystem.

## 5. CONCLUSIONS

Community evolution and ecology provides the perspective by which problems in prehistoric adaptation may be identified, formulated and analysed. This introductory chapter has attempted to outline the basis on which this statement may be made, the principles that may be derived from it, and the structure of the volume that flows from it. It has been argued that a primary research target in prehistoric archaeology and palaeoanthropology is the relationship between man and his environment. It was shown that while this has been a focus of research since the publication of the *Origin of Species*, two shortcomings have been evident. First, that it is neither possible nor useful to dichotomize hominids and their environment. Secondly, that humans should not be seen in isolation from the rest of the biological world.

The comparative framework of community evolution and ecology provided a solution to these problems, and thus their principles and fields of research were described. In turn these provided the framework for considering problems in human evolutionary studies. It is not suggested, though that such an approach can provide a monopolistic perspective. The level of the community—which is itself variable—is just one of several possible levels for examining evolutionary questions. Evolutionary problems range from the ecosystem to the genome and molecule, and each requires its own level of analysis. What they all have in common, however, is the analytical and explanatory power of the theory of natural selection, the Darwinian theory of evolution, which has equal applicability to the problems of human prehistory as to any other branch of the life sciences.

## ACKNOWLEDGEMENTS

I should like to thank Alan Turner and Clive Gamble for their helpful comments on an earlier draft of this paper.

## REFERENCES

Allison, A. C. (1954). Protection by the sickle-cell trait against subtertian malarial infection. *Brit. Med. J.* **1**: 290–292.

Butzer, K. W. (1972). *Environment and Archaeology*. London: Methuen (2nd ed).

Butzer, K. W. (1982). *Archaeology as Human Ecology: method and theory for a contextual approach*. Cambridge: Cambridge University Press.

Cavalli-Sforza, L. L. (1982). *Cultural Transmission and Evolution: a quantitative approach*. Princeton: Princeton University Press.

Chagnon, N. and Irons, W. (1979). *Evolutionary Biology and Human Social Behaviour: an anthropological perspective*. New York: Duxbury.

Clark, J. D. (1970). *The Prehistory of Africa*. London: Thames and Hudson.

Clark, J. G. D. (1952). *Prehistoric Europe: the Economic Basis*. Cambridge: Cambridge University Press.

Colinvaux, P. (1973). *Introduction to Ecology*. New York: Wiley.
Darwin, C. (1862). *On the Contrivances by which British and foreign orchids are fertilized by insects*. London: Murray.
Dawkins, R. (1975). *The Selfish Gene*. London: Oxford University Press.
Dawkins, R. (1981). *The Extended Phenotype: the Gene as the Unit of Selection*. San Francisco: Freeman.
Dunnell, R. C. (1980). Evolutionary theory and archaeology. In: M. B. Schiffer (ed) *Advances in Archaeological Theory and Method. Vol. 3*. London: Academic Press. pp. 35–99.
Ehrlich, P. R. and Raven, P. H. (1964). Butterflies and plants : a study in coevolution. *Evolution* 18: 586–608.
Eldredge, N. and Cracraft, J. (1980). *Phylogenetic Patterns and the Evolutionary Process*. New York: Columbia University Press.
Eldredge, N. and Gould, S. J. (1972). Punctuated equilibrium: an alternative to phyletic gradualism. In: T. J. M. Schopf (ed) *Models in Palaeobiology*. San Francisco: Freeman. pp. 82–115.
Forde, C. D. (1934). *Habitat, Economy and Society: a Georaphical Introduction to Ethnology*. London: Methuen.
Futuyma, D. J. (1979). *Evolutionary Biology*. Sunderland, Mass.: Sinnawer Associates.
Ghiselin, M. (1982). The intellectual path to natural selection. *New Scientist* **94** (1301): 156–159.
Gould, S. J. (1977). Eternal metaphors of palaeontology. In: A. Hallam (ed) *Patterns of Evolution as Illustrated by the Fossil Record*. Amsterdam: Elsevier. pp. 1–26.
Gould, S. J. and Eldredge, N. (1977). Punctuated equilibria: the tempo and mode of evolution reconsidered. *Palaeobiology* **3**: 115–51.
Hamilton, W. D. (1964). The genetical evolution of social behaviour. *J. Theoret. Biol.* **7**: 1–52.
Harcourt, A. H., Harvey, P. H., Larson, S. G. and Short, R. V. (1981). Testes weight, body weight and breeding systems in primates. *Nature* **293**: 55–57.
Harris, M. (1968). *The Rise of Anthropological Theory: a History of Theories of Culture*. New York: Crowell.
Harris, M. (1979). *Cultural Materialism: the Struggle for a Science of Culture*. New York: Random House.
Harrison, G. A., J. S. Weiner, J. M. Tanner and N. A. Barnicot (1964). *Human Biology: an Introduction to Human Evolution, Variation and Growth*. London: Oxford University Press.
Hoffman, A. (1979). Community palaeoecology as an epiphenominal science. *Palaeobiology* **5**(4): 357–379.
Holling, C. S. (1973). Resiliance and stability of ecological systems. *Ann. Rev. Ecol. Syst.* **4**: 1–24.
Janzen, D. H. (1966). Coevolution of mutualism between ants and acacias in Central America. *Evolution* **20**: 249–275.
Lewontin, R. (1969). The meaning of stability. *Brookhaven Symp. Biol.* **22**: 13–24.
MacArthur, R. H. (1955). Fluctuations of animal populations and a measure of community stability. *Ecology* **36**: 533–536.
MacArthur, R. H. (1962). Some generalized theories of natural selection *Proc. Nat. Acad. Sci.* **48**: 1893–197.
MacArthur, R. H. and Wilson, E. O. (1967). *The Theory of Island Biogeography*. Princeton: Princeton University Press.
Margalef, R. (1975). Diversity, stability and maturity in natural ecosystems. In: W. H. van Dobben and R. H. Lowe-McConnell (eds) *Unifying Concepts in Ecology*. The Hague: W. Junk. 151–60

Martin, R. D. (1981). Relative brain size and basal metabolic rate in terrestrial vertebrates. *Nature* **293**: 57–60.
Martin, R. D. and May, R. M. (1981). Outward signs of breeding. *Nature* **293**: 7–9.
May, R. M. (1973). *Stability and Complexity in Model Ecosystems*. Princeton: Princeton University Press.
Maynard Smith, J. (1974a). The theory of games and the evolution of animal conflict. *J. Theor. Biol.* **47**: 209–221.
Maynard Smith, J. (1974b). *Models in Ecology*. Cambridge: Cambridge University Press.
Maynard Smith, J. (1976a). Evolution and the theory of games. *Amer. Scient.* **64**: 41–5.
Maynard Smith, J. (1976b). A comment on the Red Queen. *American Naturalist* **110**: 325–330.
Mayr, E. (1970). *Populations, Species and Evolution*. Cambridge, Mass: Harvard University Press.
Mayr, E. (1982). Questions concerning speciation. *Nature* **296**: 609.
Motulsky, A. G. (1960). Metabolic polymorphisms and the role of infectious diseases in human evolution. *Human Biol.* **32**: 28–62.
Orians, G. H. (1975). Diversity, stability and maturity in natural ecosystems. In: W. H. van Dobben and R. H. Lowe-McDonnell (eds) *Unifying Concepts in Ecology*. The Hague: W. Junk. pp. 139–50.
Pianka, E. (1978). *Evolutionary Ecology*. New York: Harper and Row. (2nd ed).
Preston, F. W. (1969). Diversity and stability in the biological world. *Brookhaven Symp. Bibl.* 22: 1–12.
Pyke, G. H., H. R. Pulliam and E. L. Charnov (1977). Optimal foraging: a selective review of theory and tests. *Quat. Rev. Biol.* **52**: 137–154.
Rappaport, R. A. (1967). *Pigs for the Ancestors: ritual in the ecology of a New Guinea people,* New Haven: Yale University Press.
Raup, D. M., Gould, S. J., Schopf, T. J. M. and Simberloff, D. S. (1973). Stochastic models of phylogeny and the evolution of diversity. *J. Geol.* **81**: 525–542.
Rensch, B. (1947). *Evolution above the Species Level*. New York: Columbia University Press.
Stanley, S. M. (1977). Trends, rates and patterns of evolution in the Bivalvia. In: A. Hallam (ed) *Patterns of Evolution as Illustrated by the Fossil Record*. Amsterdam: Elsevier pp. 210–250.
Stanley, S. (1979). *Macroevolution: Pattern and Process*. San Francisco: Freeman.
Steward, J. (1955). *Theory of Culture Change: the Methodology of Multilinear Evolution*. Urbana: University of Illinois Press.
Steward, J. (1977). *Evolution and Ecology: Essays on Social Transformation*. Urbana: University of Illinois Press.
Trivers, R. L. (1971). The evolution of reciprocal altruism. *Quart. Rev. Biol.* **46**: 35–57.
Trivers, R. L. (1972). Parental investment and sexual selection. In: B. G. Campbell (ed) *Sexual Selection and the Descent of Man*. Chicago: Aldine. pp. 139–79.
Van Dobben, W. H. and R. H. Lowe-McDonnell (eds) (1975). *Unifying Concepts in Ecology*. The Hague: W. Junk.
van Valen, L. (1973). A new evolutionary law. *Evol. Theory* **1**: 1–30.
Vayda, A. P. (ed) (1969). *Evironment and Cultural Behaviour: Ecological Studies in Cultural Anthropology*. New York: Natural History Press.
Vrba, E. (1980). Evolution, species and fossils: how does life evolve? *S. African J. Sci.* **76**: 61–84.
White, L. (1959). *The Evolution of Culture*. New York: McGraw Hill.

Whittaker, R. H. (1970). *Communities and Ecosystems*. New York: Macmillan.
Williams, G. C. (1966). *Adaptation and Natural Selection: a critique of some current evolutionary thought*. Princeton: Princeton University Press.
Williamson, P. G. (1981). Palaeontological documentation of speciation in Cenozoic molluscs from Turkana basin. *Nature* **293**: 437.
Wilson, E. O. (1975). *Sociobiology: the New Synthesis*. Cambridge, Mass: Harvard University Press.
Wilson, E. O. (1978). *On Human Nature*. Cambridge, Mass: Harvard University Press.
Wilson, E. O. and C. Lumsden (1981). *Genes, Mind and Culture: the Co-evolutionary process*. Cambridge, Mass: Harvard University Press.
Woodwell, G. M. and H. Smith (1969) (eds). *Diversity and Stability in Ecological Systems*. Brookhaven Symp. Biol. 22.

# 2 *Pleistocene Environments in Time and Space*

*Neil Roberts*

## 1. INTRODUCTION

Since the 1960's there has been a revolutionary change in our understanding of the Quaternary, the geological period which roughly corresponds to the last two million years of the Earth's history. At its simplest, this change may be summarized as the recognition, not of four, but of a dozen or more glacial–interglacial cycles, and a realization that the tropics, long thought to be relatively unaffected by Quaternary climatic changes, in fact experienced environmental oscillations almost as dramatic as those in the temperate zone. In particular, the formerly held belief that high latitude glacials were accompanied by low latitude pluvials (or wet phases), is now known to be erroneous; on the contrary, most of the intertropical zone was more arid than at present during glacial maxima.

So far it has been in palaeoclimatology that these new results have had their greatest impact, but implications for other fields such as landform evolution (Street-Perrott *et al.*, in press) and island biogeography (Colinvaux and Schofield, 1976) are beginning to be appreciated. In the latter part of this chapter some implications for the twin fields of community ecology and human evolution will be considered, with discussion focusing on issues such as environmental stability versus ecological diversity, and the nature of human dispersal and adaptation during the Pleistocene. Before this the new framework will be used to describe both the sequence of climatic change over the whole of the Pleistocene, and the spatial pattern of palaeoclimates and environments within a single (the last) glacial–deglacial hemicycle. By way of introduction to these themes some consideration will be given to the background to the revolution in Quaternary science.

HOMINID EVOLUTION
*ISBN 0-12-261920-X*

## 2. BACKGROUND TO THE QUATERNARY SCIENCE REVOLUTION

The new Quaternary paradigm has emerged primarily because the methodological basis of Quaternary science has been radically restructured. The Quaternary has always rested somewhat uneasily on the top of the geological column, largely because the palaeontological method which is fundamental to Phanerozoic (i.e. post-PreCambrian) geology, is not easily applied to as short and as recent a period as the last 2 Myr. The time required for radiation of new terrestrial plant species, for instance (say 0·5 Myr), may be insignificant on a Cretaceous time scale but is certainly not geologically instantaneous in the context of the late Cenozoic. In consequence, nineteenth and early twentieth century stratigraphers relied heavily on two other methodological props to help them establish a framework for the Quaternary: first "count from the top", and secondly "inferred correlation".

The first of these is a development of the Law of Superimposition, by which younger rocks overlie older ones in undeformed sedimentary strata, so that to go down through the sediment pile is to go back through time. Counting glacials (or interglacials) downwards from the top of the pile can, of course, work successfully so long as the sedimentary sequence involved is a continuous one. Unfortunately, the areas actually studied by the early Quaternary geologists were those parts of Europe and North America which had been affected by ice sheet advance and retreat. Because later ice sheets invariably tended to destroy or at least disturb the evidence left by earlier ice advances, the record of glacial sediments in these terrestrial areas was incomplete. Counting from the top, however, makes no allowance for gaps in the stratigraphic record, and older glaciations were misattributed or even missed out altogether.

Table 2.1. Classical Penck and Brückner glacial-interglacial sequence, with regional equivalents (glacial stages in capitals)

| Alps | N. Europe | British Isles | North America | E. Africa[a] |
|---|---|---|---|---|
| WÜRM | WEICHSELIAN | DEVENSIAN | WISCONSIN | MAKALIAN/GAMBLIAN |
| R-W | Eemian | Ipswichian | Sangamon | — |
| RISS | SAALE | WOLSTONIAN | ILLINOIAN | KANJERAN |
| M-R | Holstein | Hoxnian | Yarmouth | — |
| MINDEL | ELSTER | ANGLIAN | KANSAN | KAMASIAN |
| G-M | Cromerian | Cromerian | Aftonian | — |
| GÜNZ | — | — | NEBRASKAN | KAGUERAN |

[a] Pluvials.

The classical framework for Pleistocene glaciations was first presented in 1909 by Penck and Brückner based on their studies of fluvio-glacial outwash terraces in the north Alpine foreland. In this region they identified four terraces and therefore four glaciations, which they named after the Bavarian river valleys, Günz, Mindel, Riss and Würm. Similar sequences were soon to emerge from northern Europe and North America, and like that from the Alps, they included four major Pleistocene glacial phases. Not surprisingly, it was often inferred that these glacials and interglacials could be cross-correlated with each other (Table 2.1).

Inferred correlation of this type was the second methodological prop of Quaternary research prior to the 1960's. As with "counting from the top", correlation—or the demonstration of time-equivalence across space—is not an inherently defective method. What matters is the way in which it is carried out, and in the case of the Quaternary it was carried out with reference to the nearest available glacial sequence. A good example of this process is provided by the East African pluvial–interpluvial sequence, which was built up from a hotch-potch of palaeoenvironmental evidence and ratified by the Pan-African Congress of Prehistory in 1949 (Leakey, 1952), only to be later discarded and discredited. Lacking a sound internal stratigraphy, the East African evidence was fitted into a sequence which closely resembled that from the Alps, pluvials being conveniently found to match the four "known" European glacials (Table 2.1). This not only gave some semblance of order (albeit false) to the jumble of African material, but also helped to reinforce the edifice originally constructed by Penck and Brückner. The shortcomings of these two methodological props are, of course, obvious, and they help to explain the relative backwardness of Quaternary studies during the first half of the present century.

The chain of events which eventually led to counting from the top and inferred correlation being discarded can be traced back to three technical developments which took place during the late 1940's. Two of the three originated from a group of isotope geochemists working at the University of Chicago. From this group, which included such men as Harold Urey, Willard Libby, and Samuel Epstein, came first radiocarbon dating, and secondly oxygen isotope analysis. Both methods utilize the fact that most natural elements are a mixture of several isotopes, which have identical chemical compositions but different atomic masses and physico-chemical properties. Radiocarbon and other isotopic dating techniques exploit the fact that some of these istopes, such as $^{14}C$, are unstable and decay to stable forms, in this case $^{12}C$. Because decay takes place at a fixed rate, unstable isotopes represent natural geological clocks, which can be used to date events in the past. The development of the radiocarbon dating method has been amply documented elsewhere (Libby, 1955; Berger and Suess, 1979) and need not be discussed further here; so too have the implications of the "radiocarbon revolution" for later prehistoric archaeology (e.g. Renfrew, 1973). Curiously, the implications for Quaternary environmental studies have not received anything like so much attention, perhaps because it is only now that those implications are coming to be fully appreciated (cf. Vita-Finzi, 1973).

The other method to emerge from the Chicago school—that of oxygen isotope analysis—used a different isotopic characteristic, that of the different physico-chemical behaviour of the two stable isotopes, $^{18}O$ and $^{16}O$. As Urey (1947) discovered, when these isotopes are removed from water (e.g. by evaporation or precipitation as carbonate), they are lost at different rates relative to one another as the temperature changes. This process, termed isotopic fractionation, means that the $^{18}O/^{16}O$ ratio of fossil organisms, such as molluscs, may be used to determine the temperature of the water in which they lived. If unstable isotopes represent geological clocks, it was thought stable oxygen isotopes might provide a geological thermometer. However, this method required sample material of suitable age before it could be applied to the Quaternary. Those samples were to be provided as a result of the third technological development of the late 1940's: the invention of the Kullenberg corer.

As noted above, any sequence of glacials and interglacials built up from the discontinuous sedimentary record of formerly glaciated terrestrial areas is likely to be incomplete. A complete sequence can only be obtained from environments where there has been continuous sedimentation during the Quaternary. Unfortunately, the most important such environment—the deep sea bed—is also one of the least accessible. To study the sedimentary record of the deep ocean floor requires undisturbed sediment cores many metres long, and they have to be recovered in ocean water several kilometres deep. Early coring devices were either based on gravity, in which case the cores which were recovered were no more than a metre or so long, or they deformed the sediment too badly for use. However, the development of a piston corer by Björe Kullenberg in 1947 led the way to the collection of a large number of deep sea sediment cores over 10 m long.

If the Urey–Epstein method of oxygen–isotope analysis could be applied to calcareous microfossils (e.g. foraminifera) from deep sea cores, then it should be possible to produce a continuous record of temperature changes during the Pleistocene. Precisely this reasoning led Cesare Emiliani, who was studying at Chicago during the early 1950's, to analyse the isotopic composition of planktonic foraminifera in a number of sediment cores from the Caribbean and the equatorial Atlantic. The results of his analyses were spectacularly successful, and revealed cyclical downcore variations in the $^{18}O/^{16}O$ ratio, which he interpreted as indicating multiple oscillations between cold (glacial) and warm (interglacial) oceanic and climatic conditions. What was most significant about Emiliani's data was that there were many more glacial –

interglacial cycles than the four predicted by the Alpine model. These results were to prove the first major challenge to the Penck and Brückner paradigm for the Quaternary. What occurred subsequently was, to use Kuhn's (1962) terminology, a scientific revolution in which the old paradigm was overthrown and replaced by a new one based on the deep-sea, not the continental, record of glaciation.[1]

One interesting feature of this revolution is that the new paradigm did not come to be fully accepted by the Quaternary community until the 1970's, more than 20 years after Emiliani's seminal (1955) paper. There are many reasons for this delay but one of them was a re-interpretation of oxygen–isotope analysis which took place during the 1960's. It is ironic that arguably the most significant paper published on the Quaternary this century, Emiliani's "Pleistocene temperatures", should actually have been mistitled! For, as Shackleton (1967) and others later discovered, the $^{18}O/^{16}O$ signal reflected not so much ocean temperature as ocean volume, which was reduced during glacial phases as water was locked up in expanded ice sheets. In other words, Emiliani's oxygen–isotope data represented a record of Pleistocene glaciation, not of Pleistocene temperatures.

The revolution in Quaternary science may be said to have been completed with the appearance of three key publications by the CLIMAP group in 1976 (CLIMAP Project Members, 1976; Hays *et al.*, 1976; Cline and Hays, 1976), and with the publication of the first Quaternary textbook based on the new paradigm (Bowen, 1979).[2]

## 3. PLEISTOCENE CLIMATES AND ENVIRONMENTS THROUGH TIME (LAST 2 MYR)

It is now recognized that local stratigraphies based on data agglomerated from several sites (e.g. the recommended British glacial–interglacial sequence of Mitchell *et al.*, 1973), should be fitted into frameworks provided by long, continuous records of environmental change, and not vice-versa.[3] There exist a number of depositional environments, both oceanic and terrestrial, which preserve a continuous sedimentary record suitable for palaeoenvironmental analysis, but the most important so far as the stratigraphy of the whole of the Pleistocene is concerned, is the deep ocean floor.

Deep sea sediments, often largely made up from the skeletons of micro-organisms such as foraminifera and diatoms, settle out slowly but continuously from the water column, disturbed only by the burrowing of small benthic organisms. This burrowing, or bioturbation, mixes up the layers at the sediment–water interface and blurs the quality of the palaeoenvironmental record. The problem is particularly acute at sites where sediment has been accumulating relatively slowly. Furthermore, "in lower Pleistocene sediment, oxygen–isotope stratigraphy will not be useful in cores with an accumulation rate lower than 1 cm/$10^3$ yr, because the (oxygen) isotopic changes were more frequent than in the Brunhes (palaeomagnetic) epoch and therefore more readily obscured by mixing in the sediment." (Shackleton and Opdyke 1976: 462). Yet with a 20–30 m limit to Kullenberg coring, a 2-Myr long record would require a sedimentation rate of no *more* than 1 cm/Kyr.

Table 2.2. Comparative details of three long deep-sea cores.

| Core number and reference | Core length (m) | Age of base of core (Myr) | Mean rate of sedimentation | Time interval between samples (Kyr) |
|---|---|---|---|---|
| V28–238 (Shackleton and Opdyke, 1973) | 16 | 0·85 | 1·7 cm/kyr | 6 |
| V28–239 (Shackleton and Opdyke, 1976) | 21 | 2·1 | 1·0 cm/kyr | 5 |
| V16–205 (Van Donk, 1976) | 12 | 2·4 | 0·6 cm/kyr | 20 |

This difficulty can be appreciated by reference to Table 2.2, which lists details of three of the few deep sea cores which reach back into the Lower Pleistocene (Fig. 2.1 for locations). Although equatorial Atlantic core V16–205 (van Donk, 1976) spans the entire Pleistocene (and some of the Late Pliocene), the sedimentation rate was too low to allow anything more than a very generalized climatic curve to be produced.[4] At the other extreme, core V28–238 taken northeast of New Guinea in the Pacific, has a much more detailed record, but the higher rate of sediment accumulation means that the base of the core is only 0·85 Myr old (Shackleton and Opdyke, 1973). Fortunately there is one core which combines both advantages: V28–239 covers more than 2·0 Myr, but a mean sedimentation rate of 1·0 cm/Kyr coupled with a fine sampling interval also gives it detailed time resolution. Furthermore this core was taken close to V28–238 (Fig. 2.1) which enabled a successful cross-check to be made on the reliability of the upper part of V28–239. The oxygen–isotope record for this core, which is shown in Fig. 2.2, represents the best climatostratigraphic framework presently available for the Pleistocene.

It can be seen from this that there is no evidence for a marked change in climate at the beginning of the Pleistocene.[5] The Late Cenozoic Ice Age—of which the Pleistocene forms only the latest part—began in the Miocene with ice sheet formation in Antarctica and was intensified

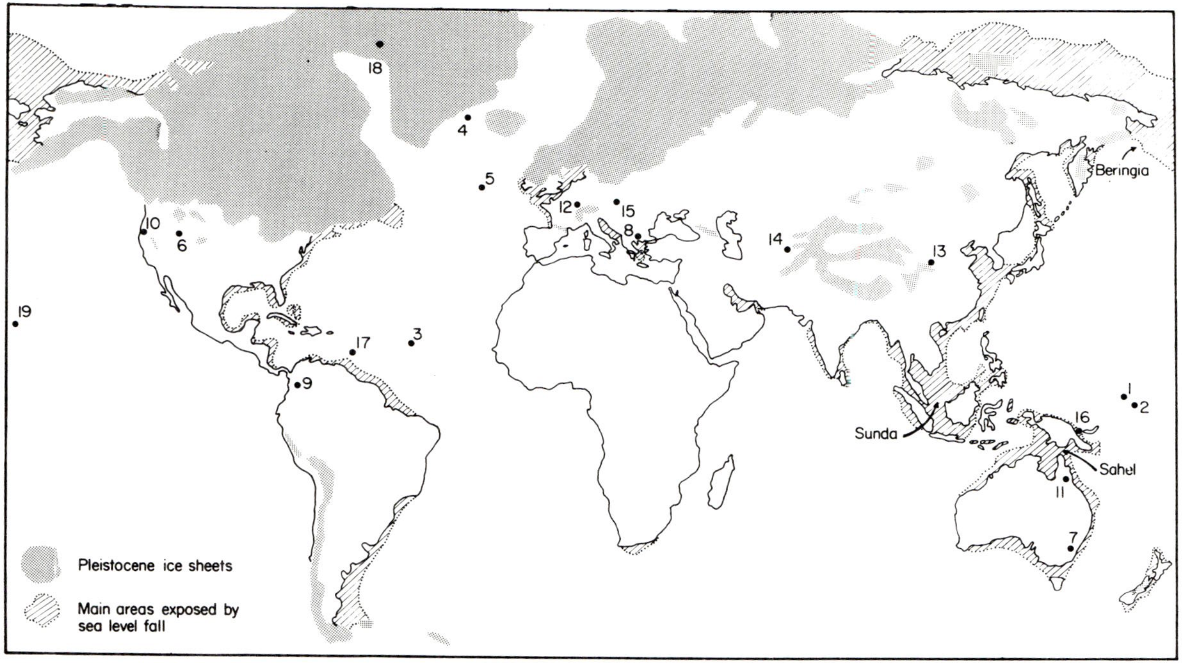


Fig. 2.1. World map showing areas affected by Pleistocene glaciation and sea level change, and locations of sites discussed in text. *Key*: DEEP SEA CORES. (1) V28-239; (2) V28-238; (3) V16-205; (4) V28-14; (5) V23-82. LAKE LEVELS. (6) Bonneville; (7) George. POLLEN (8) L. Phillipi; (9) Bogota; (10) Clear Lake; (11) Lynch's Crater; (12) Grande Pile. LOESS. (13) China; (14) Tadzhikistan; (15) Central Europe. SEA LEVELS. (16) Huon Pen; (17) Barbados. ICE CORE. (18) Camp Century. GLACIATION. (19) Hawaii.

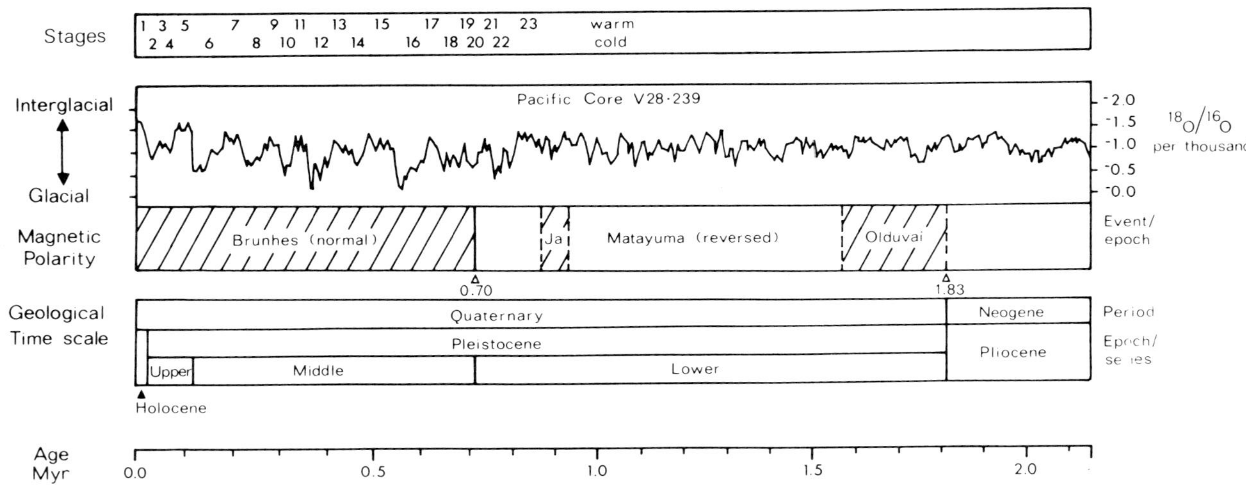


Fig. 2.2. Stratigraphic framework for the Quaternary, including the most complete sequence of Pleistocene glaciation presently available ($^{18}O/^{16}O$ record from core V28-239, after Shackleton and Opdyke, 1976).

during Pliocene times with the onset of northern hemisphere glaciation. There appears to have been a significant break in the tempo of climatic change *within* the Quaternary, however (Fig. 2.2), for after about 0·8 Myr, the amplitude of glacial–interglacial cycles increased significantly. It was almost certainly these later cycles which were responsible for the majority of the ice-sheet advances recorded on the continents (e.g. North European end moraines). The periodicity of climatic fluctuations also changed after 0·8 Myr, with major glacial–interglacial cycles occurring roughly every 100 000 years. Perhaps eight such cycles can be identified for the Middle and Upper Pleistocene; certainly more than the four originally proposed by Penck and Brückner.[6]

One interesting feature of the oxygen–isotope record for this period is that it is saw-toothed in shape (Broecker and van Donk, 1970), suggesting that ice-sheet build-up was gradual, but that ice-sheet decay was relatively rapid. A second feature is the close correlation with the temporal sequence predicted by the Croll-Milankovitch astronomical theory, from which it may be inferred that variations in the earth's orbit around the sun probably acted as the trigger for major glacial advances during the Middle and Upper Pleistocene (Hays *et al.*, 1976). The oxygen–isotope record also dispels the notion that glacial conditions were somehow atypical of the Quaternary as a whole. While it is certainly true that glacial *maxima* represent less than 20% of Middle and Upper Pleistocene time, interglacials occupied an even smaller proportion of the same period. For the bulk of the time, conditions were neither so warm as at present nor so cold as at the last glacial maximum.

If the deep ocean record provides the most complete framework for studying the Quaternary, then a key problem must be how to connect the oceanic record to the less complete but more accessible terrestrial one (Kukla, 1977). Land–sea correlation is perhaps most easily achieved using long, continuous sequences, whose climatic curves may be matched against those from deep sea cores. Correlation may also be achieved by using geochronologic dating methods alone, but for the pre-$^{14}C$ timescale this tends to be limited to the identification of specific time horizons, such as those provided by volcanic ash layers, rather than the provision of chronological sequences.

Among the terrestrial environments where successful land–sea correlation has been achieved are the following:

### 3.1. Lake Level Fluctuations

The lake bed, like the sea bed, is an environment in which sedimentation normally proceeds continuously, although accumulation rates are much higher than on the ocean floor, and sediment cores need to be an order of magnitude longer in order to encompass the same period of time. Consequently long sequences are presently limited to a few lake basins such as Lake George in Australia (Singh *et al.*, 1981) and Lake Bonneville in the United States (Eardley *et al.*, 1973) (Fig. 2.1). Both of these show multiple climatic cycles during the last million or so years, but the soil/shallow water/deepwater sediment cycles indicate that the pattern of water balance fluctuations during the Pleistocene was more complex than the sequence of northern hemisphere glaciation.[7]

### 3.2. Pollen Records

Vegetation history, as indicated by palynological analysis, has also been reconstructed from long, continuous lake sediment sequences, but the majority come from outlet lakes or peat bogs. The longest and most complete pollen record so far obtained comes from Tenaghi Philippon (or Lake Philippi) in Northern Greece (Wijmstra and van der Hammen, 1974), where a 120 m long core covering approximately the last 600 000 years has been under study. This includes half-a-dozen full glacial–interglacial cycles, which can easily be correlated with those from by deep ocean cores (see, for instance, Ruddiman and McIntyre, 1976: Fig. 14; or Street, 1980: Fig. 8.1). While the interglacial vegetation of northern Greece was dominated by oak and other tree species, glacial spectra are conspicuously lacking in arboreal pollen and instead indicate an open landscape dominated by *Artemisia*–Chenopodiaceae steppe. There was certainly no pluvial climate here during glacial stages of the Pleistocene. Other long pollen records also cover multiple glacial–interglacial cycles (e.g. Bogotá in the Colombian Andes; van der Hammen, 1974), but the correlation of pollen and deep-sea records has perhaps been most successful for the Upper Pleistocene (130–10 Kyr B.P.). Continuous pollen sequences from such disparate localities as California, France and north-eastern Australia closely parallel not only each other but also both oxygen isotope and pollen records from deep sea cores (Fig. 2.3) (e.g. Heusser and Shackleton, 1979). Land-based sequences in fact have a chronological resolution finer than that possible for deep sea cores, especially in the case of the pollen record from Grande Pile in northwest France (Woillard, 1978; Woillard and Mook, 1982), now effectively the type site for the Upper Pleistocene of North West Europe.

### 3.3. Loess Sequences

Thick deposits of this aeolian silt of glacial age, interbedded with interglacial soil horizons, are found in the non-glaciated mid-latitude belt of Eurasia. The thickest deposits occur in Tadzhikistan in Soviet Central Asia and in China,

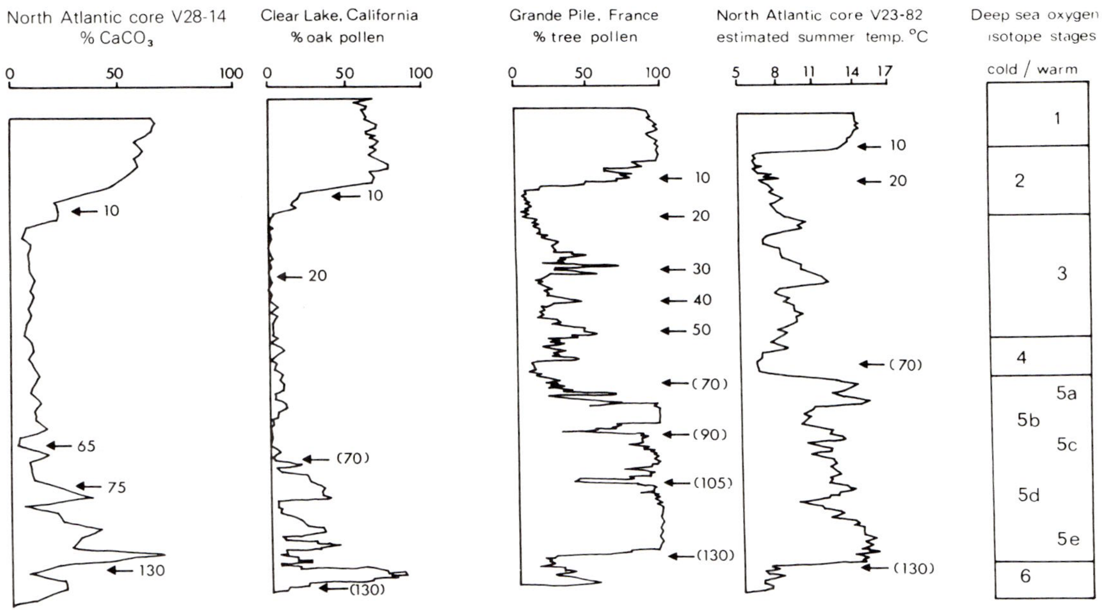


Fig. 2.3. Climatic curves from land and sea for the Upper Pleistocene and Holocene (after Emiliani, 1955; Kellogg, 1976; Adam *et al.*, 1981; Woillard and Mook, 1982; Sancetta *et al.*, 1973) (Ages in Kyr, extrapolated dates in brackets).

where 335 m of Pleistocene loess exists at a single section (Derbyshire, 1983). The Chinese loess sequence probably covers about 1·2 Myr, and includes two significant features identified in the deep ocean record: first evidence of multiple climatic cycles (more than 20 fossil soil horizons during the Middle and Upper Pleistocene), and secondly indications of a marked cooling at around 0·7–0·8 Myr. However, it is the rather more fragmentary loess deposits of Czechoslovakia and Hungary which have so far received greatest attention, and Kukla (1977) has successfully related these to the deep sea record.

### 3.4. Glaciated Environments

Direct correlation between terrestrial sedimentary and geomorphic evidence for ice-sheet advance and retreat, and ocean core sequences remains extremely poor. Apart from a smattering of inconclusive radiometric dates (e.g. Boellstorff, 1973), only the last (Wisconsinan/Weichselian/Devensian) glaciation has a sound chronologic basis (Denton and Hughes, 1981). Ironically, the best dated sequence of Pleistocene glaciation at present comes from Mauna Kea on the tropical island of Hawaii (Porter 1979)!

### 3.5. Sea Level Change

As water became locked up in Pleistocene ice sheets, so the volume of the oceans was reduced and sea level fell. The eustatic component of the sea level record must have a one-to-one correlation with changing global ice volume, and thus ought to correlate also with the oxygen–iosotope signal from the oceans. Sea level change may therefore be used as an independent test of the validity of the $^{18}O/^{16}O$ record. On the other hand, because the sea rose to similar elevations during separate interglacials, older shorelines may be disturbed by younger ones and this can lead to confusion between interglacial beaches of different ages. In coasts experiencing steady tectonic uplift, on the other hand, a series of stepped shoreline terraces have formed, with the highest being the oldest. Subsequent correction using an uplift curve provides a eustatic sea level curve which may be compared to oxygen–isotope curves from the deep sea bed. Such conditions are found in the Huon peninsula of New Guinea, where Pleistocene coral reefs and deltas form terraces which offlap downslope (Chappell, 1974). U-series dating on these terraces has given ages of *ca.* 125 Kyr, 105 Kyr and 82 Kyr for the three high sea levels associated with oxygen isotope stage 5 (Bloom *et al.*, 1974). Confirmation of the link between the last interglacial complex on land and that identified in oceanic sediment cores has been provided by oxygen–isotope analysis of molluscs from uplifted coral terraces in Barbados (Shackleton and Matthews, 1977).

### 3.6. Non-Marine Oxygen–Isotope Records

As just noted, oxygen–isotope analysis is not restricted to microfossils from deep sea cores, and such results as can be obtained from terrestrial environments ought to correlate closely with those from the oceans. One such environment is provided by the ice sheets themselves which have been gradually accumulating over thousands of years. Ice cores from Camp Century in Greenland (Dansgaard *et al.*, 1971) and from Byrd Station and Vostok in Antarctica (Epstein *et al.*, 1970; Robin, 1977) have records reaching back to oxygen isotope stage 5. These sequences agree well with the oceanic record, but it should be borne in mind that rates of ice accumulation are not constant, and that independent dating control beyond 50 Kyr is poor for ice cores. Another suitable terrestrial environment is a cave where oxygen–isotope analysis of speleothems (e.g. stalagmites) has been successfully carried out (Harmon *et al.*, 1978), often in conjunction with U-series dating (e.g. Gascoyne *et al.*, 1981).

One main conclusion which may be drawn from this brief review is that land–sea correlation has so far been relatively unsuccessful for those terrestrial environments traditionally used to build up glacial–interglacial sequences. The British and North European Pleistocene sequences, for instance (Table 2.1), can only be related with confidence to the oceanic $^{18}O/^{16}O$ record for the Upper Pleistocene. The last (Devensian/Weichselian) glaciation represents oxygenisotope stages, 2, 3 and 4, and the last (Ipswichian/Eemian) interglacial matches stage 5e, dating to *ca.* 125 Kyr. The relation between earlier European glacials and interglacials and the new global climatostratigraphic framework, however, is much less certain, and these older stages are best considered as chronologically "free-floating".

## 4. CLIMATES AND ENVIRONMENTS IN SPACE (LAST 25 000 YEARS)

Our revised knowledge of glacial–interglacial climatic cycles has tended to capture most of the headlines in the Quaternary science revolution (cf. Imbrie and Imbrie, 1979), and implications of the radiocarbon revolution have generally been relegated to back page news. Without independent chronological control, however, it has usually been assumed that Pleistocene climatic or archaeological changes have been synchronous and unidirectional the world over, although this is patently not the case. With good dating control, on the other hand, it becomes possible to analyse isochronous changes such as vegetation colonization and ice sheet growth and decay.

Although a number of other isotopic dating techniques, such as Uranium-series and Potassium-Argon, have been used successfully for the Pleistocene, it is so far just $^{14}C$ which has been applied widely enough for palaeo-environmental data to be completely freed from its former role as a provider of a chronology. Consequently it is only for the last 25 000 years that we have a sufficiently large amount of independently dated evidence to allow global patterns of climatic and ecological change to be established. We may in fact tentatively use the period covered by $^{14}C$ dating as a model for longer Pleistocene time-spans, although this procedure does raise a number of methodological questions. First, we may ask to what extent the last glacial–deglacial hemicycle is representative of earlier ones? Long deep sea and terrestrial records would suggest that its duration, intensity, and consequences for flora and fauna, were fairly typical for cycles during the last 0·8 Myr.

Secondly, does the "saw-toothed" nature of the oxygen isotope record not mean that climatic extremes, whether glacial or interglacial, have been over-represented during the last 25 Kyr, compared with the more intermediate conditions typical of the earlier part of the last glaciation? This is certainly a valid criticism, especially in such environments as the humid tropics of northern Australia, where the postglacial vegetation was complex vine forest, that around the time of the last glacial maximum (26–10 Kyr) was sclerophyll woodland, but during the early and middle part of the last glaciation (79–26 Kyr) was neither of these but was instead an araucarian vine forest (Kershaw, 1978). For much of Europe, on the other hand, vegetation cover since the last interglacial has tended to alternate between either full forest (whether mixed deciduous or coniferous) and open vegetation (steppe, tundra, etc.) (Frenzel, 1973). Finally, there remains the question of increasing human impact on ecosystems during the last 25 Kyr, as dramatically illustrated by the Late Pleistocene mammalian extinctions (Martin and Klein, 1982). Because of human impact the postglacial vegetation sequences in certain parts of the world are atypical of those of earlier interglacials.

There is no better example of the impact of $^{14}C$ dating on Quaternary research than our altered conception of Late Pleistocene and Holocene palaeoenvironments in the tropics (Livingstone, 1975; Street, 1981). Prior to 1960 it was widely believed that high latitude glacials were accompanied by low- and mid-latitude pluvials, or wet phases. One corollary of this was that the equatorial rainforests were thought to have been at least as extensive during the Pleistocene as they are today. The rainforest consequently came to be viewed as a formation of great longevity and high environmental stability (Whitmore *et al.*, 1982). One of the main lines of evidence for former pluvials came from high level shorelines around non-outlet lakes. However, when shorelines from a number of East African lake basins were $^{14}C$ dated during the 1960's, they all turned out to be of early Holocene age (Butzer *et al.*, 1972)! An increase in rainfall between 10 and 5 Kyr B.P. has now been confirmed from several parts of the tropics, including Central and South America, India, and Australia (Street–Perrott *et al.*, in press; Swain *et al.*, 1983; Bowler *et al.*, 1976); this also appears to have been true for previous interglacials (e.g. Gaven *et al.*, 1982). These wet conditions transformed deserts like the Sahara into savanna grasslands supporting perennial rivers and freshwater lakes.

But what, then, was the climate of the tropics at the last glacial maximum, *ca.* 18 Kyr B.P.?[8] Although some mid-latitude areas such as the South-West United States and the southern shore of the Mediterranean, show signs of an increase in effective precipitation at this time, most low-latitude regions were relatively dry, and this aridity intensified as the northern ice sheets began to wane. Evidence from the former extent of sand dunes (Sarnthein, 1978) and from lake level studies (Street and Grove, 1979) indicates that between 16 and 13 Kyr B.P. the climate of the tropics was both drier and cooler than at the present-day. As might be expected, this had far-reaching effect on vegetation patterns.

Tropical palynology is less developed than that in the temperate zone, and many of the existing pollen diagrams come from montane environments where the altitudinally-zoned vegetation has varied primarily in response to temperature rather than to moisture fluctuations (Flenley, 1979b). Nonetheless a significant corpus of material now exists (Flenley, 1979a; Hamilton, 1982; van der Hammen, 1974) which demonstrates that two of the world's three great rainforests, those in the Amazon and Zaire (Congo) basins, were decimated and reduced to isolated refugia during the last glaciation. One startling piece of evidence for this is the existence of fossil sand dunes beneath parts of both rainforests, indicating that there were active sand deserts here during the Pleistocene (Tricart, 1974) (Fig. 2.4).

The one part of the humid tropics to escape this great natural deforestation appears to have been South-East Asia (Maloney, 1980). This was at least partly due to the fact that this area experienced dramatic environmental changes of a different kind, those associated with sea level fluctuations. The glacial fall in sea level created the subcontinent of Sunda out of the Indonesian archipeligo (Tjia, 1980), and nourished by moisture from adjacent oceans, the Malesian rainforest was able to expand onto the new land and avoid deforestation.

The rise in sea level after the last glaciation has commanded a good deal of attention in Quaternary research. Much of this has been aimed at establishing the relative importance of eustasy (changes in ocean water volume), isostasy (compression loading and unloading by ice, water or sediment), and geoidal changes (Bloom, 1971; Mörner, 1976), or at producing a master curve for the

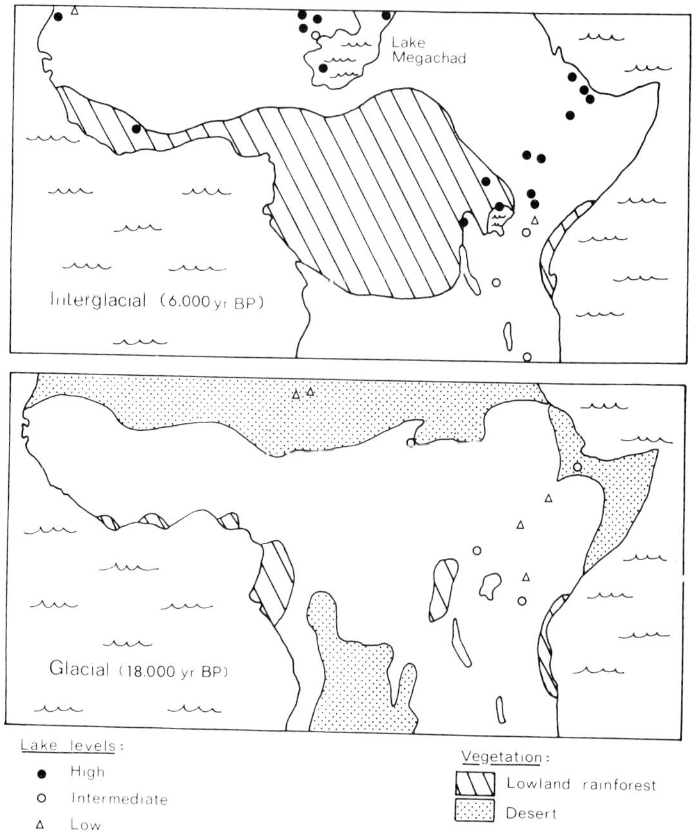


Fig. 2.4. Tropical African vegetation and lake levels at glacial and interglacial maxima (after Hamilton, 1976; van Zinderen Bakker, 1976; Sarnthein, 1978; Street-Perrott and Roberts, 1983).

marine transgression (e.g. Fairbridge, 1961). More significant in terms of palaeoecological or archaeological reconstruction is the fact that the empirical evidence collected in order to test these hypotheses has provided a worldwide data base of the position of former coastlines during the last 25 Kyr (e.g. Hopkins, 1973; Shi and Wang, 1979), and thereby also the rate of inundation and land-loss as the sea level rose (Cronin, 1983). Where offshore slopes are steep, as around most of Africa, the lowering of sea level at the last glacial maximum changed the configuration of the coastline only slightly. But in areas with extensive continental shelves, the fall of between 75 and 175 m created land bridges and new subcontinents. The linking of the New and Old Worlds via Beringia (now the Bering Strait) and of the British Isles and the European mainland, allowed the free passage of fauna, including hominids, during glacial phases of the Pleistocene (see Turner, this volume).

## 5. SOME ECOLOGICAL IMPLICATIONS

The notion that the forests of the temperate zone were massively disrupted during the Pleistocene but that those in the tropics effectively remained stable, is now known to be a gross oversimplification. It is known that the boreal, mixed deciduous and Mediterranean woodlands of Europe were reduced to isolated southern refugia during glacial stages, to be replaced by steppic and tundra-like associations (Frenzel, 1973; van der Hammen *et al.*, 1971). But in North America and East Asia the forest belts were simply displaced southwards; Japan, for instance, supported a continuous cover of mainly coniferous forest right through the last glaciation (Tsukada, 1983). Adjacent to most of the Laurentide ice sheet there was only the narrowest zone of tundra, and one of the other North American vegetation formations to suffer most during the late Wisconsin glacial was the prairie (Delcourt and Delcourt, 1983).

The pattern of vegetation change in the tropics was equally complex. While the Amazon and Zaire rainforests shrank to isolated pockets as a result of glacial aridity, that in South-East Asia remained more or less intact (Flenley, 1979a; Maloney, 1980). The high species diversity of the equatorial rainforest can therefore no longer solely be ascribed to environmental stability (cf. the palaeoclimatic model outlined in Chapter 1). On the contrary, research into, on the one hand, species distribution patterns and, on the other, Pleistocene environments, has converged towards the conclusion that differentiation of tropical elements may have been *encouraged* by alternate fragmentation and re-expansion of the rainforest (Simpson and Haffer, 1978). According to this hypothesis, any extinctions resulting from reduction in habitat size during fragmentation would have been offset by new speciation in the remaining, isolated forest communities. The resulting pattern is one of centres of endemism surrounded by zones of secondary contact. The latter were formed after the forest refugia were re-united, and they are characterized by more normal arrays of hybrids.

Disjunct populations of South American lizards, butterflies, birds and some flora, have been identified and mapped (Simpson and Haffer, 1978; Prance, 1978), and the suggested centres of endemism may be compared with the forest refuge locations independently indicated by palaeoenvironmental data. The resulting spatial correlation (Fig. 2.5) is suggestive rather than conclusive, but it should be remembered that the palaeoenvironmental data in question comes mainly from geomorphology (e.g. Tricart, 1974) and that palynological evidence for specific vegetation changes in the Amazonian lowlands remains "ridiculously slight" (Flenley, 1979a, p 71). This hypothesis also receives support from the existence of savanna pockets within the modern rainforest (Eden, 1974). Although maintained today by fire burning, floral and faunal affinities between separate savannas suggest that they are relics of a once continuous spread.

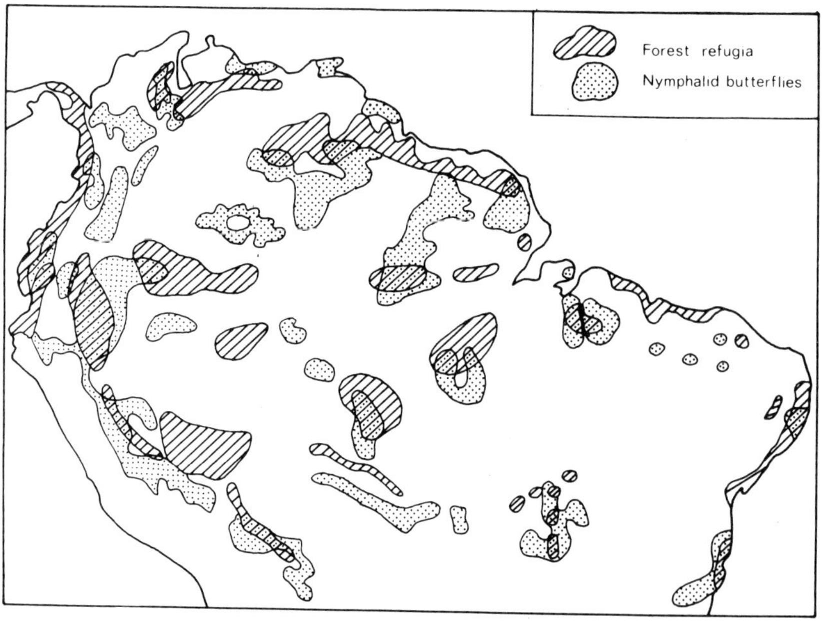


Fig. 2.5. Centres of endemism among South American nymphalid butterflies (dot shaded) and Pleistocene forest refugia (line shaded) as indicated by geomorphic evidence (after Brown and Ab'Sáber, 1979; Simpson and Haffer, 1978; following Colinvaux, 1979; Street, 1981).

In tropical Africa too, palaeoecological evidence indicates that savanna vegetation was more, not less, extensive during the Pleistocene (Hamilton, 1982), effectively putting paid to the idea that savanna grassland only exists as a vegetation type by dint of culturally-induced firing. The differentiation of African forest mammals, fishes, amphibians and reptiles can be related to glacial fragmentation of a now largely continuous rainforest (Moreau, 1969; Laurent, 1973) (Fig. 2.4). Indeed from the distribution of West African vertebrates Moreau (1969) correctly predicted that there had been a break in the forest at Baoulé within the last 20 Kyr, contrary to the palaeoclimatic data then available. In one respect, however, Africa is significantly different from South America, for its forest flora is depauperate relative to those in other humid tropical rainforests despite having experienced the same fragmentation during the Pleistocene (Flenley, 1979a, p 54). Other factors appear to have been at work here, discouraging speciation or increasing the extinction rate.

The South-East Asian rainforest was also repeatedly ruptured and re-formed during the Quaternary, not because of glacial aridity, but because high interglacial sea levels created literally, forest islands. Interestingly, a number of savanna-adapted mammals which are present on the Asian mainland, such as the horse and some bovids, are absent from the drier parts of Indonesia despite the existence of a land bridge at times of low sea level (Medway, 1972). This suggests that another barrier, presumably lowland rainforest, prevented faunal migration at such times. It may also be worthy of note that known Pleistocene forest refugia *outside* the tropics, such as the Caucasus (Gogichaishvili, 1982) are today characterized by exceptionally rich and diverse floras.

Thus almost all biotic communities had to adapt in one way or another to Pleistocene climatic change, whether the consequences were displacement, isolation or complete disintegration. The extent to which Pleistocene ecosystems were *qualitively* different from those of today is, and doubtless will remain a vexed one. On the one hand, it has been proposed (Delcourt *et al.*, 1983, p 168) that the physiognomic characteristics of many of North America's major vegetation types were maintained over the past 20 000 years, with only their distribution and areal extent being altered. On the other, pollen evidence indicates that other formations, such as the upper montane forests of New Guinea (Flenley, 1979a, p 98), were completely destroyed during glacial phases and recreated from scratch during interglacials.

A variety of ecosystem responses might perhaps be predicted on *a piori* grounds. Complex tropical rainforest communities can hardly have persisted throughout the Pleistocene unless at least some areas retained the present pattern of minimal seasonal climatic variation. Tundra vegetational communities are equally strongly adapted to the highly seasonal insolation regime of very high latitudes, but these regions—at least in North America and Europe—were occupied by ice sheets during glacial phases, so that the tundra cannot have been preserved there in its present form. The biomass of the plant communities which occupied the European periglacial zone in fact appears to have been quite different from that of the modern tundra. Furthermore Frenzel (1973) has calculated that the number of browsing mammals supported by herb-steppe communities was less than a tenth of that supported by grass-steppes during the Weichselian glaciation.[9]

One feature to have emerged from the palynological record for the last glacial–deglacial hemicycle, is that species behaviour was often strongly individualistic at times of dynamic vegetation change. During the post-glacial forest re-advance in Europe (Huntley and Birks, 1983), North America (Davis, 1976) and parts of the tropics (Flenley, 1979a, p 99) the trees migrated in as individual species, not as whole communities. This would be less significant if the Pleistocene had been characterized by long periods of climatic stability, as Penck and Brückner believed. In fact, the Pleistocene was a period of almost constant climatic change, with climatically stable phases rarely lasting more than 10 000 years. Yet many vegetation types (such as the lowland African rainforest) took up to 5000 years to achieve a balanced composition during the present interglacial (Hamilton, 1982).

In consequence, ecological communities were often unable to keep up with the pace of Pleistocene climatic change, especially during stadial and interstadial phases, whose short durations allowed only selected plants and animals time to migrate. A good example of this is the Upton Warren interstadial (*ca.* 43 Kyr B.P.) when summer temperatures in lowland England were higher than at present, but for which pollen evidence indicates a treeless landscape (Coope, 1977), presumably because the trees did not have time to migrate in during the brief climatic amelioration. Clearly, this mid-Devensian ecological community can have no modern analogue (see Chapter 9, this volume). The problem is exacerbated by the abruptness of many Pleistocene climatic and environmental changes (Flohn, 1979). Coleoptera indicate a rise in summer temperatures of more than 5°C in less than a century at 13 Kyr B.P. and again at 10 Kyr B.P. (Coope, 1977), while pollen data from Grande Pile demonstrate that the change from temperate to boreal forest at the end of the last (Eemian) interglacial took place within 150 years (Woillard, 1979). Under these circumstances, the potential for species extinctions is obviously considerable, and resilience to environmental stress must have been a prime criterion for survival in the Pleistocene.

## 6. SOME IMPLICATIONS FOR HUMAN EVOLUTION

Plio-Pleistocene hominids may be regarded as large mammals adapted to tropical and sub-tropical parkland environments (see Chapter 4, this volume). Whether or not Africa experienced progressive desiccation during the late Neogene (cf. Butzer, 1978, p 211), the contemporary rise of the graminae (Harris, 1980, p 21; Butzer, 1982, p 287) means that this period witnessed the creation of extensive grassland savannas in that continent. For early hominid populations a major advantage of the savanna biotope would have been its high herbivore biomass (Foley, 1982), but the shortage of drinking water appears to have restricted dry season occupation to riverine and lakeside locations (Harris, 1982, p 32; Butzer, 1978). It is known that fluctuations in the climate and vegetation of East Africa did occur within the period 5 to 1 Myr (Bonnefille 1981), but it is difficult to establish their magnitude or context within the wider framework of Plio-Pleistocene climatic changes (Butzer and Cooke, 1982, p 50).

On the other hand, the global climatic record for the Quaternary indicates intensified glacial–interglacial cycles after 0·8–0·9 Myr. When coupled with evidence for later Quaternary environmental change in the tropics, it is possible to predict that there would have been a significant shift in the ecological stability of Africa's intertropical zone towards the end of the Lower Pleistocene. After this time the extent and location of the savanna zone would have altered over time scales between $10^3$–$10^5$ yr. At glacial maxima savanna would have occupied much of what is now rainforest, while during interglacials it moved northward to reach beyond the tropic of Cancer (Fig. 2.4). Although never fragmented into isolated pockets as it was in South-East Asia, the African savanna was periodically split into separate regions which were later reunited.

This cyclic alternation between savanna, forest and desert presented early hominids with a number of choices. They could have responded to shifts in their habitat in the same way as other savanna mammals by migrating along with the vegetation belts. Or they could have adapted to new environments, switching their dietary emphasis from animal to plant resources. Finally there was the possibility of moving beyond the limits of tropics altogether, into habitats previously unoccupied by human populations. In practice it seems that all three strategies were utilized, with the last being especially important in terms of hominid radiation northward out of Africa. The Sahara would have acted in this regard as a pump, sucking in population during wetter savanna phases and forcing it out towards the Mediterranean as desiccation subsequently set in.

The postulated cultural adaptation to a shift in the tempo of environmental change around 0·8–0·9 Myr is consistent with existing archaeological evidence, and in particular with the spread of the Acheulian complex out of Africa at the end of the Lower Pleistocene (Clark, 1980, p 44). From this time on *Homo erectus* makes a first appearance in South-West Asia and Mediterranean Europe and is also found in a much wider range of habitats within Africa. In these extra-tropical environments *Homo erectus* encountered not only a more severe temperature regime but, more importantly, had to cope with the continual and extreme environmental fluctuations which characterized the European Middle and Upper Pleistocene. In inter-tropical Africa early hominids had been able to occupy the savanna ecotype continuously even if the savanna shifted its location through time, but in Europe, the vegetation zones were not simply displaced but were reduced to isolated refugia or eliminated completely as the climate changed. Consequently, new economic strategies had to be devised every time mixed deciduous woodland replaced steppe-tundra or vice-versa. Such environmental pressures must have favoured the better adapted Acheulian populations and helped to speed the pace of cultural development. This process is well illustrated in South-East Asia where repeated loss of land as a result of sea level change eventually prompted an adaptive strategy which included the use of water-borne craft. This in turn led to the crossing of open water at least 80–100 km wide and the Upper Pleistocene colonization of Australasia (Birdsell, 1977).

## 7. CONCLUSIONS

The recent revolution in Quaternary studies has discredited both Penck and Brückner's Alpine model of 4-fold glaciaton and the inferred correlation between high-latitude glacials and low-latitude pluvials. The new paradigm has as its basis, long, continuous sequences and radiometric dating. Global climatostratigraphy is now derived from deep-sea sediment cores, and is divided into warmer and colder stages on the basis of oxygen–isotope analysis. The deep-sea record indicates that the major break in the tempo of climatic change occurred, not at the beginning of the Pleistocene (1·83 Myr), but around 0·9–0·8 Myr, after which glacial–interglacial cycles became longer and more intense. Twenty-two oxygen–isotope stages, representing 9–10 full glacial–interglacial cycles, have been identified for this time-period. A key problem in Quaternary research has become how to relate deep-sea sequences to more accessible terrestrial ones. Land-sea correlation has so far been most successful for loess, pollen, oxygen–isotope, lake- and sea-level stratigraphies. The direct continental record of European glaciation has only been confidently related to deep-sea cores back as far as the last (Eemian/Ipswichian) interglacial, *ca.* 125 Kyr.

Contrary to previous belief, the tropics were not zones of high environmental stability during the Middle and Upper Pleistocene. $^{14}C$ dating of high lake levels, particularly in East Africa and the Sahara, has shown that the last wet (pluvial) phase in the tropics took place during the first half of the *post*-glacial period. At and after the time of the last glacial maximum, most of the tropics experienced climatic aridity which decimated the South American and African rainforests. It is therefore likely that the high diversity of tropical rainforest communities was a result of speciation in glacial refugia rather than of Pleistocene environmental stability. Some other ecological communities were also reduced to isolated refugia or even eliminated altogether during the Quaternary, but many—including the African savannas—were simply displaced geographically or altitudinally. In addition, the speed and frequency of Pleistocene climatic change often did not permit complete ecological successions. Consequently, adaptation to repeated environmental stress was a key factor in community evolution during the Pleistocene. This may have given early hominids a competitive advantage over other large mammals, especially during the period of more intense climatic fluctuations after 0·9–0·8 Myr. It is also possible that the main radiation of hominids out of Africa was related to the global environmental shift around this time.

## NOTES

1. For an eminently readable account of this "revolution" by one who was involved in it, see Imbrie and Imbrie (1979). For details of the various methods of deep sea core analysis see Shackleton (1975).
2. One spin-off of these methodological changes has been that the major role formerly played by Palaeolithic archaeology in Quaternary science as a whole has now been made redundant. Archaeological artifacts were previously used primarily as an aid to dating palaeoenvironmental sequences (or vice-versa) (e.g. Roe, 1981; Wymer, 1981). Now relieved of this role by the advent of radiometric dating methods and the creation of a global climatostratigraphic framework, archaeological and environmental evidence is instead being synthesized to produce integrated palaeoecological reconstructions (Butzer, 1982). A good example of a project where this integrated, multidisciplinary approach has been adopted is the Lake Mungo hominid site in Australia (Bowler *et al.*, 1970).
3. For the purposes of this discussion, long sequences are considered to be those spanning at least one full glacial–interglacial cycle, or approximately 100 Kyr, at a single site.
4. This may be illustrated by applying the mean sampling interval of 1 per 20 000 years to the last glacial–deglacial hemicycle: samples taken at 30 000 yr B.P. and again at 10 000 yr B.P., would miss out the last glacial maximum altogether!
5. The Plio-Pleistocene boundary is conventionally placed at 1·83 Myr (Nikiforova, 1982, p 22), and is defined by the base of: (a) the Olduvai palaeomagnetic event; (b) the *Globorotalia truncatulinoides* zone in oceanic sediment cores; (c) the marine Calabrian beds of southern Italy.
6. Twenty-two oxygen–isotope stages are identified for the last 0·8 Myr, cold stages having even numbers and warm ones odd numbers, but not all of these reach the status of full glacials or interglacials. Stage 3, for instance, represents the interstadial complex during the middle of the last glaciation.
7. The African deep drilling project (Shackleton, 1980) is likely to add to the number of long lacustrine sequences during the next few years.
8. The date used for the last glacial maximum (technically the last global ice

and snow volume maximum) is that identified by the CLIMAP Project Members (1976).

9. Frenzel's calculations suggested 0·3–0·4 animals per 100 hectares of herb-steppe, and 5 animals per 100 hectares on grass-steppe. These figures were based on the nearest living relatives of mammoth, reindeer, Taiga-antelope, horse, ass, deer, bison and rhinoceros.

## REFERENCES

Adam, D. P., Sims, J. D. and Throckmorton, C. K. (1981). 130 000-yr continuous pollen record from Clear Lake, Lake County, California. *Geology* **9**: 373–377.

Berger, R. and Suess, H. E. (eds) (1979). *Radiocarbon Dating. (Proceedings of the Ninth International Conference).* Berkeley: University of California Press.

Birdsell, J. B. (1977). The recalibration of a paradigm for the peopling of Greater Australia. In Allen, J., Golson, J. and Jones, R. (eds) *Sunda and Sahul*. London: Academic Press, pp. 113–167.

Bloom, A. L. (1971). Glacial-eustatic and isostatic controls of sea level since the last Glaciation. In Turekian, K. K. (ed) *The Late Cenozoic Glacial Ages*. New Haven: Yale University Press, pp. 355–379.

Bloom, A. L., Broecker, W. S., Chappell, J. M. A., Matthews, R. K. and Mesolella, K. J. (1974). Quaternary sea-level fluctuations on a tectonic coast. *Quat. Res.* **4**: 185–205.

Boellstorff, J. (1973). Correlating and dating some older "Pleistocene" tills in the midcontinent. *Geo. Soc. Am. Abstr.* **5**: 301.

Bonnefille, R. (1981). Palynological evidence for continental climatic changes in East Africa at 2·5 and 1·7 M.Y. *Terra Cognita* Special Issue: 102–103.

Bowen, D. Q. (1978). *Quaternary Geology*. Oxford: Pergamon Press.

Bowler, J. M., Jones, R., Allen, H. and Thorne, A. G. (1970). Pleistocene human remains from Australia: a living site and human cremation from Lake Mungo, New South Wales. *World Archaeol.* **2**: 39–60.

Bowler, J. M., Hope, G. S., Jennings, J. N., Singh, G. and Walker, R. K. (1976). Late Quaternary climates of Australia and New Guinea. *Quat. Res.* **6**: 359–394.

Broecker, W. S. and van Donk, J. (1970). Insolation changes, ice volumes and the 0–18 record in deep sea cores. *Rev. Geophys. Space Phys.* 168–198.

Brown, K. J. and Ab'Sáber, A. N. (1979). Ice-age forest refuges and evolution in the neotropics: correlation of palaeoclimatological, geomorphological and pedological data with modern biological endemism. *Palaeoclimas* 5 (Instituto de Geografia, Universidade de São Paulo).

Butzer, K. W. (1978). Geographical perspectives on early hominid evolution. In Jolly, C. (ed) *Early Hominids of Africa*. London: Duckworth, pp. 191–217.

Butzer, K. W. (1982). *Archaeology as Human Ecology*. Cambridge: Cambridge University Press.

Butzer, K. W. and Cooke, H. B. S. (1982). The palaeo-ecology of the African continent: the physical environment of Africa from the earliest geological to Later Stone Age times. In Clark, J. D. (ed) *The Cambridge History of Africa, volume 1.* Cambridge: Cambridge University Press, pp. 1–69.

Butzer, K. W. Isaac, G. L., Richardson, J. L. and Washbourn-Kamau, C. K. (1972). Radiocarbon dating of East African lake levels. *Science* **175**: 1069–1076.

CLIMAP project members (1976). The surface of the ice-age earth. *Science* **191**: 1131–1137.

Chappell, J. M. A. 1974. Geology of coral terraces, Huon Peninsula, New Guinea: a study of Quaternary tectonic movements and sea level changes. *Bull. Geol. Soc. Am.* **85**: 553–570.

Clark, J. D. (1980). Early human occupation of African savanna environments. In Harris, D. R. (ed) *Human Ecology in Savanna Environments*. London: Academic Press, pp. 41–71.

Cline, R. M. and Hays, J. D. (eds) (1976). *Investigation of Late Quaternary Palaeoceanography and Palaeoclimatology*. Geological Society of America Memoir 145.

Colinvaux, P. (1979). Ice Age Amazon. *Nature* **278**: 399–400.

Colinvaux, P. A. and Schofield, E. K. (1976). Historical ecology in the Galapagos islands. *J. Ecol.* **64**: 989–1028.

Coope, G. R. (1977). Fossil coleopteran assemblages as sensitive indicators of climatic changes during the Devensian (Last) cold stage. *Philosophical Transactions of the Royal Society of London, Series B* **280**: 313–340.

Cronin, T. M. (1983). Rapid sea level and climate change: evidence from continental and island margins. *Quat. Sci. Rev.* **1**: 177–214.

Dansgaard, W. S., Johnson, S. J., Clausen, H. B. and Langway, C. C. (1971). Climatic record revealed by the Camp Century Ice Core. In Turekian, K. K. (ed). *The Late Cenozoic Glacial Ages*. New Haven: Yale University Press, pp. 37–56.

Davis, M. B. (1976). Pleistocene biogeography of temperate deciduous forests. *Geosci. and Man* **13**: 13–26.

Davis, M. B. (1981). Quaternary history and the stability of forest communities. In West, D. C., Shugast, H. H. and Botlein, D. B. (eds) *Forest succession, Concepts and Application*. New York: Springer Verlag, pp. 132–153.

Delcourt, P. A. and H. R. (1983). Late Quaternary vegetational dynamics and community stability reconsidered. *Quat. Res.* **19**: 265–271.

Delcourt, H. R., Delcourt, P. A. and Webb, T. (1983). Dynamic plant ecology: the spectrum of vegetational change in space and time. *Quat. Sci. Rev.* **1**: 153–175.

Denton, G. H. and Hughes, T. J. (eds) (1981) *The Last Great Ice Sheets*. New York: Wiley.

Derbyshire, E. (1983) On the morphology, sediments and origin of the Loess Plateau of central China. In: Gardner, R. and Scoging, H. (eds) *Mega-geomorphology*. Oxford: Oxford University Press.

Eardley, A. J. *et al.* (1973). Lake cycles in the Bonneville Basin, Utah. *Bull. Geol. Soc. Am.* **84**: 211–216.

Eden, M. J. (1974). Palaeoclimatic influences and the development of savannas in southern Venezuela. *Jo. Biogeog.* **1**: 95–109.

Emiliani, C. (1955). Pleistocene temperatures. *J. Geol.* **63**: 538–578.

Epstein, S., Sharp, E. K. and Gow, A. J. (1970). Antarctic ice sheet isotope analysis of Byrd Station cores and interhemispheric climatic implications. *Science* **168**: 1570–1572.

Fairbridge, R. W. (1961). Eustatic changes in sea level. In Ahrens, L. H. Press, F., Rankama, K. and Runcorn, S. K. (eds) *Physics and Chemistry of the Earth* 4. Oxford: Pergamon Press, pp. 99–185.

Flenley, J. R. (1979a). *The Equatorial Rainforest: A Geological History*. London: Butterworth.

Flenley, J. R. (1979b). The Late Quaternary vegetational history of the equatorial mountains. *Prog. Phys. Geog.* **3**: 488–509.

Flohn, H. (1979). On time scales and causes of abrupt palaeoclimatic events. *Quat. Res.* **12**: 135–149.
Foley, R. (1982). A reconsideration of the role of predation on large mammals in tropical hunter-gatherer adaptation. *Man n.s.* **17**: 383–402.
Frenzel, B. (1973). *Climatic fluctuations of the Ice Age*. Cleveland.
Gascoyne, M., Currant, A. P. and Lord, T. C. (1981). Ipswichian fauna of Victoria Cave and the marine palaeoclimatic record. *Nature* **294**: 652–654.
Gaven, C., Hilaire-Marcel, C. and Petit-Maire, N. (1981). A Pleistocene lacustrine episode in southeastern Libya. *Nature* **290**: 131–133.
Gogichaishvili, L. K. (1982). Vegetational and climatic history of the western part of the Kura river basin (Georgia). In Bintliff, J. L. and van Zeist, W. (eds) *Palaeoclimates, Palaeoenvironments and Human Communities in the Eastern Mediterranean Region in Later Prehistory*. Oxford: B.A.R. Int. Series 133, pp. 325–341.
Hamilton, A. C. (1976). The significance of patterns of distribution shown by forest plants and animals in tropical Africa for the reconstruction of Upper Pleistocene palaeoenvironments: a review. *Palaeoecol. Africa* **9**: 63–97.
Hamilton, A. C. (1982). *Environmental History of East Africa*. London: Academic Press.
Harmon, R. S., Thompson, P., Schwarcz, H. P. and Ford, D. C. (1978). Late Pleistocene palaeoclimates of North America as inferred from stable isotope studies of speleothems. *Quat. Res.* **9**: 54–70.
Harris, D. R. (ed) (1980). *Human Ecology in Savanna Environments*. London: Academic Press.
Hays, J. D., Imbrie, J. and Shackleton, N. J. (1976). Variations in the earth's orbit: pacemaker of the ice ages. *Science* **194**: 1121–1132.
Heusser, L. E. and Shackleton, N. C. (1979). Direct marine-continental correlation: 150 Ka oxygen isotope pollen record from North Pacific. *Science* **204**: 837–839.
Hopkins, D. M. (1973). Sea level history in Beringia during the past 250,000 years. *Quat. Res.* **3**: 520–540.
Huntley, B. and Birks, H. J. B. 1983. *An atlas of past and present pollen maps for Europe: 0–13,000 years ago*. Cambridge: Cambridge University Press.
Imbrie, J. and Imbrie, K. P. (1979). *Ice Ages: Solving the Mystery*. London: Macmillan.
Kellogg, T. B. (1976). Late Quaternary climatic changes: evidence from deep-sea cores of Norwegian and Greenland Seas. In Cline, R. M. and Hays, J. D. (eds) *Investigation of Late Quaternary Palaeoceanography and Palaeoclimatology*. Geological Society of America Memoir 145, pp. 77–110.
Kershaw, A. P. (1978). Record of last interglacial-glacial cycle from north-eastern Queensland. *Nature* **272**: 159–161.
Kuhn, T. S. (1962). *The Structure of Scientific Revolutions*. Chicago: University of Chicago.
Kukla, G. J. (1977). Pleistocene land-sea correlations. 1. Europe. *Earth-Sci. Rev.* **13**: 307–374.
Laurent, R. F. (1973). A parallel survey of equatorial amphibians and reptiles in Africa and South America. In Meggers, B. J., Ayensu, E. S. and Duckworth, W. D. (eds) *Tropical Forest Ecosystems in Africa and South America: A Comparative Review*. Washington: Smithsonian Institution Press, pp. 259–266.
Leakey, L. S. B. (ed) (1952). *First Pan-African Congress on Prehistory*. Oxford.
Libby, W. F. (1955). *Radiocarbon Dating*. (2nd edn.) Chicago: University of Chicago Press.
Livingstone, D. A. (1975). Late Quaternary climatic changes in Africa. *Ann. Rev. Ecol. Systemat.* **6**: 249–280.

Maloney, B. K. (1980). Pollen analytical evidence for early forest clearance in North Sumatra. *Nature* **287**: 324–326.
Martin, P. S. and Klein, R. G. (eds) (1982). *Pleistocene Extinctions*. Tucson: University of Arizona Press.
Medway, Lord (1972). The Quaternary mammals of Malesia: a review. In: Ashton, P. and M. (eds) *The Quaternary Era in Malesia*. Unviersity of Hull, Department of Geography, pp. 63–83.
Mitchell, G. F., Penny, L. F., Shotton, F. W. and West, R. G. (1973). *A Correlation of Quaternary Deposits in the British Isles*. Geological Society of London, Special report No. 4.
Moreau, R. E. (1969). Climatic changes and the distribution of forest verbebrates in West Africa. *J. Zool.* **158**: 39–61.
Mörner, N. A. (1976). Eustasy and geoid changes. *J. Geol.* **84**: 125–151.
Nikiforova, K. V. (1982). Neogene-Quaternary boundary. *Geol. Correl. (IGCP report)* **10**: 22–23.
Penck, A. and Brückner, E. (1909). *Die Alpen im Eiszeitalter*. Leipzig.
Porter, S. C. (1979). Hawaiian glacial ages. *Quat. Res.* **12**: 161–187.
Prance, G. T. (1978). The origins and evolution of the Amazon flora. *Interciencia* **3**: 207–221.
Renfrew, C. (1973). *Before Civilization*. London: Jonathan Cape.
Robin, G. de Q. (1977). Ice cores and climatic change. *Philophical Transactions of the Royal Society of London, Series B*. **280**: 143–168.
Roe, D. A. (1981). *The Lower and Middle Palaeolithic Period in Britain*. London: Routledge and Kegan Paul.
Ruddiman, W. F. and McIntyre, A. (1976). Northeast Atlantic palaeoclimatic changes over the past 600,000 years. In: Cline R. M. and Hays, J. D. (eds) *Investigation of Late Quaternary Palaeoceanography and Palaeoclimatology*. Geological Society of America Memoir 145, pp. 111–146.
Sancetta, C., Imbrie, J. and Kipp, N. G. (1973). Climatic record of the past 130,000 years in North Atlantic deep-sea core V23–82: correlation with the terrestrial record. *Quaternary Research* **3**: 110–116.
Sarnthein, M. (1978). Sand deserts during glacial maximum and climatic optimum. *Nature* **272**: 43–45.
Shackleton, N. J. (1967). Oxygen isotope analyses and Pleistocene temperatures re-assessed. *Nature* **215**: 15–17.
Shackleton, N. J. (1975). The stratigraphic record of deep-sea cores and its implications for the assessment of glacials, interglacials, stadials, and interstadials in the Mid-Pleistocene. In: Butzer, K. W. and Isaac, G. L. (eds) *After the Australopithecines*. The Hague: Mouton, pp. 1–24.
Shackleton, N. J. (1980). Deep drilling in African lakes. *Nature* **288**: 211-212.
Shackleton, N. J. and Matthews, R. K. (1977). Oxygen isotope stratigraphy of Late Pleistocene coral terraces in Barbados. *Nature* **268**: 618–9.
Shackleton, N. J. and Opdyke, N. D. (1973). Oxygen Isotope and palaeomagnetic stratigraphy of equatorial Pacific core V28–238. *Quat. Res.* **3**: 39–55.
Shackleton, N. J. and Opdyke, N. D. (1976). Oxygen-isotope and palaeomagnetic stratigraphy of Pacific core V28–239. Late Pliocene to Latest Pleistocene. In: Cline, R. M. and Hays, J. D. (eds) *Investigation of Late Quaternary Palaeoceanography and Palaeoclimatology*. Geological Society of America Memoir 145, pp. 449–464.
Shi, Y. and Wang, J. (1979). The fluctuations of climate, glaciers and sea level since late Pleistocene in China. In: Allison, I. (ed) *Sea Level, Ice and Climatic Change* I. A. H. S. publication 131, pp. 281–293.

Simpson, B. B. and Haffer, J. (1978). Speciation patterns in the Amazon forest biota. *Ann. Rev. Ecol. Syst.* **9**: 497–518.
Singh, G., Opdyke, N. D. and Bowler, J. M. (1981). Late Cainozoic stratigraphy, palaeomagnetic chronology and vegetational history from Lake George, N.S.W. *J. Geol. Soc. Aust.* **28**: 435–452.
Street, F. A. (1980). Ice Age Environments. In: Sherratt, A. G. (ed) *Cambridge Encyclopedia of Archaeology*. Cambridge: Cambridge University Press, pp. 52–56.
Street, F. A. (1981). Tropical palaeoenvironments. *Prog. Phys. Geog.* 5: 157–185.
Street, F. A. and Grove, A. T. (1979). Global maps of lake-level fluctuations since 30 000 yr BP. *Quat. Res.* **12**: 83–118.
Street-Perrott, F. A. and Roberts, N. (1983). Fluctuations in closed basin lakes as an indicator of past atmospheric circulation patterns. In: Street-Perrott, F. A., Beran, M. and Ratcliffe, R. A. S. (eds) *Variations in the Global Water Budget*. Dordrecht: Reidel, pp. 331–345.
Street-Perrott, F. A., Roberts, N. and Metcalfe, S. (in press). Geomorphic implications of Late Quaternary hydrological and climatic changes in the northern hemisphere tropics. In Douglas, I. (ed) *Geomorphology and Environmental Change in the Tropics*. London: George Allen and Unwin.
Swain, A. M., Kutzbach. J. E. and Hastenrath, S. (1983). Estimates of Holocene precipitation for Rajasthan, India, based on pollen and lake-level data. *Quat. Res.* **19**: 1–17.
Tija, H. D. (1980). The Sunda shelf, southeast Asia. *Zeitschrift fur Geomorphologie* N.F. **24**: 405–427.
Tricart, J. (1974). Existence de périodes sèches au Quaternaire en Amazonie et dans les régions voisines. *Revue de Géomorphologie Dynamique* **23**: 145–158.
Tsukada, M. (1983). Vegetation and climate during the Last Glacial Maximum in Japan. *Quat. Res.* **19**: 212–235.
Urey, H. C. (1947). The thermodynamic properties of isotopic substances. *J. Chem. Soc.*, pp. 562–581.
van Donk, J. (1976). $O^{18}$ record of the Atlantic Ocean for the entire Pleistocene epoch. In: Cline R. M. and Hays, J. D. (eds) *Investigation of Late Quaternary Palaeoceanography and Palaeoclimatology*. Geological Society of America Memoir 145, pp. 147–164.
van der Hammen, T. (1974). The Pleistocene changes of vegetation and climate in tropical South America. *J. Biogeog.* **1**: 3–26.
van der Hammen, T., Wijmstra, T. A. and Zagwijn, W. H. (1971). The floral record of the Late Cenozoic of Europe. In Turekian, K. K. (ed) *The Late Cenozoic Glacial Ages*. New Haven: Yale University Press, pp. 391–424.
van Zinderen Bakker, E. M. (1976). The evolution of late Quaternary palaeoclimates of Southern Africa. *Palaeoecol. Africa* **9**: 160–202.
Vita-Finzi, C. (1973). *Recent Earth History*. London: MacMillan.
Whitmore, T. C., Flenley, J. and Harris, D. R. (1982). The tropics as the norm in biogeography? *Geogr. J.* **148**: 8–21.
Wijmstra, T. A. and van der Hammen, T. A. (1974). The Last Interglacial-Glacial cycle: state of affairs of correlation between data obtained from the land and from the ocean. *Geologie en Mijnbowe* **53**: 386–392.
Woillard, G. M. (1978). Grande Pile peat bog: a continuous pollen record for the last 140,000 years. *Quat. Res.* **9**: 1–21.
Woillard, G. (1979). Abrupt end of the last interglacial s.s. in north-east France. *Nature* **281**: 558–562.
Woillard, G. M. and Mook, W. G. (1982). Carbon-14 dates at Grande Pile: correlation

of land and sea chronologies. *Science* **215**: 159–161.
Wymer, J. J. (1981). The Palaeolithic. In: Simmons, I. and Tooley, M. (eds) *The Environment in British Prehistory*. London: Duckworth , pp. 49–81.

# 3 *Human Evolution and Biological Adaptation in the Pleistocene*

*Chris Stringer*

## 1. INTRODUCTION

Surveying previous reviews of hominid evolution is a sobering and sometimes painful exercise. The field is littered with abandoned ancestors and the theories that went with them, and it is clear that ideas put forward may be destined for only transitory relevance. Failure to realize the complexities involved in trying to interpret a few fossils scattered sparsely through space and time has characterized the approaches of even the most competent workers, resulting in naive interpretations. A major reason for this unhappy state of affairs is that neither the inherent assumptions of these workers, nor their theoretical frameworks, were consistent with the limitations of the fossil evidence and its context. Consequently whole evolutionary edifices would collapse, complete with attached ancestors and descendents, with each development in theory, investigation of an underlying assumption, or new discovery.

My task is no easier today, despite three major changes. First there has been an increase in the fossil data available, so that statements are generally based on a larger sample than was available to earlier workers. Secondly, the theoretical framework for palaeoanthropology has shifted much more towards that of evolutionary biology as a whole, towards a recognition of ecological factors and away from a dependence upon assumptions about the uniqueness of humans and consequently of their evolutionary history. Finally, the growing importance of punctuational models and cladistics in palaeontology generally is beginning to have an impact on palaeoanthropology. As a result of these changes most textbooks reflect the ideas of ten years ago, rather than the most recent research. They portray human evolution proceeding through unidirectional gradual change, with little hint of complex evolutionary mechanisms. They indicate that the main events in hominid evolution are well known and dated: the Miocene form *Ramapithecus represents* the earliest known hominid and its existence establishes that unique hominid adaptations

HOMINID EVOLUTION
*ISBN 0-12-261920-X*

had already separated our lineage from that of the African apes fifteen million years ago; *Australopithecus africanus* is probably a direct ancestor for subsequent members of the genus *Homo*; *Homo habilis* represents an evolutionary

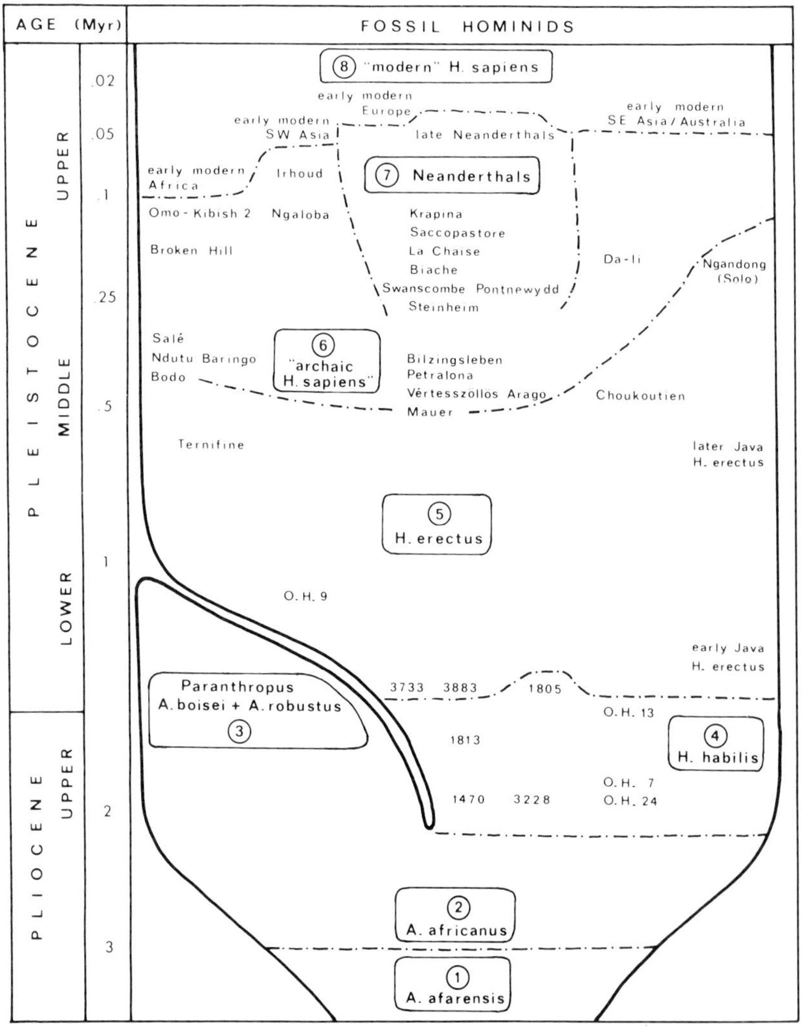


Fig. 3.1. Fossil hominids in time and space. Many of the specimens are not dated accurately. They have been divided into a grade classification, but see text for alternative classifications.

intermediate between *Australopithecus* and *Homo erectus*; the widely distributed palaeospecies *H. erectus* originated in Africa and gave rise to *Homo sapiens* in various areas of the inhabited world; the Cro-Magnons were more or less the earliest form of "modern" humans, and either simply evolved gradually from the preceding Neanderthals or swept into Europe from an adjacent area such as western Asia, replacing the inferior Neanderthals virtually overnight. Every one of these basic tenets of the anthropological textbooks must now be regarded as questionable, even if an alternative "correct" answer cannot be provided. The impact of the new data and new theoretical framework has thus far been to make us less certain than workers ten or twenty years ago or the textbooks of today. Thus this chapter is not intended to present a survey of the "facts" of hominid evolution, but of the evolutionary problems to which some of these facts relate.

## 2. REVIEW OF PLIOCENE AND PLEISTOCENE HOMINID EVOLUTION (Fig 3.1)

It is an indisputable fact that the fossil hominid record compares poorly with those of many other mammalian groups, despite the finance and effort which has been poured into palaeoanthropological fieldwork in the last 20 years. The fossil record is considerably richer than it was for the African Plio-Pleistocene, but for the late Miocene and early Pliocene, for the Plio-Pleistocene outside of Africa, and for the Middle and Upper Pleistocene of most areas, a dearth of evidence continues, particularly if we demand material which is well stratified and can be well dated, absolutely or relatively. The European area is fairly rich in later Middle and Upper Pleistocene fossil hominids, but the collapse of the old stratigraphic framework based on the Alpine glacial–interglacial sequence (see Chapter 2, this volume) has cast doubts on many of the assumptions made about European hominid evolution beyond the range of reliable radiocarbon dating (Cook *et al.*, 1982). Even where there is a reasonable fossil record within a reasonable chronological and archaeological framework, this does not guarantee agreement between workers, as the continuing "Neanderthal problem" amply demonstrates (Wolpoff, 1980, 1981; ApSimon 1981; Stringer *et al.*, 1981; Stringer, 1982). Further back in time where the record is less complete and less ordered it is not clear whether the species *H. erectus* existed at all in Europe, and if it did, what its relationship was to *H. sapiens*.

The hominid fossil record is really too poor to contribute significantly to the debate about the tempo and mode of evolution generally, although it has been cited as providing support for the model of punctuated equilibria (Gould and Eldredge, 1977; Stanley, 1979), a view already challenged (Cronin *et al.*, 1981). Much of the fossil hominid evidence used by Gould and Eldredge and

Stanley had already been invalidated by further research at the time they wrote their discussions, but Stanley has since produced additional arguments in favour of his macroevolutionary model from palaeoanthropological data (Stanley, 1982).

Since an important part of the debate concerns the *origin* of species, it must be admitted that palaeospecies recognized by most workers are to a great extent artificial constructs with an indeterminate degree of biological reality. It is impossible to know whether morphological changes recognized as determining a separate palaeospecies correspond to actual species formation. Some would argue that there is no relationship between morphological differences recognised in fossils, and genuine species differences (Cracraft, 1982). Undoubtedly the most satisfactory area in which to test the competing models of punctuated equilibria and gradualism is in living species, which have some objective reality and can be studied genetically and behaviourally as well as morphologically. The next best arca in which to test these models is in fossil groups with extensive and well stratified records (e.g. molluscs—Williamson, 1981), and the hominid record certainly does not compare favourably with these kind of data. Nevertheless if we are content (some might say unwise enough) to use the available hominid record with all its imperfections, accept dating evidence which in some cases is highly questionable, and for the moment work with one common view of hominid classification, then we can at least examine the relative strengths of the arguments in the following ways:

(i) Does the record show convincing gradual unidirectional change in the preserved fossils, with morphological intermediates (mosaics) occurring at the right time period to link generally recognised palaeospecies (gradualistic model)? or

(ii) Does the record show convincing stasis or non-directional change within palaeospecies, without intermediates between them (punctuational model)?

Further support for the punctuated equilibria model (but not proof) would follow from a demonstration that stratigraphic ranges of the various time-successive hominid species actually overlapped significantly.

## 2.1. Pliocene Hominids

It is now recognized that many of the important characters of the Pleistocene hominids, both anatomical and behavioural, have their roots in the Pliocene, and this material has already become heavily involved in the developing debate about punctuated equilibria and gradualism as modes of evolutionary change. Stanley (1979) argued that

> the fossil record now indicates that *A. africanus* survived for perhaps as long as 3

> Myr without changing enough for its lineage to be deemed a new chronospecies. In relying on phyletic evolution, a gradualistic interpretation would require that this creature, after this long period of virtual evolutionary stagnation, underwent abrupt phyletic transformation into the genus of man. The more palatable alternative is that *Homo* is not a lineal descendant of *A. africanus* but a dramatically new form that budded off by way of a small population, while the parent species persisted with little change.

There are several problems with Stanley's arguments, but the whole nature of the discussion has changed with the controversial creation of a new Pliocene species of *Australopithecus* (*A. afarensis*) by Johanson *et al.* (1978). The relatively abundant evidence from the Afar and Laetoli (Johanson and White, 1979; White *et al.*, 1981), which dates between *ca.* 4·0 Myr and 3·0 Myr has clearly demonstrated that the fundamental human adaptation of bipedalism appears to predate the first clear evidence of tool-making and expansion of the brain beyond the level found in our nearest relatives, the African apes. It seems likely, therefore, that the origin of bipedalism must be sought outside the cultural and behavioural milieu so often proposed for it. However, this would in turn imply that later Pliocene hominids were neatly preadapted to utilize their hands for the production of artifacts, once liberated from the constraints imposed by locomotor activity.

The Afar material has only recently been published in detail (see volume 57, part 4, of the *American Journal of Physical Anthropology*), and few other workers have yet had an opportunity to discuss critically published assessments of the specimens and their allocation to the single species *A. afarensis* (but see Tobias, 1980a; Olson, 1981; Senut, 1981). The phylogeny favoured by Johanson and White would displace the gracile australopithecine, *A. africanus*, from the position ancestral to *Homo* which it generally occupies in most accounts of hominid evolution. Instead it is "demoted" to an ancestor of the robust australopithecines (*A. robustus* + *A. boisei* = *Paranthropus* of some workers) only. *A. afarensis* instead represents the common ancestor of all subsequent hominids according to Johanson and White (1979), although if it is equally related to the *Homo* and *Australopithecus* lineages, it is unclear why it should be classified with *Australopithecus*. The cladogram published by White *et al.* (1981) differs from their 1979 cladogram, and although it is now consistent with their classification of the Afar and Laetoli material as *Australopithecus*, it is also now inconsistent with their proposed phylogeny.

Whatever view is taken of the *A. afarensis* material, whether it is accepted as a primitive hominid species (Delson, 1981), viewed as an early form of *A. africanus* (Tobias, 1980a), or as actually representing at least two hominid species, perhaps allied to *Homo* on the one hand and *Paranthropus* on the other (Olson, 1981; Senut, 1981), there seem to be greater resemblances to Miocene hominoids and living apes amongst the specimens than could be found in any known sample of *A. africanus*. However, the degree of dimorphism indicated by the range in size and robusticity of the material is remarkable by the standards of most other primate species, which has led some workers to doubt that only one species is represented in the material assigned to *A. afarensis*.

Models of early hominid evolution usually incorporate dietary models to account for the divergence of robust australopithecines from the lineage leading to *Homo*, although it should be noted that Walker's (1981) recent analyses of tooth surfaces indicate that all hominids prior to *H. erectus* may have had diets sharing much in common with those of modern apes. Indeed one of the main reasons why Johanson and White eliminated *A. africanus* from the lineage of *Homo* is that they perceive in its gnathic morphology deviation in the direction of increased grinding, a deviation which they believe culminated in the morphology of the robust australopithecines. Other workers have accepted this fact (Rak, 1981; Grine, 1981) but not all accept that *A. africanus* was therefore too specialized to represent an ancestor for *Homo* also (e.g. Tobias, 1980a, 1980b). Some workers see in *A. africanus* derived characters linking it exclusively to *Homo*, such that the gracile australopithecines should be incorporated into that genus as *H. africanus* (Robinson, 1972; Olson, 1981).

So while there is already a large measure of disagreement between workers concerning the status of *A. afarensis* and its relationship to *A. africanus*, and whether *A. afarensis* is better qualified to represent the common ancestor for the lineages leading to *Homo* and the robust australopithecines, there is little support for Stanley's view of a three million year stasis in *A. africanus*. Once all the material referred to *A. afarensis* has been widely studied, we will be in a better position to establish whether the main cladogenetic event in hominid evolution (following the split from the African apes) occurred prior to, or after, about 3·00 Myr. The recognition of the timing of this split between the lineages of the robust australopithecines (*Paranthropus*) and *Homo* will have a critical effect on the phylogenetic position assigned to *A. africanus*, and whatever else is decided it is likely that the days of the grade term "australopithecine" to encompass all this material are surely numbered. Furthermore, doubts about the dating of the Hadar fossils (Brown, 1982), will have to be resolved before the phylogenetic scheme of White *et al.* (1981) can finally be accepted.

## 2.2. *Homo Habilis*

In the late Pliocene-early Pleistocene there are a number of specimens attributed by various workers to the species *H. habilis* which have been regarded as actual or potential ancestors for the Pleistocene species *H. erectus* and *H. sapiens*. Most of these specimens are from the sites of Olduvai Gorge, Tanzania, and Koobi Fora, Kenya, and they have been interpreted in various ways. The view that the entire sample considered here (including Olduvai hominids 7, 13 and 24, and Koobi Fora hominids KNM-ER 1470, 1802, 1805 and 1813) represents one species, temporally and morphologically intermediate between *A. africanus* and *H. erectus*, is perhaps the prevailing one amongst palaeoanthropologists. But this view requires that the species is highly variable, with some of its early fossil representatives possessing large brain sizes, but also large dentitions, and some of the later specimens possessing smaller *Homo*-like dentitions, but also small brain sizes (see later discussions). Additionally, there is the problem of possible temporal overlap between later representatives of this *H. habilis* grouping (e.g. 1805, and perhaps 1813) and early representatives of *H. erectus*. It is true that certain characters of this assemblage can be taken to be intermediate between *A. africanus* and *H. erectus*, and that there are some traits in the specimens which suggest a progressive trend towards a *H. erectus* morphology (Cronin *et al.*, 1981;). But this view is not entirely persuasive, given the short time range of the group (*ca.* 0·4 Myr) and the large morphological changes still necessary to convert any of its members into *H. erectus* over the time period available. Other views of the material include the recognition of at least two groups within it, one large brained (including 1470), the other a late form of *A. africanus* (including O.H. 13 and 1813) (Walker and Leakey, 1978). Yet another view sees the material as representing an earlier "australopithecine" portion (including 1470) and a later *H. erectus* portion (including 1805) (Wolpoff, 1980). The latter view demands an extreme and surely unacceptable range of variability in early *H. erectus* populations. The former view would remove the need to accept a large range of brain size in the *H. habilis* sample, but at the expense of greatly increasing the morphological range of *A. africanus*.

A particularly important fossil which has received a wealth of attention is now generally, but not universally, classified as *H. habilis*. This specimen, KNM-ER 1470, was utilized by Gould and Eldredge (1977) to enhance their claims for a punctuation event at the origin of the genus *Homo*. They claimed that the specimen represented a large-brained member of the genus *Homo*, unrelated to australopithecines, with which it lived sympatrically as long as 3 Myr ago. Even in 1977, it was apparent that there were serious problems over the dating of this fossil, and these have now been resolved by repeat potassium-argon, argon-argon, fission-track, palaeomagnetic, mineralogical, stratigraphic and faunal analyses (see e.g. Drake *et al.*, 1980; McDougall *et al.*, 1980). These converge on a best date of *ca.* 1·9 Myr (old half-life constant for K-Ar decay) for the tuff and the hominid specimens (including 1470) which lay below it.

The 1470 cranium can therefore be said to be contemporaneous with, or earlier than, *robust* australopithecines (*A. boisei* or *A. robustus* = genus *Paranthropus*) but not gracile australopithecines (*A. africanus*), unless controversial fossils such as KNM-ER 992 and 1813 are regarded as belonging to this species, or the much criticized redating of the Taung specimen is accepted (Howell, 1978; Tobias, 1980b; Grine, 1981). Even if late forms of *A. africanus* were found to overlap the time range of *H. habilis* (used here to include 1470), this would not necessarily exclude earlier forms of *A. africanus* (probably dated 2·5–3·0 Myr) or specimens classified as *A. afarensis* (dated *ca.* 3·0–4·0 Myr) from a possible ancestral status. However, while a later date for the 1470 cranium contradicts Gould and Eldredge's views, it also creates another problem for a gradualistic model by accentuating the morphological differences of the 1470 specimen from *H. erectus* and *H. sapiens*, to which it becomes much closer in time. Some of the characteristics of 1470 are now seen to resemble *Australopithecus*, leading to doubts about its classification as *Homo* rather than *Australopithecus* (Walker and Leakey, 1978; Reader, 1981).

## 2.3. The Origin of *H. erectus*

One of the major problems facing students of human evolution is an explanation for the origin of the *H. erectus* fossils which appear in the African record at or slightly before 1·5 Myr. Certain *H. erectus* characters (e.g. larger supraorbital and occipital torus, larger brain size, longer and lower cranial vault, nasal structure, and robust post-cranial morphology) are foreshadowed in the *H. habilis* sample already discussed, but convincing directional trends in the sample are difficult to establish, despite attempts (Cronin *et al.*, 1981). There are certainly important differences in robusticity and cranial thickness between typical *H. erectus* fossils and African Pliocene hominids, although it is possible that the early *H. erectus* fossils from Koobi Fora may not themselves be very "typical" in this respect, perhaps because they are rather primitive. On the other hand, the robust *H. erectus* postcranial morphology, exemplified by the O.H. 28 pelvic and femoral fragments from Olduvai, seems to have extended in time not only into the later Middle Pleistocene (Arago 44) but back into the Pliocene if the partial pelvis KNM-ER 3228 is genuinely stratified below the KBS Tuff dated *ca.* 1·8–1·9 Myr (Drake *et al.*, 1980; McDougall *et al.*, 1980; Day, 1982). This Koobi Fora fossil would then be older than any *H. erectus* specimens identified from crania or mandibles, and might indicate that some specimens referred to *H. habilis* were already enlarged in body size with a greater postcranial robusticity. Perhaps the *H. erectus* cranial morphology is an allometric consequence of an enlarged body size (see p 71 and following), but this alone is hardly a convincing explanation for that morphology.

An alternative explanation for the origin of *H. erectus* which I believe still merits consideration, is that the species developed outside of the area with the best Plio-Pleistocene fossil record (East Africa). Evidence for the existence of hominids in Asia during the Pliocene is still poor, but claims for Yuanmou, China (Zhou Mingzhen *et al.*, 1982), Java (Ninkovich and Burckle, 1978; Siesser and Orchiston, 1978) and 'Ubeidiya, Israel (Repenning and Fejfar, 1982) deserve further examination. K-Ar determinations for Modjokerto which pointed to a late Pliocene age for the Perning fossil (Curtis, 1981) may not in fact date the hominid (Sartono, 1982). But the hominids which *are* most satisfactorily associated with the Pucangan beds in Java are all robust, and these include recent discoveries of facial or cranial parts assigned to "*Meganthropus*": Sangiran 27 (Jacob, 1980) and 31 (Sartono, 1982). The latter specimen is remarkably robust and, although distorted, shows either an extreme development of a sagittal keel, or even a sagittal crest. Thus while probable early *H. erectus* specimens from Africa (such as KNM-ER 992, 3733 and 3883) display less robust crania or mandibles than later *H. erectus* fossils, those from Indonesia are remarkably robust. This may be because the Indonesian fossils are, in fact, later in date, but another possibility is that they had evolved from a Pliocene or early Pleistocene ancestor already more robust than Pliocene "advanced hominids" known from African sites. Now evidence supporting a younger age for the Asian sample is discussed in Matsu'usa (1982) and Pope (1983).

### 2.4. Variation in *H. erectus*

Only recently has it become possible to assess geographical variation in early *H. erectus* fossils, and although much of the material remains to be fully described, it certainly points to a greater morphological stability in *H. erectus* characteristics in time and space than seemed likely when Olduvai hominid 9 was the sole reasonably complete early African *H. erectus* fossil. Some of the evidence can be taken to reinforce the views of Howells (1980) and Rightmire (1981) that *H. erectus* was a relatively stable taxon in time and space, and if the known African Pliocene hominids are taken as the ancestral model, considerable cranial changes (but perhaps not postcranial changes, as discussed earlier) must have occurred in the period between 2–1·5 Myr. But at present, comparisons of the various fossils to establish or deny the existence of evolutionary trends in the cranium and dentition (Rightmire, 1981) are bound to founder on the scarcity of well-dated material from any one area, allowing too much flexibility in sample composition and chronological ordering to make a convincing case either way. For endocranial volume increase, my analyses provide somewhat better support for gradual change than did Rightmire's (1981) study (see p 77). But there is certainly little evidence to support Gould and Eldredge's (1977) claim for stasis in absolute brain size for over a million years in *H. erectus*, nor is there much justification for regarding early specimens (such as KNM-ER 3733 and 3883) as virtually identical morphologically to late forms (such as the Choukoutien or Ngandong fossils). Finally, while I have stated above that there is a *general* stability of characters in the *H. erectus* group, the fact that certain typical *erectus* characters (e.g. occipital angulation, pelvic structure) are found outside the group while supposedly "typical" characters (e.g. cranial thickness, sagittal keeling) are not universally present within it, leads to problems in defining the species morphologically by its own within-group characteristics.

## 2.5. The Transition from *H. erectus* to *H. sapiens*

While the supposed transition from *Australopithecus* to *H. erectus* via *H. habilis* provides only very arguable support for a gradual evolution between Pliocene and Pleistocene hominids, the transition from *H. erectus* to *H. sapiens* (*sensu lato*) seems to be represented by genuine mosaics which occur at the right time and place to support the model by establishing that evolutionary intermediates do occur over a major part of the known Pleistocene range of the two species. Fossils from Europe (e.g. Arago, Vértesszöllös, Petralona), Africa (e.g. Salé, Omo-Kibish 2, Bodo, Broken Hill, Ndutu) and Asia (e.g. Ngandong and Da-Li) have proved difficult to classify using the conventional morphological definitions of *H. erectus* and *H. sapiens (s. lato)*, even to the extent of suggesting to some workers that the existence of such mosaics should lead to the incorporation of the palaeospecies *H. erectus* into *H. sapiens* (Jelínek, 1980). Below, I will argue that the taxon name *H. sapiens* may already be used too broadly when classifying fossils, and I therefore do not support such suggestions. But the available evidence certainly contradicts Gould and Eldredge's (1977) view that early and late representatives of hominid species are little different from each other and therefore indicate evolutionary stasis. The classification of the Petralona cranium, for example, is controversial but whether we choose to classify the specimen as a late *H. erectus* or an early *H. sapiens* (maintaining current usage for the present) it is clearly considerably different from both early *H. erectus* (e.g. KNM-ER 3733) and late *H. sapiens* (e.g. Neanderthal or modern human) crania. The relatively frequent occurence of "mosaic" fossils in different Continents would not be expected if change from one species to the next was a very restricted event in time and space as claimed by the proponents of punctuated equilibria. The only alternative explanations for these widely distributed mosaics would appear to be less plausible than regarding them as morphological intermediates. They might represent another lineage of hominids living in the same areas as *H. erectus* and *H. sapiens*, displaying mixed characters, or they might be hybrids between the two species, although the extent of temporal or geographic overlap between the species is difficult to establish.

## 2.6. The Origin of Anatomically Modern (a.m.) *H. sapiens*

Two extreme views of the origin of a.m. *H. sapiens* have been distinguished by Howells (1976) and Wolpoff (1980). One model suggests that a.m. *H. sapiens* evolved in a geographically restricted part of the Pleistocene range of *Homo* and that this morphology then radiated from that centre by migration or gene flow to replace the more archaic forms of *H. sapiens* (e.g. the Neanderthals of Eurasia). In this model, modern geographical variation ("races") would have evolved relatively recently (probably during the last 50 000 years or so) during the Upper Pleistocene spread of a.m. *H. sapiens* through the Old World, Australasia and the Americas. There would be no morphological continuity between the archaic and a.m. *H. sapiens* forms in each area, except in the place of origin of a.m. *H. sapiens*, or where there was gene flow between them. At the other extreme is the local continuity model first proposed in detail by Weidenreich, developed by Coon, and revived recently in a different guise by workers such as Thorne and Wolpoff (1981). Here there is morphological continuity in each area where archaic *H. sapiens* forms are succeeded by a.m. forms and for Thorne and Wolpoff, local "clade" characters are considered to be traceable back to the Middle Pleistocene across perhaps half a million years. Different workers allow for different degrees of gene flow between separate areas during the evolution of *H. sapiens*, and in the Thorne/Wolpoff model, the "centre and edge hypothesis", there is a balancing effect between gene flow and local selection. In their test case area they perceive "clade" characters which link modern Australoids and *H. erectus* precursors in Indonesia. The model could apply universally, however, but Wolpoff (e.g. 1981) has repeatedly argued that the European Neanderthals of the early Upper Pleistocene are the ancestors of modern Europeans, with no necessity to invoke gene flow or migration at the time of appearance of a.m. *H. sapiens* in Europe (*ca.* 30–35 kyr). From this model, modern human variation has its roots in the Middle rather than the Upper Pleistocene.

The argument about the Neanderthals is one in which I have become embroiled, and it seems to me that this is one of the few areas in palaeoanthropology where the fossil evidence and available chronology should be sufficient to allow a resolution of a phylogenetic problem (Trinkaus and Howells, 1979; Stringer and Burleigh, 1981). Because of the relative precision of dating available at the time of the transition between Neanderthals and a.m. humans in Eurasia it should be possible to document either co-existence of Neanderthals and a.m. *H. sapiens* followed by the replacement of the former group by the latter, or the gradual transformation of Neanderthals into a.m. humans. The recent discovery of the Saint-Césaire Neanderthal (Lévèque and Vandermeersch, 1981) provisionally dated by archaeological correlation to *ca.* 31–35 kyr, would seem to provide clear evidence that at least *some* Neanderthals were clearly too distinct and too late in time to have evolved into a.m. *H. sapiens*, with whom they overlapped temporally and by whom they were replaced. To convert even the earlier last glaciation Neanderthals into the earliest a.m. Europeans would require a tremendous acceleration in evolutionary change over a few thousand years, preceded and followed by relative (but not absolute) morphological stasis, i.e. a punctuation event. Distinctive Neanderthal characters, some of which had become established through at least the previous 100 000 years, were replaced by characters entirely within the present human range in less (perhaps much less) than a tenth of that period of time. To me, a more likely explanation than punctuational change is population replacement, although there are still problems for a total replacement model. For example, there do appear to be scattered indications of local continuity in the skeletal remains and in the archaeological record in eastern Europe and southwest Asia. These may be explicable by postulating a staggered rather than synchronous spread of a.m. humans, probably from Africa, with some degree of cultural and genetic diffusion. It can even be argued that the reverse process of replacement occurred in southwest Asia, with Neanderthals spreading into the area during the last glaciation, temporarily displacing early a.m. populations already established there (Vandermeersch, 1981; Jelinek, 1982). However, I do not find this argument convincing, which in turn does lead to problems in matching the archaeological record with the demands of population replacement if we reject the model of *in situ* evolution of a.m. *H. sapiens* from Neanderthal precursors in an area such as southwest Asia.

My own view about the origin of "modern" humans is that the Middle/early Upper Pleistocene hominids of Africa are more plausible "ancestors" for a.m. *H. sapiens* because they lack what I perceive to be "specialized" (derived) characters found in the Neanderthals of Eurasia. Furthermore a number of probable early Upper Pleistocene African fossils (e.g. Border Cave, Klasies Cave, Florisbad, Omo-Kibish 1, Irhoud 2) display derived characters shared with a.m. humans, suggesting that "modern" *H. sapiens* had evolved, or was evolving, in Africa fully 50 000 years earlier than the first occurrence in Europe. But accepting this model has its problems too, since a.m. *H. sapiens* seems to have arrived in Australia at least as early as in Europe, and a newly discovered fossil (the Da-li cranium) from the Chinese Middle or early Upper Pleistocene (Wu Rukang and Wu Xinzhi, 1982) seems to show resemblances to both modern and more ancient populations of the area, once again suggesting local continuity. The interpretation based on the replacement hypothesis would be that such similarities between modern populations and "archaic" predecessors in these areas are due either to parallel evolution, some degree of gene flow at the time of population replacement, or to different combinations of retention of "primitive" characters in various a.m. populations. One real test of such a model would be an assessment of whether the characters regarded as specifically linking Australoids and Indonesian *H. erectus* fossils in the local continuity model are, in fact, merely primitive characters for *H. sapiens*, retained to a greater degree in Australoids compared with other modern human groups.

## 3. CLIMATIC ADAPTATION

It was apparently only during the Middle Pleistocene that humans began to adapt to lower temperature—higher latitude environments, and there is insufficient skeletal evidence to indicate if this adaptation was purely behavioural, or was morphological (and physiological?) as well. Some signs of differences in nasal form between the Petralona cranium and the related Middle Pleistocene hominids of Africa may be indicative of morphological adaptation to cooler environments in Europe, but with the pitifully small samples available, this possibility cannot be tested at present. It is not until the Upper Pleistocene that the evidence becomes sufficient for a real test of the idea that adaptation to life in higher latitudes was being reflected in the skeleton. The idea that the Neanderthals of Eurasia may have been cold adapted has a long history. Coon (1962) proposed that their large projecting noses, maxillary sinuses and infraorbital foramina all reflected adaptation to a periglacial environment, while Brose and Wolpoff (1971) added frontal sinus development and degree of occipital bunning to the list of Neanderthal cranial characters which could be so explained. Coon also believed that the body proportions of the Neanderthals were like those of modern cold-adapted peoples. In order to examine these claims it is necessary to study evidence from recent populations where we have a better chance to assess the reality of physical adaptation to climate in humans. Parallels may then cautiously be drawn with Pleistocene hominids where appropriate.

Recent human variation in features such as skin colour, hair form, nasal shape and physique are often explained through the agency of environmental, and specifically climatic, factors, acting via the genotype (perhaps reflecting inherited variation selected for in previous generations), phenotype (environmental effects on growth, etc.), or a combination of these mechanisms. Such explanations are often difficult to support from experimental and observational data. However, variation in nasal form, and in body size and shape, are two of the more convincing cases in which correlation seems to exist between morphology and climate in recent human populations, and from our point of view this is especially fortunate since, unlike superficial features of the skin, hair or eyes, these can be assessed in fossil hominids where the skeletal evidence is complete enough.

In the case of the nasal aperture, it has long been suspected that the shape of the nose in different human groups is in some way related to climate. Relative nose width seems to vary in relation to absolute and relative humidity, and temperature, and it appears that relatively narrow noses can be expected amongst populations in dry, cold and high altitude environments (Carey and Steegman, 1981). The nose seems to function as both a humidifier and temperature regulator for inspired air, and while a narrow, projecting nose would appear most effective in dry conditions, a flat broad nose seems most effective in climates with a seasonally high absolute humidity. Low humidity during the coldest months, even more than temperature, seems to be reflected by degree of nasal projection.

For most fossil hominids where nasal form is preserved, the typical morphology is of a flat and fairly broad nasal opening, as would be expected for relatively moist and warm environments at low latitudes and altitudes. But there is a notable exception to this generalization since Neanderthal skulls from Eurasia are unusual, as has already been mentioned, for their extremely voluminous and projecting nasal openings. The projecting form of the Neanderthal nose could be explained through adaptation to conditions of low humidity and temperature, but since the Neanderthal nose was also *broad* rather than narrow, there would appear to be no obvious analogy to it among recent populations. This might be a reflection of the fact that no present day populations are adapted to an environment which closely resembles that of Europe during the last glaciation, where there may have been unique combinations of temperature and humidity (see Chapter 9, this volume). Alternatively there may have been skeletal constraints on the form which Neanderthal facial adaptation could take, since such modification would have been made on an archaic rather than "modern" cranial form. Additionally, the Neanderthal nose may have been only one part of a facial complex giving protection against frost bite and damage to internal structures (Steegman, 1972).

Variation in physique in modern humans appears to conform fairly well to Bergmann's (1847) and Allen's (1877) "rules" developed for living organisms generally. As predicted by these "rules", humans who are indigenous to cold areas do tend to have, respectively, greater body masses and shorter extremities relative to those who live in warmer climates. By these means, the surface area through which body heat can be lost is minimized compared to body volume. Conversely, the contrasting morphologies, maximizing the surface area relative to volume, and hence the ability to lose surplus heat, are generally found in human populations indigenous to hot environments. Given fairly complete skeletal remains it is possible to assess body size and shape, and hence compare the physique of past populations with those of the present day. Unfortunately the fossil record for Pliocene and early Pleistocene hominids is not adequate for this kind of comparison, and only two moderately complete skeletons are known, for *A. afarensis* (AL-288-1—"Lucy") and for *H. erectus* (KNM-ER 1808). Body size estimates for early hominids will be discussed later, but it is not until the Upper Pleistocene, apparently related to the introduction of burial practices, that body proportion reconstructions can also be attempted using fairly complete skeletons (Trinkaus, 1981).

At various times in the past (e.g. Vallois, 1958; Coon, 1962) it has been pointed out that Neanderthal skeletons are unusual in their limb segment proportions, in that the distal limb segments (radius and ulna, tibia and fibula) are short in comparison with the proximal segments (humerus, femur). Coon (1962) recognized the possible adaptive significance of this feature and various researchers have demonstrated that in other animals as well as humans, such limb segment differences could be indicative of selection operating to minimize the surface areas of extremities, or of repeated vasoconstriction slowing growth, in a cold environment. Either or both of these factors could have been expected to operate on the human populations of the last glaciation in Europe, but only recently has enough data been gathered to present a comprehensive analysis of the fossil evidence. This study (Trinkaus, 1981) has yielded some surprising results (Fig. 3.2). When recent populations from different climatic backgrounds are compared with each other in values of the brachial (radius/humerus length) and crural (tibia/femur length) indices, there is a reasonable correlation between high temperature environments and high values of the indices (i.e. longer distal limb segments) as would be predicted by Allen's rule. When the best available Neanderthal samples (still consisting of only eight or nine individuals) are analysed in the same way, they very closely match the limb segment proportions of the recent population samples from cold environments (especially the Lapps), and are quite distinct from values expected in recent populations from warm environments. This conforms perfectly to the cold-adaptation model for the Neanderthals, but the surprise of the analyses comes in the results for the slightly larger samples of the early a.m. hominid skeletons associated with the European Upper Palaeolithic. Their ranges barely overlap the Neanderthal range for crural indices and the mean values are amongst the highest for a.m. *H. sapiens* groups, while the Neanderthal values compare with the lowest. The pattern is similar for arm and leg bones, although that illustrated (Fig. 3.2) shows the clearest separation. If one accepts the results at face value, it would suggest a long-term cold adaptation for the Neanderthals, but an origin or genetic influence for the earliest a.m. populations of Europe not merely from outside the area, but specifically from a high temperature environment.

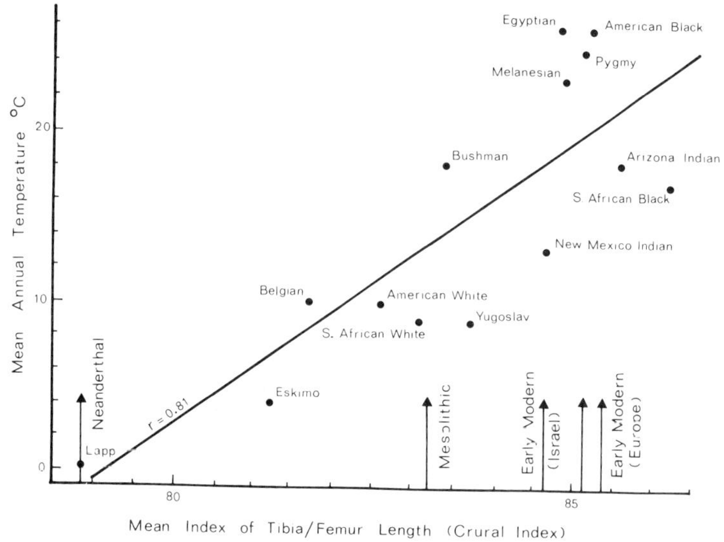


Fig. 3.2. Limb segment proportions and environmental temperature plotted for a range of recent populations of *Homo sapiens*. Fossil samples are placed by their mean crural index, but no attempt has been made to fit them to the chosen regression line of crural index to temperature based on the recent samples. Samples of recent migrant groups (e.g. South African whites) are plotted against a temperature based on their "ancestral" place of origin (i.e. Western Europe) rather than their present location as they appear to approximate more closely their ancestral body proportions after a few generations. All data from Trinkaus (1981).

The fact that the "late" Upper Palaeolithic sample analysed by Trinkaus (1981) maintains the distinctively high brachial and crural indices of the "early" sample is not even suggestive of significant environmental effects during the coldest stage of the last glaciation. This could be explained by assuming that the Neanderthals (and recent Lapps and Eskimos) suffered more prolonged or more severe environmental exposure to cold, while the Upper Palaeolithic populations were better shielded culturally or behaviourally. Another problem with the results is that the European and southwest Asian Neanderthals show the same "cold adapted" limb segment proportions, even though southwest Asia was never such a low temperature environment as Europe during the last glaciation. The similarity could be accounted for by gene flow or by descent from a common ancestor, provided that the "cold adapted" limb proportions were not selectively disadvantageous in the environment of southwest Asia. The only other explanation apart from cold adaptation proposed for the limb proportion data, relates the Neanderthal morphology to their greater robusticity and muscularity, since shorter distal limb segments might be expected to give greater power to distal limb movements. However, this would not explain modern variation in limb segment proportions, and the apparent relationship to environmental temperature. Nor would it explain the distinctive early a.m. values, since the Upper Palaeolithic samples and Mousterian-associated samples from Skhūl and Qafzeh (Israel) analysed by Trinkaus are more robust than most modern populations, yet have limb proportions at the opposite end of the scale from the Neanderthals, among the highest known for a.m. *H. sapiens*. As Trinkaus (1981) concludes, the most reasonable interpretation is that the Neanderthals were cold adapted, and that the distinctiveness of early a.m. limb proportions is a reflection of gene flow or population movement from warmer environments at the transition to "modern" *H. sapiens* in Eurasia.

## 4. BODY SIZE IN EARLY HOMINIDS

Estimates of body size (generally synonomous with body *weight*) in early hominids are useful for several reasons. First, body weight may give information about possible locomotor patterns and general behaviour e.g. mobility and agility. Differences in body weight between males and females within a species may be a reflection of competition for mates and the type of social organization, such as role specialization. Body size may be related to climatic adaptation (see p 68) or nutritional limitations. Finally, body size is an important factor in allometric analyses which investigate the relationships of changes in shape and size for different morphological characters, and one particularly relevant character is brain size, which is directly or indirectly related to body size in other mammals, including Primates (Martin, 1981).

Body size estimates for fossil hominids are usually made from dimensions of parts of the postcranial skeleton such as the femur or vertebrae. Fairly complete skeletons are rare, especially for early hominids, so the degree of compatibility of body size estimates from various parts of the skeleton cannot often be tested. The oldest moderately complete skeleton is that of "Lucy" (AL-288-1) from Hadar, assigned to *A. afarensis*, as discussed earlier. This was obviously a small individual, even compared with the average size for *Pan*, and several workers using different techniques have estimated her weight as less than 30 kg (Johanson and Edey, 1981; Aiello, 1981; Jungers, 1982). However, the degree of size variation described in the entire Hadar sample is very marked (*American Journal of Physical Anthropology* Volume 57, No. 4, 1982) since other individuals were considerably larger or more robust than "Lucy" in various skeletal characters. Her body size is comparable with typical estimates for *A. africanus* (18–35 kg: Tobias, 1971; Robinson, 1972; Pilbeam and Gould, 1974; McHenry, 1975, 1976; Holloway, 1978; Steudel, 1980), but other individuals from Hadar must have had body sizes comparable with robust australopithecines (*A. robustus* + *A. boisei* = *Paranthropus*). Typical estimates for body size in these hominids range from 36–80 + kg (references as for *A. africanus* estimates). If the *afarensis* material does indeed span a very large range of body sizes (perhaps 27–67·5 kg—Johanson and Edey, 1981), the debate about the extent of sexual dimorphism in early hominids, and the possible existence of more than one hominid species at Hadar will be further fuelled.

For the body sizes of other hominids, typical estimates for *H. habilis* range from 31 to 53 kg, for *H. erectus* from 42 to 53 kg, and for *H. sapiens* 52 to 60 kg (Tobias, 1971; Pilbeam and Gould, 1974; Holloway, 1978). Taking mean values for the estimates, a gradual increase in body size from the supposed ancestral form *A. africanus*, to *H. sapiens*, could be proposed (Cronin *et al.*, 1981). However, averaging dates and estimates for body size might disguise a true punctuational pattern, and for modern *H. sapiens* alone there is a range of *average* adult body size in different human groups of *at least* 40–70 kg (Eveleth and Tanner, 1976). We might not expect variation on quite this scale in Pleistocene hunter-gatherers with a more restricted geographical and environmental range, but widely dispersed groups would certainly have been expected to vary in body size following Bergmann's rule (see p 68). Because of the lack of postcranial material, there are few estimates of body size published for Middle and early Upper Pleistocene specimens, so I would like to conclude this discussion with some exploratory results for such hominids.

A new approach to estimating body size for fossil hominids was proposed by Steudel (1980). Regression analyses for ten Old World primate species and modern humans were used to examine the relationship between body size (measured indirectly as log. partial skeletal weight) and various measurements of the cranium and postcranium. Relationships between partial skeletal weight and palate breadth ($r = 0{\cdot}96$), bizygomatic breadth ($r = 0{\cdot}98$), orbital width ($r = 0{\cdot}97$) and femoral circumference ($r = 0{\cdot}98$) were considered to provide the best data base for predicting body weights of extinct forms such as fossil hominids. Using this approach, Steudel estimated an average body weight of *ca.* 36 kg for gracile early hominids (= *A. africanus*) and *ca.* 59 kg for the robust forms (= *Paranthropus*), with a much greater size variation in the latter group, probably related to a level of sexual dimorphism comparable to that of modern *Pongo*.

Although Steudel did not exactly define how she took the cranial measurements, assuming bizygomatic and orbit breadths are taken in a standardized way (Howells, 1973) and palate breadth is taken internally at the $P^4/M^1$ alveolus, it is possible to apply Steudel's approach to cranial data of other fossil

Table 3.1. Estimated partial skeletal weights for fossil hominids using Steudel's (1980) technique.

| | Actual Log. partial skeletal weight (kg) (Steudel 1980) | | | Mean value for fossils (no. estimators) |
|---|---|---|---|---|
| 3·00 | | | | |
| | ♂ *Gorilla* max. | | | [Bodo (1)] |
| 2·90 | | | | |
| | ♂ *Gorilla* mean | | | Ferrassie 1 (3) Petralona (3) Irhoud 1 (3) |
| 2·80 | | | *Pongo* range (22) and probable *Paranthropus* range (Steudel) | Broken Hill (3) |
| | | | | La Chapelle (2) Qafzeh 6 (2) Shanidar 1 (3) |
| 2·70 | | *H. sapiens* range (31) | | Circeo 1 (2) Cro-Magnon 1 (2) |
| | ♀ *Gorilla* ♂ *H. sapiens* mean | | | Arago 21 (2) Saccopastore 1 (2) and 2 (3) |
| 2·60 | | | | Sangiran 17 (2) Qafzeh 9 (3) Gibraltar 1 (3) |
| | ♀ *H. sapiens* mean | | | KNM-ER 3883 (2) |
| | | | | KNM-ER 3733 (3) |
| 2·50 | ♂ *Pan* mean | | | Abri Pataud (3) |
| | | | | [Steinheim (1)] [KNM-ER 1470 (1)] |
| | ♀ *Pan* mean | | | KNM-ER 1813 (3) *A. africanus* mean (Steudel) |
| 2·40 | | | | |

hominids (Table 3.1). As can be seen, the results are somewhat surprising. While in Steudel's study the size of the gracile and robust australopithecines appeared comparable with that of modern *Pan* and *Pongo*, respectively, as found in some earlier studies, many of the new estimates for fossil hominids of the Middle and Upper Pleistocene equal or exceed the largest values for modern *Homo*, suggesting that some fossil hominids could have been like large specimens of *Gorilla* in body size (as much as 150 kg in weight). The early *H. erectus* specimens from Koobi Fora (KNM-ER 3733 and 3883) appear comparable in body size to modern human females and males respectively. But variation in the Middle Pleistocene of Europe and Africa appears to have been much larger, although estimates for the most extreme specimens are based on orbital breadth measures only and are accordingly bracketed. Male Neanderthals, based on more complete data, would still have equalled or exceeded the largest modern male values, while even probable female specimens (Saccopastore 1 and Gibraltar 1) would have exceeded the average modern male figures. Early anatomically modern specimens such as Qafzeh and Cro-Magnon 1 are also estimated to have been very large-bodied, while the later Upper Palaeolithic female individual from Abri Pataud is estimated to be comparable to modern female values. These last results are reasonable considering the surviving postcranial material.

This method needs further development and testing on modern and fossil skeletons where comparisons can be made between values on the living, skeletal weights and body size estimates from the cranium and various parts of the postcranium. In particular a check should be made as to whether cranial size increase in hominids could itself be affecting the selected breadth measurements leading to overestimations of body weight. But the approach is promising and gives fairly consistent results for specimens where all three relevant cranial measurements can be taken. Biorbital breadth tended to give the largest estimates (perhaps explaining the very large value for the Bodo cranium, but not the very small value for Steinheim), while bizygomatic breadth tended to give the smallest estimates, in contrast to Steudel's findings in her australopithecine samples.

If these indirect body weight estimates for Pleistocene hominids are even approximately correct, they have various implications for studies of human evolution. The robusticity of Middle and early Upper Pleistocene hominids could be related to a very large body mass compared to recent and earlier hominids. The increase in absolute endocranial volume clearly documented during Pleistocene human evolution and its apparent decline at the end of the Pleistocene (in Europe at least) (see p 77) might be related to an initial increase in body size compared to Plio-Pleistocene hominids, followed by a more recent decline (see Chapter 4, this volume). Cultural and behavioural factors might well have been responsible for such changes in body size (see e.g. Brues,

Table 3.2. Data on endocranial volume and age.

| Fossil | Endocranial volume (ml) | | Source | Assigned age (Myr) |
|---|---|---|---|---|
| Sterkfontein Sts 5 | 485 | | Holloway (1981a) | 2·75 |
| 60 | 436 | | Holloway (1981a) | 2·75 |
| 19/58 | 436 | | Holloway (1975) | 2·75 |
| KNM-ER 1470 | 752 | | Holloway (1980a) | 1·90 |
| Olduvai O.H.7 | 725 | (mean value) | Holloway (1980a) | 1·80 |
| O.H.24 | 590 | | Holloway (1975) | 1·80 |
| KNM-ER 1813 | 515 | | Holloway (pers. comm.) | 1·70 |
| O.H.13 | 650 | | Holloway (1980a) | 1·60 |
| KNM-ER 1805 | 580 | | Holloway (pers. comm.) | 1·50 |
| KNM-ER 3733 | 848 | | Holloway (pers. comm.) | 1·50 |
| KNM-ER 3883 | 804 | | Holloway (pers. comm.) | 1·50 |
| O.H.9 | 1067 | | Holloway (1975) | 1·20 |
| Sangiran 4 | 908 | | Holloway (1981c) | 1·00 |
| Trinil 2 | 940 | | Holloway (1981c) | 0·70 |
| Sangiran 2 | 813 | | Holloway (1981c) | 0·70 |
| Sangiran 10 | 855 | | Holloway (1981c) | 0·70 |
| Sangiran 12 | 1059 | | Holloway (1981c) | 0·70 |
| Sangiran 17 | 1004 | | Holloway (1981c) | 0·70 |
| O.H.12 | 750 | | Holloway (1980a) | 0·70 |
| Choukoutien II | 1030 | | Weidenreich (1943) | 0·40 |
| Choukoutien III | 915 | | Weidenreich (1943) | 0·40 |
| Choukoutien X | 1225 | | Weidenreich (1943) | 0·40 |
| Choukoutien XI | 1015 | | Weidenreich (1943) | 0·40 |
| Choukoutien XXI | 1030 | | Weidenreich (1943) | 0·40 |
| Petralona | 1230 | | Protsch (pers. comm.) | 0·40 |
| Salé 1 | 850 | | Holloway (1981b) | 0·40 |
| Ngandong 1 | 1172 | | Holloway (1980b) | 0·20 |
| Ngandong 6 | 1251 | | Holloway (1980b) | 0·20 |
| Ngandong 7 | 1013 | | Holloway (1980b) | 0·20 |
| Ngandong 11 | 1231 | | Holloway (1980b) | 0·20 |
| Ngandong 12 | 1090 | | Holloway (1980b) | 0·20 |
| Broken Hill 1 | 1285 | | Holloway (1981a) | 0·20 |
| Laetoli L.H.18 | 1367 | | Holloway (pers. comm.) | 0·13 |
| Djebel Irhoud 1 | 1305 | | Holloway (1981b) | 0·05 |
| Spy 1 | 1305 | | Holloway (1981b) | 0·05 |
| Spy 2 | 1553 | | Holloway (1981b) | 0·05 |
| La Quina 5 | 1350 | | Boule (quoted Holloway 1981b) | 0·05 |
| La Chapelle | 1626 | | Boule (quoted Holloway 1981b) | 0·05 |
| La Ferrassie 1 | 1689 | | Heim (1976) | 0·05 |
| Amud 1 | 1750 | | Ogawa *et al.* (1970) | 0·05 |
| Modern *H. sapiens* (13) | Mean 1475 | | Holloway (pers. comm.) | 0·00 |
| minimum | 1166 | | Holloway (pers. comm.) | 0·00 |
| maximum | 1659 | | Holloway (pers. comm.) | 0·00 |

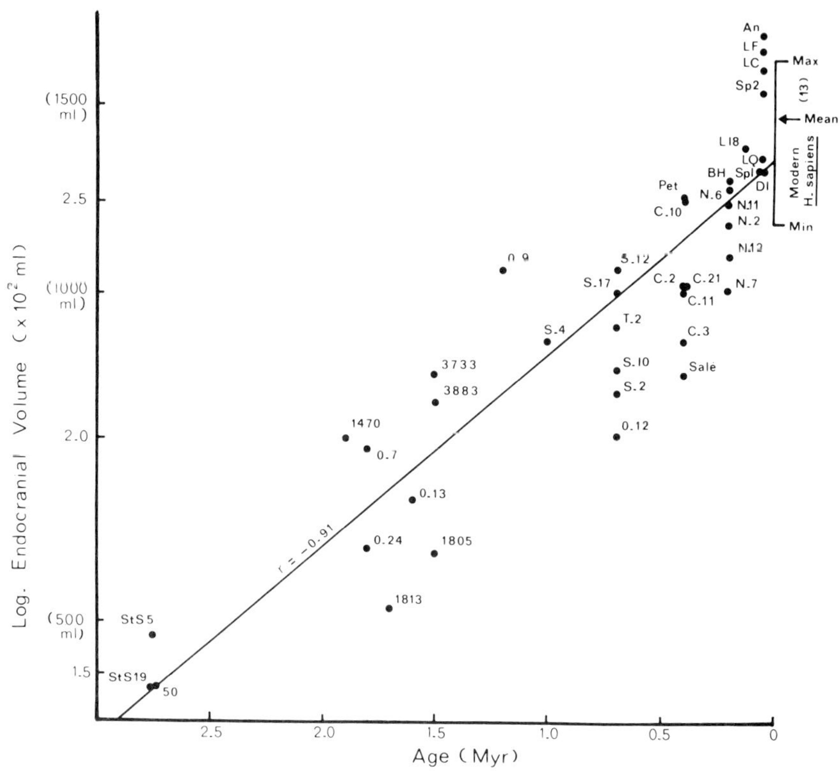


Fig. 3.3. A plot of log endocranial volume against actual or estimated geological age for fossil and living hominids. The least squares regression line shown is the regression for endocranial volume to age, but the regression for age to volume and the reduced major axis were also calculated. They have not been plotted as there is little difference between them and the line shown. The equation for the regression of age to log volume is $y = -0{\cdot}50x + 2{\cdot}66$, and the reduced major axis equation is $y = -0{\cdot}45 + 2{\cdot}62$.

1959; Frayer, 1981), but nutritional and social factors (e.g. changes in reproductive patterns) could also have been involved. However, given the present data, while the increase in brain size from early *H. erectus* to "archaic *H. sapiens*" specimens might be attributed wholly to body size changes, the even larger brain sizes of Neanderthal and a.m. specimens are not so easily explained since body size estimates seem comparable between the Middle and Upper Pleistocene fossils, whereas absolute brain volume is significantly increased. A collaborative study in which I am currently involved, using skeletal measurements to estimate body weights, suggests that the Neanderthals did not differ markedly from modern humans in body weight.

## 5. BRAIN SIZE AND TIME

It has been a commonly-held assumption that hominid evolution was accompanied by a steady increase in brain size (see e.g. Tobias, 1971; Lestrel, 1976). Some workers have argued for a strictly linear relationship between time and brain size (Lestrel, 1976; Cronin *et al.*, 1981), others for an autocatalytic model of change, where the rate of change accelerates through time (see e.g. Tobias, 1971; Holloway, 1972). More recently, proponents of the punctuated equilibria model have argued that a stepped pattern of brain increase, followed by stasis, may be more appropriate (Gould and Eldredge, 1977; Zindler, 1978; Godfrey and Jacobs, 1981). In this view the apparent gradual trend for increasing hominid brain size through time might, in reality, be due to combining values for successive separate taxa, each with fairly static brain sizes through time. In Table 3.2 and Figure 3.3 I present data for endocranial volume and age determined for fossil hominids and a recent human sample. Endocranial volume can be measured or calculated in various ways, often giving disparate values for the same specimens. The endocranial data presented here are mainly based on one method, that of direct endocast volume determinations, with the minimum of reconstruction. These data are probably the most accurate and consistent available at present since most determinations derive from one experienced worker, Ralph Holloway. Only adult or sub-adult specimens have been used, but the dates quoted are based on my own assessment of published absolute or relative dates. So as in other recent investigations (e.g. Godfrey and Jacobs, 1981; Rightmire, 1981), the dating of the samples is far less certain than the endocranial volume determinations. For this reason I have presented the data in full so that other workers can make their own modifications where appropriate. The early and late portions of the data are actually more reliably dated than the later Lower Pleistocene/Middle Pleistocene part, where there is a dearth of reliable absolute dates. For Indonesia in particular, chronological control is very poor, and even the relative ages of the most important specimens are subject to doubt. Some of the sample derives from the Pucangan Beds, most from the Kabuh Beds, and some from the Notopuro Beds. Whereas some data would place the oldest Indonesian fossils in the earliest Pleistocene or even Pliocene (Ninkovich and Burckle, 1978; Siesser and Orchiston, 1978; Curtis, 1981; Sartono, 1982), the majority of finds would postdate 0·73 Myr if we accept Sartono's (1982) identification of the Brunhes–Matuyama boundary at the top of the Pucangan Beds, while the opposite would be the case if we accept the Itihara *et al.* (1982) identification of a tektite bed dated at 0·70 ky nearer the top of the Kabuh Beds. Furthermore there is a fundamental disagreement about the dating of the Ngandong specimens. Sartono (1982) would place these in the Upper Pleistocene, Jacob (1981) in the Middle Pleistocene, close to other

Sangiran specimens. Finally, the faunal divisions on which the old Indonesian biostratigraphic sequence was erected are now also under reexamination (de Vos *et al.*, 1982) and are likely to be fundamentally modified in the future. So the Indonesian sample, which is so important a part of studies on brain size in fossil hominids, cannot be definitively dated at present [for further discussion see Matsu'ura (1982) and Pope (1983)]. As a result I have, with no great conviction, assigned the main Indonesian sample a date of 0·70 Myr, while Sangiran 4, generally agreed to be of true lower Pleistocene age (Sartono, 1982), is assigned an age of 1·00 Myr. The Ngandong fossils are dated here at 0·20 Myr. Caveats about these dates must apply equally to other workers utilizing this material in studies of brain size changes in fossil hominids (e.g. Godfrey and Jacobs, 1981; Rightmire, 1981).

Two graphs of endocranial volume against time were prepared. The first used actual volume determinations, and the second (Fig. 3.3) shows a logarithmic transformation of the endocranial data. The first linear regression analysis of volume to age fitted the data less satisfactorily ($y = -3{\cdot}60x + 13{\cdot}1$, where $y$ = endocranial volume in $10^2$ml, $x$ = age in Myr, $r = -0{\cdot}85$), rather underestimating the capacity of the earliest and latest hominid specimens. Not surprisingly, logarithmic transformation of the endocranial values provides a much better fit for the linear regression line of volume to age ($y = -0{\cdot}41x + 2{\cdot}59$, where $y$ = log. endocranial volume in $10^2$ ml, $x$ = age in Myr, $r = -0{\cdot}91$), although the Upper Pleistocene increase in endocranial volume is still remarkable, as Godfrey and Jacobs (1981) also note. From this data, the autocatalytic model of endocranial volume increase seems most appropriate since there is an increasing rate of change until the late Pleistocene, when endocranial capacity values stabilize or even decline. Alternatively, interpreting the data by the punctuational model, three main separate segments each with *relative* (but *not* absolute) stasis might be recognized—an early *Australopithecus*—*H. habilis* portion, a *H. erectus*—"early archaic *H. sapiens*" portion, and a Neanderthal—a.m. *H. sapiens* portion. However, data (at present unavailable) from later Pleistocene non-Neanderthal specimens such as Omo 2 and Da-li might well provide a greater overlap between the later Middle Pleistocene and Upper Pleistocene groups.

## 6. CONCLUSIONS

This paper has been more concerned with identifying problem areas in the study of fossil hominids than providing conclusions about the course of human evolution. Larger samples and better dating should improve our knowledge of these areas, such as the origins of the genus *Homo*, of "archaic *H. sapiens*", and of a.m. *H. sapiens*. But new analytical approaches will be needed too, and these must include the wider application of cladistic analysis and the use of cladistic principles in hominid classification. I believe that this approach could clarify whether *A. afarensis* is a genuine single species, whether *A. africanus* is a sister group to the genus *Paranthropus* or to the genus *Homo*, whether *H. habilis* and *H. erectus* are real morphological species, and whether the "archaic *H. sapiens*" group in fact consists of separate subgroups allied cladistically to (at least) a.m. *H. sapiens* and Neanderthals. In the most commonly accepted hominid classification (Fig. 3.1) the main taxonomic division in the later Pleistocene is made between late *H. erectus* and *H. sapiens* (including "archaic *H. sapiens*", the Neanderthals, and a.m. *H. sapiens*). In common with other palaeoanthropologists, I find many difficulties in clearly distinguishing between *H. erectus* and "*H. sapiens*" in the Middle Pleistocene but I find no such problems in dividing the Upper Pleistocene material into at least three groups—a.m. *H. sapiens*, Neanderthals and "others". Such difficulties may be a reflection of different modes of evolution. Perhaps the *erectus*—"early *sapiens*" transition was gradual and widespread, characterized by mosaicism, whereas the appearance of Neanderthals and a.m. *H. sapiens* was punctuational and localized, leading to easy recognition of taxonomic boundaries. But alternatively, is it possible that the main later Pleistocene taxonomic divisions are drawn in the wrong place and should instead come between a.m. *H. sapiens*, the Neanderthals, and all the other Middle Pleistocene fossils? If so, are we justified in extending the term *H. sapiens* beyond the group which we can define most satisfactorily on derived, within-group characteristics—a.m. *H. sapiens* (Day and Stringer, 1982). If the answer is no, then the Neanderthals might be recognized as a distinct species once again (*H. neanderthalensis*), and the "archaic *H. sapiens*" group could either be allocated a single specific name (such as *H. heidelbergensis* or *H. rhodesiensis*) if it can be characteristized by its own distinct within-group characteristics (a very difficult task), or it could be divided into separate subgroups related, by shared derived characteristics, to (at least) *H. erectus*, *H. sapiens* or *H. neanderthalensis*. If, on the other hand, this approach is considered by other workers to lead to excessive splitting, retention of the previous approach of applying subspecific names would only be justifiable, in my opinion, if these groups could also be formulated on cladistic, as well as geographic and chronological grounds (see Bonde, 1981; Day and Stringer, 1982).

Taking available genetic evidence into account, the origin of "modern" *H. sapiens* appears to have been quite recent, of the order of 100 000 years ago. (Jones, 1981). This independent evidence can be used to support the proposal, based on the fossils, that "true" *H. sapiens* appeared well after the time of *H. erectus sensu stricto*, and hominid classification may have to take account of this fact.

# REFERENCES


Aiello, L. (1981). Locomotion in the Miocene Hominoidea. In C. B. Stringer (ed) *Aspects of human evolution*. London: Taylor and Francis. pp. 63–97.

Allen, J. A. (1877). The influence of physical conditions in the genesis of species. *Radical Rev*. **1**: 108–140.

ApSimon, A. P. (1981). Matters arising. *Nature* **289**: 823–824.

Bergmann, C. (1847). Über die Verhältnisse der Wärmeökonomie der Thiere zu ihrer Grösse. *Göttinger Stud*. **8**.

Bonde, N. (1981). Problems of species concepts in palaeontology. *International Symposium of Concepts and Methods in Palaeontology*. Barcelona. 1981.

Brose, D. S. and Wolpoff, M. H. (1971). Early Upper Palaeolithic man and late Middle Palaeolithic tools. *Am. Anthrop*. **73**: 1156–1194.

Brown, F. H. (1982). Tulu Bor Tuff at Koobi Fora correlated with the Sidi Hakoma Tuff at Hadar. *Nature* **300:** 631–633.

Brues, A. M. (1959). The spearman and the archer. *Am. Anthrop*. **72**: 1073–1078.

Carey, J. W. and Steegman, A. T. Jr. (1981). Human nasal protrusion, latitude, and climate. *Am. J. phys. Anthrop*. **56**: 313–319.

Cook, J., Stringer, C. B., Currant, A. P., Schwarcz, H. P. and Wintle, A. G. (1982). A review of the chronology of the European Middle Pleistocene hominid record. *Yb. phys. Anthrop*. **25:** 19–65.

Coon, C. S. (1962). *The Origin of Races*. New York: Knopf.

Cracraft, J. (1982). Pattern and process in palaeobiology: the role of cladistic analysis in systematic palaeontology. *Palaeobiol*. 7: 456–468.

Cronin, J. E., Boaz, N. T., Stringer, C. B. and Rak, Y. (1981). Tempo and mode in hominid evolution. *Nature* **292**: 113–122.

Curtis, G. H. (1981). Establishing a relevant time-scale in anthropological and archaeological research. *Phil. Trans. R. Soc. Lond*. B. **292**: 7–20.

Day, M. H. (1982). The *Homo erectus* pelvis: punctuation or gradualism? *Congrès International de Paléontologie Humaine. Nice*. Prétirage: pp. 411–421.

Day, M. H. and Stringer, C. B. (1982). A reconsideration of the Omo Kibish remains and the *erectus-sapiens* transition. *Congrès International de Paléontologie Humaine. Nice*. Prétirage: pp. 814–846.

Delson, E. (1981). Palaeoanthropology: Pliocene and Pleistocene human evolution. *Palaeobiol*. 7: 298–305.

De Vos, J., Sartono, S., Hardja-Sasmita, S. and Sondaar, P. Y. (1982). The fauna from Trinil, type locality of *Homo erectus*; a reinterpretation. *Geologie Mijnb*. **61**: 207–211.

Drake, R. E., Curtis, G. H., Cerling, T. E., Cerling, B. W. and Hampel, J. (1980). KBS Tuff dating and geochronology of tuffaceous sediments in the Koobi Fora and Shungura Formations, East Africa. *Nature* **283**: 368–372.

Eveleth, P. B. and Tanner, J. M. (1976). *Worldwide Variation in Human Growth*. Cambridge: Cambridge University Press.

Frayer, D. (1981). Body size, weapon use, and natural selection in the European Upper Palaeolithic and Mesoliothic. *Am. Anthrop*. **83**: 57–73.

Godfrey, L. and Jacobs, K. H. (1981). Gradual, autocatalytic and punctuational models of hominid brain evolution: a cautionary tale. *J. hum. Evol*. **10**: 254–272.

Gould, S. J. and Eldredge, N. (1977). Punctuated equilibria: the tempo and mode of evolution reconsidered. *Palaeobiol*. **3**: 115–151.

Grine, F. E. (1981). Trophic differences between "gracile" and "robust" Australopithecines: a scanning electron microscope analysis of occlusal events. *S. Afr. J. Sci.* **77**: 203–230.

Heim, J.-L. (1976). Les hommes fossiles de la Ferrassie. *Arch. Inst. Paléont. Hum.* Mémoir **35**: 1–331.

Holloway, R. L. (1972). Australopithecine endocasts, brain evolution in the Hominoidea, and a model of hominid evolution. In: Tuttle R. H. (ed). *The Functional and Evolutionary Biology of Primates*. Chicago: Aldine-Atherton.

Holloway, R. L. (1975). Early hominid endocasts: volumes, morphology and significance for hominid evolution. In: Tuttle, R. H. (ed). *Primate Functional Morphology and Evolution*. The Hague: Mouton.

Holloway, R. L. (1978). Problems of brain endocast interpretation and African hominid evolution. In: Jolly, C. J. (ed) *Early Hominids of Africa*. London: Duckworth.

Holloway, R. L. (1980a). The O.H.7 (Olduvai Gorge, Tanzania) hominid partial brain endocast revisited. *Am. J. phys. Anthrop.* **53**: 267–274.

Holloway, R. L. (1980b). Indonesian "Solo" (Ngandong) endocranial reconstructions: some preliminary observations and comparisons with Neanderthal and *Homo erectus* groups. *Am. J. phys. Anthrop.* **53**: 285–295.

Holloway, R. L. (1981a). Exploring the dorsal surface of hominid brain endocasts by stereoplotter and discriminant analysis. *Phil. Trans. R. Soc. Lond.* B.**292**: 155–166.

Holloway, R. L. (1981b). Volumetric and asymmetry determinations on recent hominid endocasts: Spy I and II, Djebel Irhoud 1, and the Salé *Homo erectus* specimens, with some notes on Neanderthal brain size. *Am. J. phys. Anthrop.* **55**: 385–393.

Holloway, R. L. (1981c). The Indonesian *Homo erectus* brain endocasts revisited. *Am. J. phys. Anthrop.* **55**: 503–521.

Howell, F. C. (1978). Hominidae. In: Maglio, V. J. and Cooke, H. B. S. (ed) *Evolution of African Mammals*. Cambridge, Mass: Harvard University Press.

Howells, W. W. (1973). Cranial variation in man: a study by multivariate analysis of patterns of difference among recent human populations. *Pap. Peabody Mus.* **67**: 1–259 (whole volume).

Howells, W. W. (1976). Explaining modern man: evolutionists *versus* migrationists. *J. hum. Evol.* **5**: 477–495.

Howells, W. W. (1980). *Homo erectus*—who, when and where: a survey. *Yb. phys. Anthrop.* **23**: 1–23.

Itihara, M., Kadar, D., Sudijono, Shibasaki, T., Kumai, H. and Yoshikawa, S. (1982). *Abstracts XI INQUA Congress, Moscow* Vol. **I**: 139.

Jacob, T. (1980). The *Pithecanthropus* of Indonesia: phenotype, genetics and ecology. In: Königsson, L.-K. (ed.) *Current Argument on Early Man*. Oxford: Pergamon Press.

Jelinek, A. (1982). The Tabūn cave and Palaeolithic man in the Levant. *Science*, N.Y. **216**: 1369–1375.

Jelínek, J. (1980). European *Homo erectus* and the origin of *Homo sapiens*. In Königsson, L.-K. (ed). *Current Argument on Early Man*. Oxford: Pergamon Press.

Johanson, D. C. and Edey, M. A. (1981). *Lucy: The Beginnings of Mankind*. London: Granada.

Johanson, D. C. and White, T. D. (1979). A systematic assessment of early African hominids. *Science*, N.Y. **203**: 321–330.

Johanson, D. C., White, T. D. and Coppens, Y. (1978). A new species of the genus *Australopithecus* (Primates: Hominidae) from the Pliocene of Eastern Africa.

*Kirtlandia*. **28**: 1–14.
Jones, J. S. (1981). How different are human races? *Nature* **293**: 188–190.
Jungers, W. L. (1982). Lucy's limbs: skeletal allometry and locomotion in *Australopithecus afarensis*. *Nature* **297**: 676–678.
Lestrel, P. E. (1976). Hominid brain size versus time: revised regression estimates. *J. hum. Evol.* 5: 207–212.
Lévèque, F., and Vandermeersch, B. V. (1981). Le Néandertalien de Saint-Césaire. *Recherche* **12**: 242–244.
Martin, R. D. (1981). Relative brain size and basal metabolic rate in terrestrial vertebrates. *Nature* **293**: 57–60.
Matsu'ura, S. (1982). A chronological framing for the Sangiran hominids. *Bull. Nat. Sci. Mus. Tokyo*. D. **8**: 1–53.
McDougall, I., Maier, R., Sutherland-Hawkes, P. and Gleadow, A. J. W. (1980). K-Ar age estimate for the KBS Tuff, East Turkana, Kenya. *Nature* **284**: 230–234.
McHenry, H. M. (1975). Fossil hominid body weight and brain size. *Nature* **254**: 686–687.
McHenry, H. M. (1976). Early hominid body weight and encephalization. *Am. J. phys. Anthrop*. **45**: 77–84.
Ninkovich, E. and Burckle, L. H. (1978). Absolute age of the base of the hominid-bearing beds in Eastern Java. *Nature* **275**: 306–308.
Ogawa, T., Kamiya, T., Sakai, S. and Hosokawa, H. (1970). Some observations on the endocranial cast of the Amud Man. In: Suzuki, H. and Takai, F. (eds) *The Amud Man and his Cave Site*. Tokyo: Academic Press of Japan.
Olson, T. R. (1981). Basicranial morphology of the extant hominoids and Pliocene hominids: the new material from the Hadar formation, Ethiopia, and its significance in early human evolution and taxonomy. In: Stringer, C. B. (ed) *Aspects of Human Evolution*. London: Taylor and Francis.
Pilbeam, D. R. and Gould, S. J. (1974). Size and Scaling in human evolution. *Science, N.Y.* **186**: 892–901.
Pope, G. G. (1983). Evidence on the age of the Asian hominidae. *Proc. Natl. Acad. Sci. USA* **80**: 4988–4992.
Rak, Y. (1981). *The Morphology and Architecture of the Australopithecine Face*. Ph.D. thesis, University of California, Berkeley.
Reader, J. (1981). *Missing Links*. London: Collins.
Repenning, C. A. and Fejfar, O. (1982). Evidence for earlier date of 'Ubeidiya, Israel, hominid site. *Nature* **299**: 344–347.
Rightmire, G. P. (1981). Patterns in the evolution of *Homo erectus*. *Palaeobiol*. 7: 241–246.
Robinson, J. T. (1972). *Early Hominid Posture and Locomotion*. Chicago: University of Chicago Press.
Sartono, S. (1982). Characteristics and chronology of early men in Java. *Congrès International de Paléontologie Humaine. Nice*. Prétirage: 491–533.
Siesser, W. G. and Orchiston, D. W. (1978). Micropalaeontological reassessment of the age of *Pithecanthropus* mandible C from Sangiran, Indonesia. *Mod. Quat. Res. SE Asia* **4**: 25–30.
Senut, B. (1981). L'humerus et ses articulations chez les hominidés pliopléistocènes. Paris: Centre National de la Recherche Scientifique.
Stanley, S. M. (1979). *Macroevolution*. San Francisco: Freeman.
Stanley, S. M. (1982). *The New Evolutionary Timetable*. New York: Basic Books.
Steegman, A. T. Jr. (1972). Cold response, body form and craniofacial shape in two racial groups of Hawaii. *Am. J. phys. Anthrop*. **37**: 193–222.

Steudel, K. (1980). New estimates of early hominid body size. *Am. J. phys. Anthrop.* **52**: 63–70.
Stringer, C. B. (1982). Towards a solution to the Neanderthal problem. *J. hum. Evol.* **11**: 431–438.
Stringer, C. B. and Burleigh, R. (1981). The Neanderthal problem and the prospects for direct dating of Neanderthal remains. *Bull. Br. Mus. nat. Hist. (Geol.).* **35**: 225–241.
Stringer, C. B., Kruszynski, R. G. and Jacobi, R. M. (1981). Matters arising. *Nature* **289**: 823–824.
Thorne, A. G. and Wolpoff, M. H. (1981). Regional continuity in Australasian Pleistocene hominid evolution. *Am. J. phys. Anthrop.* **55**: 337–349.
Tobias, P. V. (1971). *The Brain in Hominid Evolution.* New York: Columbia University Press.
Tobias, P. V. (1980a). "*Australopithecus afarensis*" and *A. africanus*: critique and alternative hypothesis. *Palaeont. afr.* **23**: 1–17.
Tobias, P. V. (1980b). A survey and synthesis of the African hominids of the late Tertiary and early Quaternary periods. In: Königsson, L.-K. (ed.) *Current Argument on Early Man.* Oxford: Pergamon Press.
Trinkaus, E. (1981). Neanderthal limb proportions and cold adaptation. In: Stringer, C. B. (ed.) *Aspects of Human Evolution.* London: Taylor and Francis.
Trinkaus E. and Howells, W. W. (1979). The Neanderthals. *Scient. Am.* **241**: 118–133.
Vallois. H. V. (1958). La Grotte de Fontéchevade. II: Anthropologie. *Arch. Inst. Paléont. Hum.* Mémoir **29**: 1–164.
Vandermeersch, B. V. (1981). Les premiers *Homo sapiens* au Proche-Orient. In: *Les Processus de l'Hominisation.* Paris: Centre National de la Recherche Scientifique.
Walker, A. (1981). Dietary hypotheses and human evolution. *Phil. Trans. R. Soc. Lond.* **B.292**: 57–64.
Walker, A. C. and Leakey, R. E. F. (1978). The hominids of East Turkana. *Scient. Am.* **239**: 44–56.
Weidenreich, F. (1943). The skull of *Sinanthropus pekinensis*: a comparative study on a primitive hominid skull. *Palaeont. Sinica.* **D.10**: 1–292.
White, T. D., Johanson, D. C. and Kimbel, W. H. (1981). *Australopithecus africanus:* its phylogenetic position reconsidered. *S. Afr. J. Sci.* **77**: 445–470.
Williamson, P. G. (1981). Palaeontological documentation of speciation in Cenozoic molluscs from Turkana Basin. *Nature* **293**: 437–443.
Wolpoff, M. H. (1980). *Palaeoanthropology.* New York: Knopf.
Wolpoff, M. H. (1981). Matters arising. *Nature* **289**: 823.
Wu Rukang and Wu Zinzhi (1982). Comparison of Tautavel man with *Homo erectus* and early *Homo sapiens* in China. *Congrès International de Paléontologic Humaine. Nice.* Prétirage: 605–616.
Zhou Mingzhen, Li Yanxian and Wanglinghong (1982). Chronology of the Chinese fossil hominids. *Congrès International de Paléontologie Humaine. Nice.* Prétirage: 593–604.
Zindler, R. E. (1978). On the increase of cranial capacity in mankind's lineage: augments and elaborations. *J. hum. Evol.* **7**: 295–305.

# 4 *Early Man and the Red Queen: Tropical African Community Evolution and Hominid Adaptation*

*Robert Foley*

## 1. INTRODUCTION

Adaptation is the process of survival. Adaptation may be genetically determined or environmentally induced; it may be physiological, anatomical, biochemical, or it may be behavioural. Indeed, the continuities between these categories are all too obvious. The implications, are, though, that both morphology, as Stringer (see Chapter 3, this volume) has shown, and behaviour are fields of palaeoanthropological study. Behavioural characteristics should be examined in terms of the extent to which they enhance survival and reproductive success. Applying this to the early hominids, it indicates that an evolutionary approach should be applied as much to the observed behavioural patterns of prehistoric populations as to their fossilized bones. The study of human evolution must incorporate the selection for types of behaviour and subsistence activity.

The evolutionary ecology of early hominid behaviour patterns is thus the subject of this paper, which will examine in this context early hominid adaptation in Africa during the Plio-Pleistocene. The starting point for such an analysis must be the hunter-gatherer. Since the publication of Lee and DeVore's influential book (1968), *Man the Hunter*, it has become a truism that the evolution of the hominids has been the evolution of the hunter-gatherer way of life, and that hunting and gathering has been the basis for survival throughout the Pleistocene. The recognition of the relationship between hunting/gathering and human evolution has had a profound effect on palaeoanthropology, directing researchers' interests towards more behavioural questions. Problems with such an approach can however arise. First, we are faced with the possibility of imposing the way of life of the modern hunter-

HOMINID EVOLUTION
*ISBN 0-12-261920-X*

gatherer onto the archaeological record, rather than testing for differences between the two. At the very least this has the effect of rendering invisible any patterns of temporal variation in the behaviour of early man. Secondly, it has removed the study of early man from the context of the particular environment in which he evolved. The hunter-gatherer model is a global model; as such it must account for hunter-gatherer activities in universal terms. Once this is done, though, there seems little point or possibility in relating the characteristics of the early hominids to the specifics of their environment.

It is these, among other more strictly empirical problems, that have lead to the current air of uncertainty in early hominid studies, as well as considerable debate (see Chapters 5 and 6, this volume, for a discussion of these issues). This paper will address some of these difficulties, and, more specifically, try to place characteristics of early hominid subsistence behaviour into the context of Later Cenozoic tropical African evolution and ecology. First I shall describe briefly the "global hunter-gatherer model" and the congruence between this and the Plio-Pleistocene archaeological evidence. Secondly, I shall look at the nature of adaptive and evolutionary processes, to provide a means for relating the characteristics of hominids to their environment. Thirdly, I shall examine the ecological problems faced by the early hominids and the selective forces this may lead to. Finally, I shall discuss the relationship between the adaptive problems faced by the early hominids and the patterns of community evolution and ecology in Africa.

## 2. HUNTER-GATHERERS AND EARLY HOMINIDS

The hunter-gatherer adaptation is the term given to those humans who survive by exploiting resources as they occur in the wild, and over the reproduction, behaviour, and distribution of which they exert little control. Understanding of these adaptations is advancing rapidly, but regrettably not as rapidly as the people themselves are disappearing. Currently hunter-gatherers are limited in distribution to environments that are marginal to agriculture, and they exist in a world dominated by agriculturalists and are often closely connected with more settled peoples. During the Pleistocene, though, some form of hunting and gathering was the universal way of life (Lee and DeVore, 1968).

Certain characteristics seem general to living hunter-gatherers. They live at relatively low population densities (Hassan, 1975), and their home ranges are large (Foley, 1978), larger than that expected for equivalent sized mammals and primates (McNab, 1963; Milton and May, 1975; Harested and Bunnell, 1979). They live in social groups, often referred to as bands and based on ties of reproductive purposes (Birdsell, 1953, 1968). The hunter-gatherer bands have a home base or settlement, which will act as a spatial focus for many activities. It is, however, seldom permanent, and most hunter-gatherer bands are mobile, moving several times a year. This movement might either take the form of a fixed annual cycle or a less patterned transhumance, depending upon the structure of resource availability (Binford, 1980). Hunter-gatherers are largely omnivorous, although the proportion of vegetable foods, animal foods, and aquatic foods varies with availability (Lee, 1968; Binford, 1980; Foley, 1982). There is usually a pattern of division of labour, with men responsible for acquiring meat, women plant foods, and both (depending upon circumstances) aquatic resources. Less specifically, a key characteristic of hunter-gatherer adaptation is its flexibility—in relation to resource base, in relation to spatial location, and in relation to group size.

The above description is at a very general level, and furthermore, presents a picture of uniformity where variation exists. The pattern of variability has recently been described and analysed by Binford (1979, 1980), but for the purposes of this paper a generalized model is sufficient at this stage. Rather, the central concern here is whether it is possible to identify any of these characteristics in the fossil and archaeological evidence of early hominids in Plio-Pleistocene Africa.

The earliest evidence for the appearance of hominid-like forms comes from the Late Miocene, and while it is not clear whether *Ramapithecus*, found widely in the tropical and sub-tropical Old World between 15 and 8 Myr, is actually a hominid or simply indicates similar trends among a range of primates, it represents patterns of primate evolution that would be expected to occur at the base of hominid ancestry (Pilbeam, 1979). There is, however, very little to link these early traces to the now abundant fossil remains of the Plio-Pleistocene. These later remains, though, are confined to sub-Saharan Africa (Howell, 1978). These archaeological and palaeontological data may be used to test for the occurrence of hunter-gatherer attributes among the early hominids.

### 2.1. Bipedalism

On the basis of the footprints discovered at Laetoli (Leakey, 1979) it may be stated with some degree of certainty that hominids were habitually bipedal (Charteris *et al.*, 1981). This is broadly speaking confirmed by anatomical studies, although it has recently been suggested that the bipedalism may have differed substantially from that of modern man (Jungers, 1982).

## 2.2. Body Size

Throughout the Pleistocene there has been a gradual increase in body size among hominids, from the diminutive *Australopithecus afarensis* at 25 kg (Johanson and Edey, 1981) to the more substantial *Homo sapiens sapiens* at approximately 60 kg (Pilbeam and Gould, 1974) (Fig. 4.1). Increase in size will result in increased metabolic costs and food requirements.

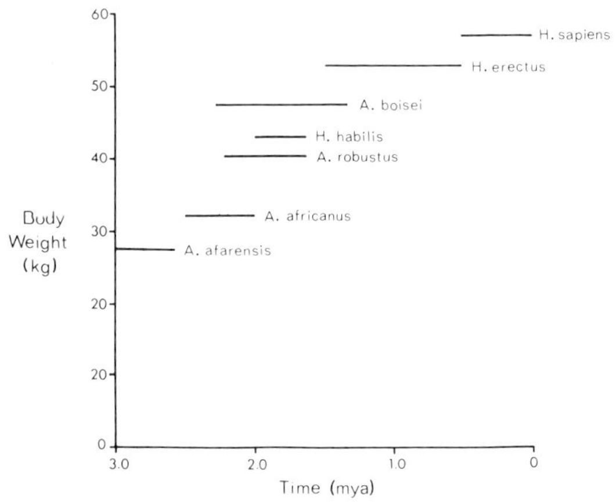


Fig. 4.1. Body weights of hominid species (data from Pilbeam and Gould, 1974; Johanson and Edey, 1981).

## 2.3. Population Density

Estimates of the density of populations in the past are notoriously tenuous, but various attempts have been made for early hominids. Boaz (1979), using relative trophic abundance has estimated an early hominid population density of between 0.006 and 1·7/km$^2$. R. A. Martin (1981) used the relationship between population density and body size (calculated from bicondylar width) known for some species (Harested and Bunnell, 1979) to calculate early hominid density. His estimates depend on diet, and range from 0·0009/km$^2$ for a carnivore to 1·61/km$^2$ for a herbivore. Apart from the range of the estimates, the method is further affected by group size considerations. Population density of early hominids is at present unknown.

### 2.4. Home Range Size

Patterns of home range size have been established for mammals (McNab, 1963; Harested and Bunnell, 1979), and also for primates (Milton and May, 1975; Clutton-Brock and Harvey, 1977). Using Milton and May's equation for primates total home range ($HR_t$), if the early hominids are behaving as ordinary primates they should have a home range size of 4·43 $km^2$. This would differ markedly from modern hunter-gatherer home range size (Foley, 1978). Direct evidence of home range size, independent of assumptions about behaviour, is difficult to obtain. Estimates of distances travelled, although not necessarily directly related to home range, have been estimated by distances from raw material sources. Isaac (1976, p 561) has stated that hominids at East Turkana during the Plio-Pleistocene were transporting rocks at least 4 km. Treating this as the diameter of a home range, a home range of 12·56 $km^2$ is indicated; as a radius the home range would be 50·2 $km^2$. Gowlett (see Chapter 7, this volume) has shown that hominids at Olduvai covered distances of 12 kilometres to obtain raw materials—i.e. a home range size of between 113·1 $km^2$ and 452·3 $km^2$. Even allowing for the innaccuracy and dangers of this method, the calculated home ranges are smaller than those of modern hunter-gatherers but larger than other primates (Foley, 1978).

### 2.5. Group Size

No method is available for calculating group size for early hominids. The only known potential method is that of Cook and Heizer (1951), but this is based on ethnographic data and the behaviour of fully modern man. Furthermore, in order to use it one would have to make assumptions about home bases, one of the variables we are trying to test.

### 2.6. Home Base

Early assumptions about all archaeological localities representing settlements have given way to a more healthy scepticism (Binford, 1981; Isaac and Crader, 1981). Non-hominid processes of accumulation have been recognized (see Chapters 5, 6, 8 and 10, this volume). However, it is clear that hominids *are* responsible for some accumulations of stones and bones (Isaac, 1983; Chapter 6, this volume). Despite hominid authorship of these localities, though, the inference that these represent home bases is not justified. What is justified is the inference that spatial foci of activity were part of the hominid ranging behaviour by the Plio-Pleistocene.

## 2.7. Mobility

Equal difficulties arise from trying to establish the degree of mobility of early hominids. On *a priori* grounds, and on account of the absence of evidence for a home base, mobility would seem likely, but its pattern is unknown.

## 2.8. Vegetable Diet

Tropical hunter-gatherers are known to have a substantial proportion of vegetable foods in their diet (Lee, 1968), and primates in general are essentially dependent upon plant foods (Harding, 1981). It would thus be expected that early hominids obtained a high proportion of their food from plant foods. Evidence for processing of plant foods has been presented by Keeley and Toth (1981), on the basis of polish on stone artifacts. More direct evidence comes from tooth morphology and wear. All Plio-Pleistocene hominids have enlarged posterior teeth (Howell, 1978), suggesting a diet requiring heavy mastication (Jolly, 1970). This would be expected on the basis of the character of plants in the dry tropics. Such a view is substantiated by hominid thickened tooth enamel (Howell, 1978), and by the pattern of wear on the teeth (Walker, 1981).

## 2.9. Meat Eating

The importance of meat in the diet of early hominids has always been contentious (see Chapter 5, this volume), and is particularly so at the present time [(Binford, 1981); see Chapter 6, this volume, for a full discussion of this issue]. Binford (1981) has claimed that there is no evidence for significant carnivory among early hominids. Bunn (1981) and Potts and Shipman (1981), on the other hand, have demonstrated on the basis of cutmarks on fossil bone that hominids were handling meat. Keeley and Toth (1981) have suggested that patterns of polish associated with meat processing can be identified on stone tools from the Plio-Pleistocene. Isaac and Crader (1981) have also assembled more conventional evidence for the early occurrence of meat-eating. The occurrence of predatory behaviour among other primates may also be significant (Harding, 1981). Thus, while it seems certain that hominids were involved in animal foods, the character and frequency of it is a matter of considerable debate.

## 2.10 Fishing

Secure evidence for the use of aquatic resources in Africa does not occur until the Upper Pleistocene, although some fish bones have been found in Olduvai Beds I and II (Leakey, 1971). Fluctuations in sea level and lake levels have possibly obscured much of the evidence, but dependence upon aquatic resources is probably a Late Pleistocene event.

## 2.11. Division of Labour and Food Sharing

Isaac (1978a,b) has argued that the occurrence of a home base and meat-eating is an indication of food sharing and a division of labour. Problems arise from his model on account of the ambiguity of the evidence for these characteristics (see above), and the fact that it is based on an ethnographic model (see also Chapters 5 and 7, this volume). However, on independent grounds, Lovejoy (1981) has also argued for the early appearance of these characteristics as they are related to a change in reproductive strategy and the evolution of bipedalism, for which there is sound evidence (see above).

## 2.12. Environment

Taphonomic problems are particularly acute in considering the evidence for the environment of Plio-Pleistocene Africa (Behrensmeyer, 1978). Broadly speaking all the hominid fossil and archaeological localities occur in the relatively dry regions of high Africa, rather than the wet forested lowlands; the environment of this area is dry woodland, bushland and grassland, generically known as the savanna. Bonnefille (1976, p 425) has suggested on the basis of pollen remains that the Omo Basin between 2·0 and 2·4 Myr had an open grassland/woodland mosaic, although the density and distribution of trees would have varied through time. Despite the evidence for broad habitat characteristics (Jolly, 1978), little is known of how hominids were distributed in relation to microhabitats (Hay, 1976).

What do these observations of the Plio-Pleistocene fossil and archaeological record add up to? They certainly cannot be said to provide evidence that the hunter-gatherer adaptive strategy, as known from living populations, was present. To some extent the evidence is inconclusive rather than inconsistent. Key elements—bipedalism, meat-eating, enlargement of home range, etc.—were probably present, but, as Hill (Chapter 5, this volume) points out, there is no reason for assuming that these were articulated in the same manner. Given these difficulties the most fruitful approach would seem to be to examine the evolutionary and ecological context in which early hominids were living, and try to establish whether these characteristics, or others, are expected outcomes. That is, we should look at the early hominids in relation to their biological community.

## 3. PROCESSES OF EVOLUTIONARY ECOLOGY

In order to proceed with this appoach it is necessary to consider the nature of the evolutionary process and its relationship to adaptation. The evolutionary process is based on selection. There are, though, several ways of modelling selection. One such model would see adaptive and thus evolutionary changes as "improvements". According to this model organisms are for the most part in balance with their environment. Occasionally, however, a mutation will occur that leads to an adaptive improvement, placing an individual at a selective advantage within a population. As a result of the reproductive benefits that this confers the mutant form will then spread through the population, and consequently an improvement may be said to have occurred. If this model is applied to the behavioural evolution of the hominids the picture that emerges is a group of terrestrial primates living in the increasingly dry environments of the tropics. Among one of these groups mutations occurred that lead to adaptive shifts—bipedalism, tool-making, meat-eating, etc.—that placed the protohominids at an advantage over the other terrestrial primates.

This model probably describes in very general terms what happened in the past, and yet it is unsatisfactory. It is descriptive rather than analytical; it argues on the basis of hindsight, explaining the emergence of the hominids by reference to the characteristics it is known arose at some stage of human evolution; and furthermore, it makes no reference to the context in which these changes took place—ecological, spatial or temporal—and thus reduces an evolutionary problem to a series of historical accidents. A further problem still lies in determining the criteria that should be used to measure an evolutionary improvement or advance. For example, if population density is used as an indication of evolutionary success—the number of reproductive units in an area—and then from a comparison of tropical hunter-gatherers with the group on which they are supposed to represent an improvement (the terrestrial primates), it can be seen that, after correcting for resource availability and body size differences, that in fact the non-human terrestrial primates have a higher biomass than the hunter-gatherers (Fig. 4.2). There are several ways in which the relationship might be explained—for example, in terms of the effects of hunter-gatherers occupying a higher position on the trophic pyramid—but it highlights the inadequacy of viewing the evolutionary process as a progressive one, or one that involves improvements.

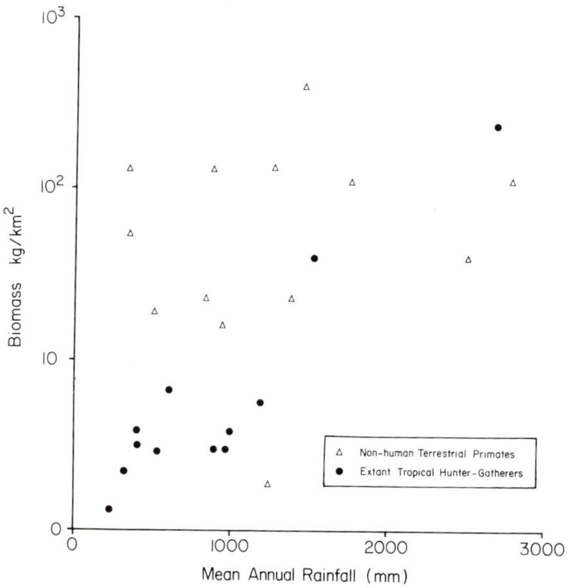


Fig. 4.2. Relationship between biomass and mean annual rainfall of tropical hunter-gatherers and terrestrial primates (data from Silberbauer, 1972; Hassan, 1975; Clutton-Brock and Harvey, 1977; Harris, 1978; Jones, 1980).

An alternative to this model of the evolutionary process is one that views evolutionary changes arise as solutions to problems faced by the organism in its environment. Adaptation is a problem solving process (Dobzhansky, 1974). New adaptations are selected for if they solve new problems or problems previously unsolved. Organisms thus survive, and where necessary undergo change, by solving the problems thrown up by their physical and social environment. Failure to solve these problems results in extinction. Selective agents are thus the organism's environmental problems. According to this model the characteristics of a hunter-gatherer adaptation would appear among the early hominids if they were solutions to problems faced by the hominids in their environment. In this case the reduced biomass of tropical hunter-gatherers would be referrable to problems not faced by the other terrestrial primates.

Where, though, do these environmental problems come from? An answer to this question has been provided by van Valen (1973). He argued that the selective agents operating on a species—i.e. its problems—are those other species which it eats, with which it competes, and which eat it. Thus, as

Maynard Smith (1976, p 325) put it, "each evolutionary advance made by one species in an ecosystem is experienced as a deterioration of the environment by other species". In other words, the problems faced by an organism are those caused by evolutionary changes among groups with which the organism interacts. The environmental problems referred to above are therefore not inanimate forces, but dynamic evolving creatures. From this model of selection van Valen (1973) deduced that to survive an organism would have to evolve as rapidly as possible simply to maintain its ecological *status quo*. He called this the Red Queen model, referring to the Red Queen in *Alice in Wonderland*, who remarked "Now here, you see it takes all the running you can do to stay in the same place".

Although van Valen's model has been criticized (Maynard Smith, 1976) for some of its assumptions, the basic model of the character of selective agencies is appropriate. This model may therefore act as the theoretical framework for studying the behavioural evolution of the early hominids. Populations undergo adaptive changes through natural selection to solve the problems of the environment, and the source of those problems is the broader ecological community. For this analysis, therefore, two things are required. First, identification of the ecological problems faced by the early hominids; and secondly, the relationship between these problems and the evolution of the tropical savanna community as a whole.

## 4. ADAPTIVE PROBLEMS

Three major ecological problems may be identified: (1) the problems of being a large mammal; (2) the problems of being a terrestrial primate; and (3) the problems of living in a tropical savanna environment.

### 4.1. Problems of Being a Large Mammal

The first mammals, appearing in the Permian more than 250 Myr, were very small, and indeed the majority of mammals today are small. According to Eisenberg (1981, p 234) their modal size is 32 cm. There has, however, been a trend towards body size increase throughout their evolution. Although hominids are not in the same class as the large sea mammals or *Baluchitherium*, the largest land mammal (Romer, 1966), they are nonetheless large mammals relative to the class as a whole. As such it is worth looking at what being large and warm blooded involves.

It has long been known that size influences many life history parameters (Kleiber, 1961; McNab, 1963; Jarman, 1974; Western, 1979; Eisenberg,

1981; R. D. Martin, 1981; Hennemann, 1983). Body size sets constraints on adaptation, enabling the prediction and description of the evolutionary and ecological changes that will occur with shifts in body size. Broadly speaking these constraints derive from the allometric relationship between linear measurements of body size and the exponential character of metabolic features (Kleiber, 1961).

From Table 4.1 the ecological and evolutionary consequences of becoming larger, as the hominids have done (see Fig. 4.1, and Chapter 3, this volume), can be seen. One effect will be an increase in total food requirements, because of higher metabolic rates. Each individual will therefore have to range over a larger area to obtain food, and as home range and population density are reciprocally related there will be a decrease in population density. The increased food requirements will not, though, be in a linear relationship with body size (Kleiber, 1961), and relative to body size there will be a decrease in metabolic requirements. An effect of this will be an increase in dietary breadth,

Table 4.1. Primary consequences of increases in body size for mammals (based on Eisenberg, 1981).

| Size effects | Ecological implications |
|---|---|
| Increase in absolute metabolic costs | 1. Greater food requirements<br>2. Larger home range<br>3. Lower population density |
| Decrease in relative metabolic costs | 1. Ability to survive on low quality food |
| Greater mobility | 1. Ability to exploit spatially and temporally patchy resources |
| Greater heat retention | 1. Ability to adapt to cooler climates + increased activity<br>2. Increased water dependence<br>3. Heat stress in tropical environments |
| Increased strength/speed | 1. Predator resistance/avoidance |
| Enlarged brain | 1. Exploitation of complex resource patches<br>2. Flexibility in behaviour<br>3. Increased sociality<br>4. Increased metabolic costs |
| Increased longevity | 1. Increased reproductive span<br>2. Increased developmental period |
| Longer pre-natal period | 1. Higher reproductive costs<br>2. Lower birth rate<br>3. Greater birth interval<br>4. Precocial young |

as indeed is the case with modern hunter-gatherers relative to non-human primates, as lower quality foods may be utilized. Larger mammals also tend to be more mobile than small mammals—they are physically capable of covering larger areas. This too has important ecological implications. An animal that can move around freely is capable of subsisting off resources that are patchily or unevenly distributed in space and time. A small mammal could only afford to depend on such a resource if the patch could sustain it for a long or indefinite period of time, while the larger mammal can move from patch to patch. This will often allow the larger bodied species to incorporate unpredictable, rare and ephemeral resources, often of very high nutritional quality, into its diet.

Greater heat retention is another consequence of increased size (Bergmann's Rule) (see Chapter 3, this volume). Large mammals can cope with extremes of cold better than small mammals, which may be a contributory factor in the expansion of the hominids out of the tropics during the Pleistocene (see Chapter 8, this volume). More importantly from the perspective of human evolution and adaptation within the tropics during the Pliocene and Lower Pleistocene, increased body size will lead to greater thermoregulatory problems. The evolution of the distinctively human sweating apparatus, consequent extreme water dependence, and reduced activity periods may be solutions to thermoregulatory problems enhanced by increasing body size. Clearly it can be seen that selective conflicts arise between reduced activity periods and increased food requirements, stressing the fact that these problems do not occur in isolation.

Changes in body size will also affect predator–prey relationships. Predators, especially those that hunt alone, are limited to certain prey sizes (Bourliere, 1963), and thus an increase in size will release an organism from the dangers of predation, and, if it is omnivorous or carnivorous, increase the range of potential prey. The effect of this will be to select for yet broader dietary niches.

It has already been stated that the basis for the influence on life history of body size is the non-linear relationship between different physical features. This is the case with brain size, which has classically been thought to vary in relation to body surface area (Jerison, 1973). Analysis of further data (R. D. Martin, 1981) has suggested that a closer relationship exists between brain weight and female metabolic turnover. The main consequence of Martin's hypothesis is that brain enlargement imposes very high metabolic costs, and thus brain enlargement, a consequence of both body size increases and changes in other life history characteristics, energy maintenance problems arise (Martin, 1982). However, this means that brain enlargement will only occur where there is considerable selective pressure for the advantages that it provides, and it is worthwhile considering those circumstances in which increased body size and the benefits of brain enlargement are linked.

A relationship exists between longevity and body size (Western, 1979). This is to be expected—large mammals represent a considerable reproductive and metabolic investment and thus are unlikely to be ephemeral—and several consequences follow (Trivers, 1972). First, a long-lived mammal is likely to encounter environmental changes, and thus it must be sufficiently flexible to track and respond to shifting conditions. Secondly, an animal that lives a long time would clearly be at an advantage if it were capable of benefitting from previous experience, and thus learning capacity would be at a premium. Thirdly, a long-lived mammal would need to deal with repeated encounters with the same individuals during its life time, and thus increased social complexity. All these characteristics are ones that would benefit from brain enlargement.

Finally, with respect to the problems that arise from increased body size, reproductive parameters are altered. Large mammals have longer gestation periods, prolonged development and growth, and a greater birth interval. Thus although large mammals may increase their reproductive span through increased longevity, their actual costs of reproduction may rise and reproductive output may fall. Large mammals must therefore cope with problems relating to low reproductive output and ensuring maximum offspring survival, through such behaviours as increased parental care (Lovejoy, 1981).

From viewing adaptation as problem-solving it has been possible to identify a series of adaptive problems that would have faced the earlier hominids as a result of the observable increase in body size. Some implications for adaptation, drawn from comparative biology, can thus be identified — extensive ranging behaviour, complex resource scheduling, broad dietary niches, high mobility, behavioural flexibility, intelligence and learning capacity, sociality and parental care—as potential solutions.

## 4.2. Problems of Being a Terrestrial Primate

The hominids are members of the Order Primates, and although this group is primarily an arboreally adapted taxon, some 40 or more extinct and living species, including the hominids, have secondarily adapted to a terrestrial existence (Napier and Napier, 1967; Simons, 1972). Two basic reasons can be suggested for the radiation of ground-dwelling primates in the Later Cenozoic. First, the fall in global temperatures from the Late Miocene (Kennett, 1977) resulted in increasingly open (woodland, bushland and grassland) environments in tropical latitudes (Badgeley and Behrensmeyer, 1980; Shipman *et al.*, 1980). With a reduction in forested environments from the Middle Miocene selection would be expected to result in terrestrially adapted species. Secondly, the main constraint operating on primates in their arboreal environment is that of size; for primates to be part of the general trend of increased body size to be found among the mammals they must become at least partially terrestrial.

Having "come down from the trees" three major adaptive difficulties would face the primate—locomotion, shelter and competition for resources. With regard to locomotion it is clear that the primate locomotor systems evolved to move through the three dimensional arboreal environment (Le Gros Clark, 1964); consequently on the ground they are less efficient than specialized ungulates. All terrestrial primates have undergone some locomotor modifications—cercopithecine quadrupedalism, pongid modified brachiation and hominid bipedalism. This diversity of solutions to the same adaptive problem partly reflects the consequences of historical constraints, and partly differing levels of selection and differing requirements once on the ground [see Aiello (1981) for a discussion of locomotor patterns]. That selection involving locomotor patterns was considerable is indicated by the early date by which at least some hominids had achieved bipedalism (Leakey, 1979).

The problem of shelter is a more acute one. Virtually all primates have special sleeping areas, as well as a general pattern of movement into trees to avoid predators. Even those primates that have adapted to a terrestrial way of life will return to trees at night. Where this is not possible they will sleep on cliff faces or in caves if available. The patas monkey (*Erythrocebus patas*), which lives in very open environments will disperse widely and sleep individually in trees rather than spend the night on the ground (Napier and Napier, 1967). Only the large hominoids, generally safe from predators, will make ground nests (Napier and Napier, 1967). Yet a terrestrial primate, especially one with a specialized terrestrial locomotor system, must employ some type of ground nest or base. Solving the problems of nocturnal shelter were critical ones facing terrestrial primates, as would be diurnal shelter from high temperatures. These are factors that need to be considered in relation to the development of a home base, the early evidence for which is so contentious (see above).

The third set of problems are those relating to competition for food in relation to more fully adapted terrestrial species. Primates require high quality plant food, and they may be expected to have problems in acquiring these on the ground. Solutions to the problems of feeding in the context of a biological community of specialized terrestrial species might either be to become highly specialized and thus increase competitive efficiency, or alternatively to maintain a broad dietary niche and expand the range of foods taken. A broad-based niche is one of the characteristics of later hunter-gatherer adaptation in tropical environments, and expansion of the range of foods taken to include meat is a characteristic of human evolution (see above p 90, and Chapter 6, this volume). It is also a feature shared with other terrestrial primates (Harding, 1981).

The problems of being a terrestrial primate can thus be seen to have major adaptive consequences that may relate to the development of characteristics observed in later hunter-gatherers. Leaving aside the locomotor changes, the constraints for sleeping areas and competitive interactions with other species, including other terrestrial primates, would lead to distinctive ecological and spatial patterning. It is, however, difficult to distinguish the problems associated with being a terrestrial primate from those of living in an open environment.

## 4.3. Problems of Living in a Savanna Environment

Terrestriality enables populations to live in drier grasslands and woodlands, and several adaptive problems can be recognized as a result (Table 4.2). First, there will be a restricted and patchy distribution of water, which, given the thermoregulatory problems and solutions mentioned above (see page 00) is required by hominids on a regular basis. It may be expected that hominids would be restricted to those areas where water is available. This would be combined with the problems of high temperatures and lack of shade, putting a higher premium on water availability.

Table 4.2. Ecological problems and consequences of the tropical savanna African environment.

| Attribute | Ecological implications |
|---|---|
| Low rainfall | 1. Fewer resources<br>2. Limited surface water distribution<br>3. Patchy resource distribution |
| Highly seasonal | 1. Complex and flexible foraging strategy required<br>2. Broad dietary niche<br>3. High mobility |
| Low quality plant food | 1. High foraging costs<br>2. Intense competition for high quality plant foods<br>3. Selective advantage for meat-eating |
| Abundant animal resources | 1. Competition for plant foods<br>2. High potential for hunting |
| Large numbers of predators | 1. Competition for animal resources<br>2. Predator avoidance |
| Reduced tree cover | 1. Competition for shelter/shade<br>2. Restricted distribution *or* complete ground-dwelling |
| High temperature | 1. Reduced activity<br>2. Water dependence<br>3. Selection for efficient cooling adaptation |

Another problem facing a savanna-dwelling mammal or primate is that of the low quality of the plant food. Low species diversity, low productivity, seasonal fluctuations and intense competition all mean that plant food will require high search time and/or processing time (see Hawkes *et al.*, 1982). Marked seasonality in availability of plant foods that are edible would result in a whole series of species being needed to ensure an adequate diet throughout the year. The limited nature of the savanna flora would be expected to exert considerable influence on hominid adaptation.

Although plant diversity is low, those species that do occur are ones with rapid turnover and growth rates, principally the grasses, which although largely unsuitable for primates, are nonetheless the reason for the very high secondary biomass that occurs in savanna environments. Ability to tap these animal resources would represent a major solution to the problems of low quality plant food availability. However, the abundance of diversity of prey species has also resulted in a rich carnivorous fauna, and a major adaptive problem is predator avoidance or resistance, and, if meat-eating is adopted, a means of competing with them and gaining access to animal resources.

The savanna would therefore pose a series of major adaptive problems to large terrestrial primates that would result in major selective forces. Solutions to those problems—i.e. the outcome of selection—would be expected to result in an adaptation involving water dependence and yet wide ranging behaviour; a broad resource base involving extensive subsistence time and a complex extractive strategy from a patchy resource base; and incorporation of meat-eating as a means of coping with periods and places of plant food scarcity. Among other species these characteristics are associated with a high encephalization quotient or relative brain enlargement (Eisenberg, 1981).

## 4.4. Combined Adaptive Problems

Characteristics that relate to the adaptive behaviour of later hunter-gatherers may thus be seen also to relate directly to the specific problems faced by early hominids. Hunter-gatherer characteristics, although not demonstrable, may be considered expected ecological and evolutionary outcomes. However, none of the problems described on their own are unique. They are shared by many other organisms, none of which have achieved the dubious status of being human. The reason for this is that it is unreasonable to treat adaptive problems independently. The uniqueness of every species derives from having a unique combination of problems requiring adaptive solutions. Thus while hominids might share the problems of savanna dwelling with elephants, and those of ground-dwelling with gorillas, their suite of problems differs from each (Fig. 4.3). Uniqueness lies not in specific adaptive traits, but

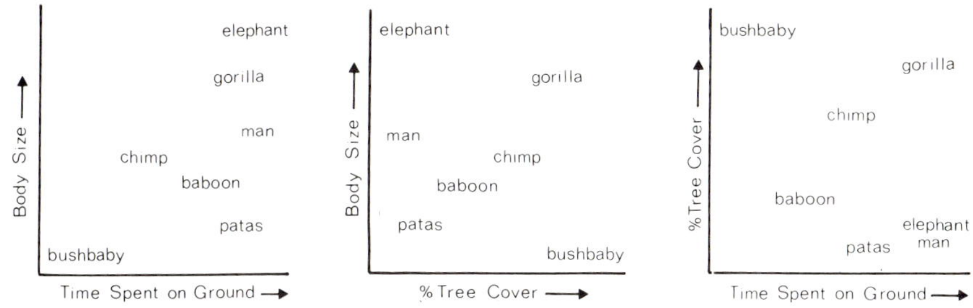


Fig. 4.3. Relationship between the "ecological problems" of various African species (see text for explanation).

in their combinations.

This may be illustrated with reference to the development of meat-eating. In this case, instead of starting with adaptive problems and working towards solutions, it is possible to examine a characteristic that occurs in later hominid adaptation and see what problems it solves (Fig. 4.4). Meat-eating can be related to a series of adaptive problems and opportunities. Carnivory is a means

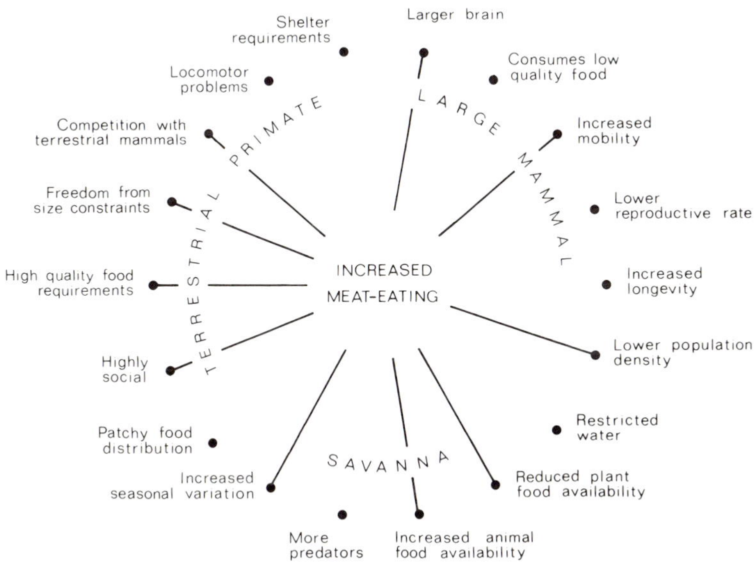


Fig. 4.4. Problems of survival influencing the development of meat eating among hominids.

Table 4.3. Genera of African large mammals contemporary with the hominids during the Pliocene and Early Pleistocene. Numbers in parenthesis refer to the number of species per genus. Data from Maglio and Cooke (1978).

| | |
|---|---|
| Herbivorous Competitors | Competitors for Roots |
| and Potential Prey | *Nyanzochoerus* (6) |
| *Equus* (8) | *Notochoerus* (3) |
| *Taurotragus* | *Metridiochoerus* (3) |
| *Tregalaphus* | *Kolpochoerus* (6) |
| *Eotragus* | *Stylochoerus* (1) |
| *Ugandax* | *Potamochoroides* (1) |
| *Pelorovis* | *Phacochoerus* (4) |
| *Simatherium* | *Sus* (1) |
| *Hemibos* | *Potamochoerus* (4) |
| *Sylvicapra* | *Hylochoerus* (1) |
| *Cephalophes* | |
| *Redunca* | Megafauna |
| *Kobus* | (Scavanged Food) |
| *Henelikia* | *Anancus* (2) |
| *Thalerocerus* | *Stegodon* (1) |
| *Hippotragus* | *Stegotetrabelodon* (1) |
| *Oryx* | *Primelephas* (1) |
| *Megalotragus* | *Mammuthus* (1) |
| *Connochaetes* | *Loxodonta* (2) |
| *Oreonagor* | *Deinotherium* (1) |
| *Dameliscus* | *Chalicotherium* (1) |
| *Parmularius* | *Diceros* (1) |
| *Beatragus* | *Dicerorhinus* (1) |
| *Aepycerus* | *Brachypotherium* (1) |
| *Raphicerus* | *Hipparion* (2) |
| *Oreotragus* | *Hexaprotodon* (2) |
| *Ourebia* | *Hippopotamous* (3) |
| *Madoqua* | *Giraffa* (3) |
| *Antidorcas* | *Sivatherium* (1) |
| *Gazella* | |
| *Tossunoria* | Terrestrial Primates |
| *Numidocapra* | *Macaca* (2) |
| | *Parapapio* (5) |
| Predators and | *Papio* (9) |
| Hunting Competitors | *Dinopithecus* (1) |
| *Lycaeon* (1) | *Gorgopithecus* (1) |
| *Leecyaena* (1) | *Theropithecus* (4) |
| *Euryboas* (2) | *Cercopithecus* (1) |
| *Hyaena* (4) | *Australopithecus* (2) |
| *Crocuta* (2) | *Gorilla* (1) |
| *Machairodus* (3) | *Pan* (2) |
| *Homotherium* (2) | |
| *Panthera* (4) | |

of overcoming the paucity of plant foods in their savanna, their seasonal distribution, and the competition involved in their extraction. It takes advantage of the high abundance of animals in savanna environments. As terrestrial primates the high protein content of meat will solve the problems of their high quality food requirements, and the costs of brain and body size enlargement. Increased body size will further enhance speed, strength, and most important mobility. Meat-eating, whether scavanging or hunting, may therefore be seen as an integrated adaptation and a probable outcome of the normal problem-solving processes of evolution. Similar analyses could also be carried out relating other hominid characteristics to environmental and phylogenetic problems.

## 5. COMMUNITY EVOLUTION

Viewing adaptation as a problem-solving process and seeking to identify the problems faced by the early hominids is a useful means of investigating the origins of some hunter-gatherer adaptive characteristics. It remains incomplete, though, because it has treated adaptive problems in terms of abstract entities, rather than in the context of a specific biological community. For example, the "savanna environment" is not a real biological entity, but a descriptive term used to describe the set of coevolving plants and animals that live in certain rainfall regimes of Africa; and the term "hominid" refers to more than one species (Walker and Leakey, 1979; Johanson and White, 1979), and therefore more than one set of problems and solutions. To overcome these problems it is necessary to examine the patterns of evolution and ecology of the animals with which hominids coexisted, and in particular patterns of competitive interaction.

Table 4.3 shows the large mammals that coexisted with the hominids in the Later Cenozoic in Africa. It is not a comprehensive list, but one based on ecological categories with which competitive interactions are likely to take place.

(a) *Large predators*: predators large enough to prey on hominids and also to compete with them for prey.

(b) *Terrestrial primates*: species with similar ecological requirements to the hominids, and with which competition may be intense.

(c) *Pigs*: large mammals specialized towards high quality plant foods, including underground resources.

(d) *Ungulates*: main consumers of the plant base, and potential prey species and scavanged food.

(e) *Megafauna*: very large, usually pachydermic species of herbivore—a potential source of scavanged food.

Among these groups, and among the hominids themselves, certain patterns can be discerned, relating to two important ecological parameters, diversity and abundance. With regard to the former, diversity reached maximum levels for the Later Cenozoic, during the Lower Pleistocene (Fig. 4.5). This is true for total diversity, for diversity of bovid genera, and also for all but one of the ecological groups. The one exception are the Suidae, which have a maximum diversity in the Pliocene.

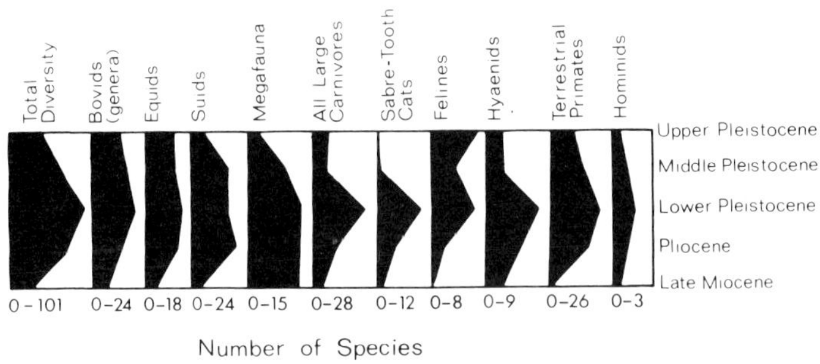


Fig. 4.5. Patterns of species diversity among the Later Cenozoic large mammal community of Africa (data from Maglio and Cooke, 1978).

The radiation of the Bovidae and Suidae during the Pliocene and Lower Pleistocene indicates evolutionary and ecological dynamism in relation to the above and below ground plant resources. Further evidence for this is provided by the fact that there are some 20 species of terrestrial primate in sub-Saharan Africa during this period. Expansion of these herbivorous forms would have imposed severe competitive constraints on the archaic megafauna, which decline rapidly during the Pleistocene. All these changes at the secondary biomass level would have consequences for the predators. As a group they reach maximum diversity during the Lower Pleistocene (Savage, 1978), and after this the sabre tooth cats decline markedly and become extinct, while the modern cats (felines) maintained their position. The hominids also reached maximum diversity during this period.

Turning to abundance, *Theropithecus* is the dominant group among the terrestrial primates during the Lower Pleistocene. This contrasts with the present situation where the genus *Papio* is the most abundant non-human terrestrial primate in Africa. And within the Hominidae, *Australopithecus* is more abundant than *Homo* in the fossil record (Howell, 1978, Tables 10·1–10·4).

Clearly there are major taphonomic problems in interpreting these observations on the palaeontological record, and the data need to be considered in relation to fossil robusticity and the relative lengths of the geological periods. However, two general conclusions may be drawn. First, that the expanding open environments were closely related to the radiation of large mammal species that were partitioning the herbivorous trophic level. Secondly, among the terrestrial primates it was those that it has been suggested are the most specialized herbivores (Jolly, 1970) that were most abundant. Although

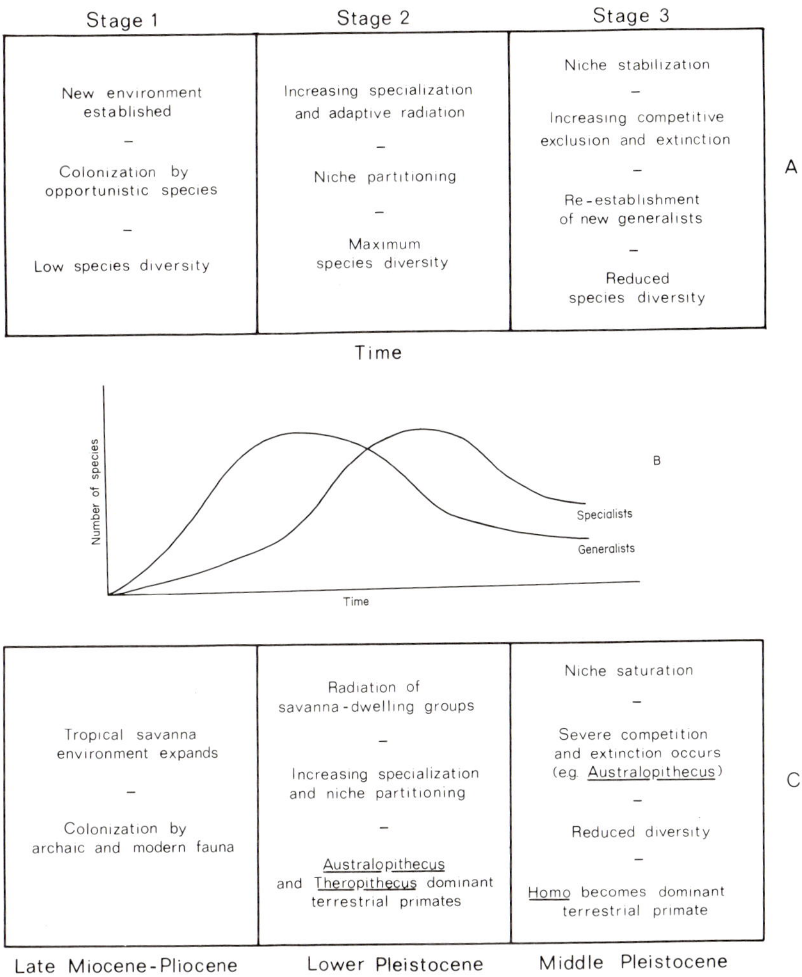


Fig. 4.6. (a) Stadial model of community evolution following the establishment of a new environment. (b) Model of changing frequencies of specialist and generalist species following establishment of a new environment. (c) Community evolution in the savanna environments of Africa during the Later Cenozoic, with special reference to the hominids.

adapted to very different niches, these two genera appear to have responded to the problems associated with the expanding savanna community through dietary specialization, and were thus part of the savanna radiations. In other words, the most successful terrestrial primates at this time were those that were most specialized.

However, as is all too apparent, *Australopithecus* is now extinct, and *Theropithecus* is now confined to a relict population in the Ethiopian highlands—i.e. the patterns of abundance among terrestrial primates shifts through the Pleistocene. This shift may be related to the patterns of diversity. With the increasing partitioning of the herbivore niche among the terrestrial mammals, primates, with their higher quality food requirements, came under considerable competitive pressure, such that the niches became too finely partitioned for a primate to survive as a dietary specialist. *Homo*, it may be argued, adopted an alternative strategy in which niche breadth was expanded to include a broad range of plants and animals, thus minimizing the problems of competitive exclusion. *Papio* would also stand in the same relationship to *Theropithecus*. The pattern of community evolution is summarized in Fig. 4·6.

## 6. CONCLUSIONS

The primary purpose of this paper has been to show that the evolutionary success of the human species is based on the solutions that selection produced to the problems faced by early hominids as part of an evolving biological community, and these solutions may in turn be related to characteristics found in the adaptive strategies of later hunter-gatherers. The important analytical steps involved were: (1) discussion of the evolutionary process as a problem-solving one; (2) recognition of species as reciprocal selective agencies, and thus the placing of evolutionary problems within a community framework; (3) identification of early hominid adaptive problems; (4) placing adaptive problems and potential solutions into the context of Later Cenozoic large mammal evolution and ecology in Africa.

This analysis goes some way towards explaining aspects of the emergence of hominid behavioural and foraging strategy. Clearly the proposed model suffers from limitations, some of which are examined in later chapters of this book. As Hill (see Chapter 5, this volume) points out, the formation of the fossil record may lead to oversimplifications and spuriously convergent patterns. Potts (see Chapter 6, this volume) discusses in more detail the evidence for meat eating among early hominids. And Gowlett (see Chapter 7, this volume) looks at the evidence for the mental apparatus by which adaptive solutions are absorbed within the hominid behavioural repertoire. However, even allowing for these modifications, two crucial points can be stated wih some degree of certainty.

First, the problems of human evolution may be analysed from a perspective of community evolution, and that the emergence of hominids as hunters and gatherers was an expected and probable outcome of ordinary biological processes. Secondly, that looking at hominids and their adaptive strategies in the context of the species with which they coexisted is a response to what Maynard Smith (1982, p 128) argues is currently the central problem in evolutionary theory—that is, the need for a "theory that says something about selection, and hence about the environment. Since the major component of most species environment consists of other species, it follows that we need a theory of ecosystems in which the component species are evolving by natural selection."

## ACKNOWLEDGEMENTS

A version of this paper was presented as the Lister Lecture for 1982 to the British Association for the Advancement for Science at Liverpool.

## REFERENCES

Aiello, L. (1981). Locomotion in the Miocene Hominoidea. In: C. B. Stringer (ed) *Aspects of Human Evolution*. London: Taylor and Francis, pp. 63–97.

Badgeley, C. and Behrensmeyer, A. K. (1980). Palaeoecology of Middle Siwalik sediments and faunas, North Pakistan. *Palaeogeog. Palaeoclimat. Palaeoecol.* **30**: 133–155.

Behrensmeyer, A. K. (1978). The habitat of Plio-Pleistocene hominids in East Africa: taphonomic and microstratigraphic evidence. In: C. Jolly (ed) *Early Hominids of Africa*. London: Duckworth, pp. 165–190.

Binford, L. R. (1979). Organisation and formation processes: looking at curated technologies. *J. Anthropol. Res.* **35**(3): 255–273.

Binford, L. R. (1980). Willow smoke and dogs' tails: hunter-gatherer settlement systems and archaeological formation processes. *Am. Antiquity* **45**(1): 4–20.

Binford, L. R. (1981). *Bones: Ancient Men and Modern Myths*. London: Academic Press.

Birdsell, J. B. (1953). Some environmental and cultural factors influencing the structuring of Australian aboriginal populations. *Am. Naturalist* **87**: 171–207.

Birdsell, J. B. (1968). Some predictions for the Pleistocene based on equilibrium systems among recent hunter-gatherers. In: R. B. Lee and I. DeVore (eds) *Man the Hunter*. Chicago: Aldine, pp. 229–240.

Boaz, N. (1979). Early hominid population densities: new estimates. *Science* **206**: 592–4.

Bonnefille, R. (1976). Palynological evidence for an important change in the vegetation of the Omo Basin between 2·5 and 2·0 million years. In: Y. Coppens, F. C. Howell, G. Isaac and R. Leakey (eds) *Earliest Man and Environments in the Lake Rudolf basin: Stratigraphy, Palaeoecology, Evolution*. Chicago: University of Chicago Press, pp. 421–431.

Bourliere, F. (1963). Specific feeding habits of African carnivores. *Afr. Wildlife* **17**: 21–27.

Bunn, H. T. (1981). Archaeological evidence for meat-eating by Plio-Pleistocene hominids from Koobi Fora and Olduvai Gorge. *Nature* **291**: 574–577.

Charteris J., Wall, J. C. and Nottrodt, J. W. (1981). Functional reconstruction of gait from the Pliocene hominid footprints at Laetoli, Tanzania. *Nature* **290**: 496–8.

Clark, W. E. Le Gros (1964). *Fossil Evidence for Human Evolution*. Chicago: University of Chicago Press (2nd edn).

Clutton-Brock, T. H. and Harvey, P. H. (1977). Primate ecology and social organisation. *Proc. Zool. Soc., Lond.* **183**: 1–39.

Cook, S. F. and Heizer, R. F. (1951). The physical analysis of nine Indian mounds of the Lower Sacramento Valley. *Univ. Calif. Publ. Am. Archaeol. Ethnol.* **40**: 281–312.

Dobzhansky, T. (1974). Chance and creativity in evolution. In: F. J. Ayala and T. Dobzhansky (eds) *Studies in the Philosophy of Biology*. London: Macmillan, pp. 309–339.

Eisenberg, J. F. (1981). *The Mammalian Radiations*. London: Athlone Press.

Foley, R. (1978). Incorporating sampling into initial research design: some aspects of spatial archaeology. In: J. F. Cherry, C. Gamble, and S. Shennan (eds) *Sampling in Contemporary British Archaeology*. B. A. R. (British Series) 50, pp. 49–66.

Foley, R. (1982). A reconsideration of the role of predation on large mammals in tropical hunter-gatherer adaptation. *Man* (N.S.) **17**: 393–402.

Harding, R. S. O. (1981). An Order of omnivores: non-human primate diets in the wild. In: R. S. O. Harding and G. Teleki (eds) *Omnivorous Primates*. New York: Columbia University Press, pp. 344–421.

Harested, A. S. and Bunnell, F. L. (1979). Home range and body weight: a re-evaluation. *Ecology* **60**(2): 389–402.

Harris, D. R. (1978). Adaptation to a tropical rain forest environment: aboriginal subsistence in north-eastern Queensland. In: N. G. Blurton-Jones and V. Reynolds (eds) *Human Behaviour and Adaptation*. London: Taylor and Francis, pp. 113–134.

Hassan, F. (1975). Determination of the size, density and growth rate of hunter-gatherer populations. In: Polgar, S. (ed) *Population, Ecology and Social Evolution*. The Hague: Mouton, pp. 27–52.

Hawkes, K., Hill, K. and O'Connell, J. F. (1982). Why hunters gather: optimal foraging and the Aché of Eastern Paraguay. *Am. Ethnol.* **9**(2): 379–398.

Hay, R. L. (1976). *Geology of the Olduvai Gorge: a study of sedimentation in a semiarid basin*. Berkeley: University of California Press.

Hennemann, W. W. III (1983). Relationship among body mass, metabolic rate and the intrinsic rate of natural increase in mammals. *Oecologia* **56**: 104–108.

Howell, F. C. (1978). Hominidae. In: V. J. Maglio and H. B. S. Cooke (eds) *Evolution of African Mammals*. Cambridge, Mass.: Harvard University Press, pp. 154–248.

Isaac, G. L. (1976). Plio-Pleistocene artefact assemblages from East Rudolf. In: Y. Coppens, F. C. Howell, G. Isaac, and R. Leakey (eds) *Earliest Man and Environments in the Lake Rudolf Basin: Stratigraphy, Palaeoecology, Evolution*. Chicago: University of Chicago Press, pp. 552–564.

Isaac, G. L. (1978a). The food sharing behavior of proto-human hominids. *Sci. Am.* **238**(4): 90–106.

Isaac, G. L. (1978b). Food sharing and human evolution: archaeological evidence from the Plio-Pleistocene of East Africa. *J. Anthropol. Res.* **34**(3): 311–325.

Isaac, G. L. (1983). Bones in contention: competing explanations for the juxtaposition of early Pleistocene artefacts and faunal remains. In: J. Clutton-Brock and C. Grigson (eds) *Animals and Archaeology. Volume 1: Hunters and their Prey*. B. A. R. (International Series), pp. 3–20.

Isaac, G. L. and Crader, D. (1981). To what extent were the early hominids carnivorous? an archaeological perspective. In: R. S. O. Harding and G. Teleki (eds) *Omnivorous Primates*. New York: Columbia University Press, pp. 37–103.

Jarman, P. J. (1974). The social organisation of antelopes in relation to their ecology. *Behaviour* **58**: 215–267.

Jerison, H. J. (1973). *Evolution of the Brain and Intelligence*. London: Academic Press.

Johanson, D. C. and Edey, M. A. (1981). *Lucy: the Beginnings of Mankind*. London: Granada.

Johanson, D. C. and White, T. (1979). A systematic assessment of early African hominids. *Science* **203**: 321–330.

Jolly, C. J. (1970). The seed-eaters: a new model of hominid differentiation based on a baboon analogy. *Man* (N.S.) **5**: 5–26.

Jolly, C. J. (ed). (1978). *Early Hominids of Africa*. London: Duckworth.

Jones, R. (1980). Hunter-gatherers in the coastal Australian savanna. In: D. R. Harris (ed) *Human Ecology in Savanna Environments*. London: Academic Press, pp- 107–146.

Jungers, W. L. (1982). Lucy's limbs: skeletal allometry and locomotion in *Australopithecus afarensis*. *Nature* **297**: 676–8.

Keeley, L. and Toth, N. (1981). Microwear polishes on early stone tools from Koobi Fora, Kenya. *Nature* **293**: 464–465.

Kennett, J. P. (1977). Cenozoic evolution of Antarctic glaciation, the circum-Antarctic ocean, and their impact on global oceanography. *J. Geophys. Res.* **82**: 3843–3860.

Kleiber, M. (1961). *The Fire of Life*. New York: Wiley.

Leakey, M. D. (1971). *Olduvai Gorge. Volume 3: Excavations in Beds I and II, 1960–63*. Cambridge: Cambridge University Press.

Leakey, M. D. (1979). Footprints in the ashes of time. *Nat. Geograph.* **155**(4): 446–457.

Lee, R. B. (1968). What hunters do for a living, or, how to make out on scarce resources. In: R. B. Lee and I. DeVore (eds) *Man the Hunter*. Chicago: Aldine, pp. 30–48.

Lee, R. B. and DeVore, I. (eds) (1968). *Man the Hunter*. Chicago: Aldine.

Lovejoy, O. (1981). The origins of man. *Science* **211**: 340–350.

McNab, B. K. (1963). Bioenergetics and the determination of home range size. *Am. Naturalist* **97**: 130–40.

Maglio, V. and Cooke, H. B. S. (eds) (1978) *Evolution of African Mammals*. Cambridge, Mass.: Harvard University Press.

Martin, R. A. (1981). On extinct hominid population densities. *J. Hum. Evol.* **10**: 427–8.

Martin, R. D. (1981). Relative brain size and basal metabolic rate in terrestrial vertebrates. *Nature* **293**: 57–60.

Martin, R. D. (1982). *Human Brain Evolution in an Ecological Context*. James Arthur Lecture, American Museum of Natural History.

Maynard Smith, J. (1976). A comment on the Red Queen. *Am. Naturalist* **110**: 325–330.

Maynard Smith, J. (1982). *Evolution Now*. London: Macmillan.

Milton, K. and May, M. L. (1975). Body weight, diet and home range area in primates. *Nature* **259**: 459–462.

Napier, J. R. and Napier, P. H. (1967). *A Handbook of Living Primates*. London: Academic Press.

Pilbeam, D. (1979). Recent finds and interpretations of Miocene hominoids. *Ann. Rev. Anthrop.* **8**: 333–52.
Pilbeam, D. and Gould, S. J. (1974). Size and scaling in human evolution. *Science* **186**: 892–901.
Potts, R. and Shipman, P. (1981). Cutmarks made by stone tools on bones from Olduvai Gorge, Tanzania. *Nature* **291**: 577–580.
Romer, A. S. (1966). *Vertebrate Palaeontology*. Chicago: University of Chicago Press.
Savage, R. G. (1978). Carnivora. In: V. J. Maglio and H. B. S. Cooke (eds) *Evolution of African Mammals*. Cambridge, Mass.: Harvard University Press, pp. 249–267.
Shipman, P., Walker, A., van Couvering, J. A., Hooker, P. J. and Miller, J. A. (1980). The Fort Ternan hominoid site, Kenya: geology, age, taphonomy and palaeo-ecology. *J. Hum. Evol.* **10**: 49–72.
Silberbauer, G. S. (1972). The G/wi Bushman. In: M. G. Bicchieri (ed) *Hunters and Gatherers Today*. New York: Holt, Rinehart and Winston, pp. 271–325.
Simons, E. (1972). *Primate Evolution: An Introduction to Man's Place in Nature*. New York: Macmillan.
Trivers, R. L. (1972). Parental investment and sexual selection. In: B. G. Campbell (ed) *Sexual Selection and the Descent of Man*. Chicago: Aldine, pp. 156–179.
van Valen, L. (1973). A new evolutionary law. *Evol. Theory* **1**: 1–30.
Walker, A. C. (1981). Dietary hypotheses and human evolution. *Phil. Trans. Roy. Soc. Lond.* **292**: 56–64.
Walker, A. C. and Leakey, R. E. (1979). The hominids of East Turkana. *Sci. Am.* **239**: 54–66.
Western, D. (1979). Size, life history and ecology in mammals. *Afr. J. Ecol.* **17**: 185–205.

# 5 *Hyaenas and Hominids: Taphonomy and Hypothesis Testing*

*Andrew Hill*

## 1. INTRODUCTION

Bone remains in archaeological contexts have often been used as the starting point for sequences of reasoning leading to behavioural and social inferences about early hominids. This, along with more general palaeoecological considerations, has been part of the stimulus to much recent work on contemporary bone assemblages in various environments and situations. Some of this has concerned the nature of whole assemblages found on different kinds of land surface in areas where the ecosystem is more or less known (Hill, 1975, 1979, 1980a; Behrensmeyer *et al.*, 1979). Others deal with more specific collectors and modifiers of bone (Brain, 1980, 1981; Binford, 1981). With regard to archaeological sites, the rationale of this work, which forms part of taphonomy, is that a detailed documentation of various assemblages formed where humans have not been involved may serve as a comparative standard against which archaeological accumulations can be compared, and any distinctive features they have might be assessed in terms of behaviour. More generally, taphonomy constitues a means of judging the kind of sample an accumulation provides of the ecosystem from which it is derived, and which forms the basis of palaeoecological interpretation.

This chapter is a review of some of the problems involved in interpreting the very early archaeological accumulations, particularly in behavioural terms. First I look at the kind of evidence that is available in the form of bone at archaeological sites by describing briefly as an example the situation at Olduvai FLK, and then referring to a classification of this and other kinds of sites. Makapansgat is another example that is considered. The following section gives an account of the behavioural and social inferences that have been made from these kinds of faunal evidence.

The rest of the chapter is devoted to a criticism of the inferences. They can

HOMINID EVOLUTION
*ISBN 0-12-261920-X*

be criticized in a number of ways. One takes the material evidence itself and by incorporating more extensive taphonomical information from other bone assemblages judges how diagnostic it is likely to be of the postulated hominid behaviours. This is done by describing the features of a bone assemblage produced by spotted hyaenas and comparing it with the archaeological examples. As well as forming a criticism of suggested models of lower Pleistocene hominid behaviour, this highlights other more general difficulties with the interpretation of such sites in palaeoecological and behavioural terms. The following section digresses into a short discussion of these.

These sections make it clear that the material evidence does not on its own warrant the extensive conclusions made about early hominid behaviour. The theories are presented as being justified by the nature of the bones, but in fact seem to originate in the practice of phylogenetic analogy. This incorporates the idea that animals close phylogenetically are more likely to have the same behaviour than are those more distantly related. Simplicity is often invoked as a defense of theories which incorporate elements of phylogenetic analogy. Criticisms of these residual arguments are presented; first, some general methodological difficulties with the ethnographic approach, and secondly some of the pervasive confusions in the application of notions of simplicity to these problems, which indeed apply equally to many other areas of archaeology and palaeoanthropology.

The conclusion summarizes these arguments and suggests that more fruitful theories about early hominid behaviour would come from treating the early hominids as the separate species that they are, distinct from *Homo sapiens*, possessing distinct morphologies and possibly distinct behaviours. Ecological and adaptive principles derived from a far wider range of species are available for this analysis, and their use would obviate the anthropocentric presentism that tends to dominate the ethnographic approach.

## 2. THE NATURE OF THE ARCHAEOLOGICAL EVIDENCE

The material evidence for the behaviour of early African hominids consists of rare fragments of the hominids themselves, the slightly more common accumulations of bones of other animals with artifacts, and certain information about their context. More exactly it consists in what a particular observer chooses to report and how he chooses to report it. This selection of what constitutes data mostly reflects those features that are interesting or necessary for the justification of the explanations and inferences that are to be drawn from them. In addition to this material evidence there is also a body of information in the form of adaptational theory, which derives from knowledge of the principles governing the morphology, behaviour and ecology of other creatures, including contemporary humans.

The general nature of the material evidence may best be appreciated by considering the example of a site, not chosen because it demonstrates a particular relative paucity of data and all the problems that entails, but because if it is at all atypical it is because it provides more information than do most archaeological sites of the time period. This is FLK, in Middle Bed 1 of Olduvai Gorge, Tanzania, dated at about 1·8 Myr.

On the main level of FLK, level $2^2$, were found some 2470 artifacts and 3510 pieces of bone, not counting microvertebrates and birds. The total excavated area was 290 $m^2$ (Potts, 1982). This represents a bone concentration greatly in excess of background densities reported for African grassland environments (Hill, 1975; Behrensmeyer *et al.*, 1979). Over 600 of the bones could be identified to taxon, and they belong to a large number of genera and species. Most come from the upper few centimetres of the level. They show a degree of spatial patterning; and more than 1000 of the finds were concentrated within an area of 21 × 15 feet (6·4 × 4·6 m) (Leakey M. D., 1971; see also Chapter 6, this volume). These bones are for the most part broken and damaged in a consistent manner, and different parts of the skeleton occur in anomalous proportions when compared to the numerical proportions in a single skeleton. The discovery which led to the excavation of this site was that of the cranium of *Australopithecus boisei*, "Zinjanthropus". Additional hominid material was later recovered at this site, at the same level, and this is generally held to represent a distinct taxon (see Howell, 1978). The site is thought to have been situated on the margin of the lake which occupied the Olduvai area at this time (Hay, 1976).

To summarize the principal features; this site is a lake marginal accumulation of bones and artifacts, that are relatively highly concentrated, that show some form of patterning, occurring at a single horizon. The bones represent parts of many species, the parts are anomalous in their representation compared with their proportions in a single skeleton, they are broken in consistent and repeated ways.

These characteristics are typical of many sites, but there are other permutations of bones and stones, and classifications have been devised to accommodate them. Isaac (1978) for example, has invented one which is useful to the extent that it removes one kind of inference from the description. He refers to associations such as FLK as Type C, where there are many artifacts along with the remains of many animals. His Type B accumulations are those with artifacts and the bones of a single animal, and artifacts occurring mainly without any bones are referred to as Type A. Isaac and Crader (1981) add a further three categories of occurrence to include those where material is not restricted to a single horizon, but is distributed through a thickness of sediment (Type D); others where there is evidence of transportation (presumably by water) and the material is out of primary context (Type G); and other sites that contain osteological material only (Type O). A site which might be classified as Type O is the large assemblage of bones at Makapansgat, South Africa, which included remains of australopithecines. This was described in a number of publications by Dart (1957, for example), and amongst the features he selected as being interesting were that skeletal parts were represented in rather odd proportions compared to those in a single skeleton, and that the different bones were broken in consistent repeated ways.

## 3. INFERENCES FROM THE ARCHAEOLOGICAL MATERIAL

What sort of inferences have been drawn from this archaeological material? I will be looking in particular at the conjectures derived from the bone remains. Other things may be inferred from the presence and nature of the artifact assemblages, but as far as the questions I shall be discussing here are concerned, they are mainly used to establish hominid involvement in accumulating the faunal material.

Various general inferences have been made about the different configurations of stones and bones discussed above. Sites of Type A are generally thought of as being factory sites; Type B are butchery sites; Type C are living or occupation sites.

Makapansgat provides an example of more detailed inference. Dart believed that the bone assemblage was due to its collection by the hominids, and that it partially represented the remains of their food, but the bones were also their weapons. He invented the term osteodontokeratic culture to describe this. A very detailed analysis of this material led Dart to suggest such social consequences for the australopithecines as the following. They were:

> . . . carnivorous creatures, that seized living quarries by violence, battered them to death, tore apart their broken bodies, disembodied them limb from limb, slaking their ravenous thirst with the hot blood of victims and greedily devouring livid writhing flesh. (Dart 1953; cited by Walker, 1981)

This may be a little unfair, in that it is an extreme quotation, but a more moderate chain of reasoning has led to this extreme point. Dart's main criteria for these kinds of conclusions were that the skeletal parts in the assemblage were represented in rather unusual proportions compared with that in a single skeleton, and that the bones were broken in a consistent fashion, producing repeated forms to which plausible implemental functions could easily be attributed.

Taking a site of Type C, such as Olduvai FLK, other sets of conclusions have been reached from the existence and nature of the bone material. Since Isaac has articulated these conclusions most elaborately, although in a tentative and conditional manner, his arguments provide a most complete example.

Basically, consideration of sites such as these leads to the notion that they represent "home bases". This implies a social focus where some members of the hominid group stay while others forage, scavenge or hunt. Food is brought back to these bases and shared amongst other members of the group. These ideas stem from the fact that bones occur in patterned high concentrations which are thought to have been produced by hominids because of the artifacts associated with them, and a high degree of contemporary ethnographic phylogenetic analogy is also involved. Such modern hunter gatherers as are known, like the !Kung, do similar things. Additional elements in the model involve division of labour, with females and males ranging in separate groups; the females gathering and the males contributing hunted food. This leads to an enhanced ability to exchange information, the development of reciprocal social obligations, to male bonding in family groups, and in general to a picture of social structure very human in character [see for example Isaac (1971, 1976, 1978, 1980) although in more recent publications (in press 1983) Isaac is modifying his earlier hypotheses, or explaining ways in which they have been misunderstood].

## 4. TAPHONOMY AND PROBLEMS WITH THE INFERENCES

There are a number of problems with these kinds of inferences. Some stem from the nature of the material evidence, and others, perhaps less obvious and more disputable, from the way this evidence is considered. In this section I will discuss in particular those that arise from the confrontation of the archaeological bone material with the results of taphonomical investigations of other bone assemblages.

Taking the basic view that sites of Types A, B and C, are factory sites, butchery sites and living floors, then clearly these kinds of site could plausibly look like these configurations, but the arrangements could equally well be produced by other means, and this consideration is what led Isaac to formulate his classificatory scheme.

In looking at more specific examples, I chose Makapansgat, because by looking at an example about which there is now relatively little controversy, it may be easier to appreciate some of the factors involved in more current arguments. At the time the osteodontokeratic culture was proposed, it was an imaginative and plausible idea. No bone assemblage, at least not in Africa, had been studied and analysed in this detail, with a great deal of attention being devoted to the proportions of different skeletal parts, and to the overall breakage pattern of the bones. Almost no comparative information at all was available on the condition of modern assemblages of bone produced by other processes, nor even about what processes produced them. At the time some others (e.g. Washburn, 1957) tried to implicate hyaenas in the formation of the assemblage, despite Dart (1954) and Hughes (1954, 1958) having asserted that hyaenas did not collect bones. Sutcliffe (1970) showed that they did, at least some of them and at least in small quantities, and my investigations in Amboseli, described below, show that significant amounts of material can be accumulated. This incidentally mirrors a much earlier controversy between Buckland (1822) and Knox (1822). Buckland suggested hyaenas were responsible for an assemblage at Kirkdale Cave, England, and Knox denied that hyaenas were bone collectors.

Hyaenas apart, however, work by Brain (1967, 1969) began to show that other kinds of assemblages such as those produced by humans and their dogs simply in the course of feeding, reveal a pattern of breakage and skeletal representation more or less indistinguishable from that of the Makapansgat collection. Later work on a variety of assemblages where humans have not been involved (Hill, 1975, 1976, 1980a), and on a range of specific animals (Brain, 1980, 1981), show a similar consistency of general damage and selective preservation. This can be related to a number of fairly basic factors having little to do with the behaviour of the animals concerned. These involve, in the case of consistency of bone damage, the relative strength and structure of different parts of each kind of bone, and in the case of selective skeletal representation, the relative density and strength of different kinds of bone as a whole.

Hence Dart's main objective criteria, based on the bone remains, for his osteodontokeratic culture and for the superstructure of predatory and social behaviour he built on it, are no longer valid. A large number of different kinds of bone assemblages share the characteristics he thought were diagnostic of hominid implemental activity. A few of the specimens from Makapansgat remain problematical, but the great majority are mimicked by those in other situations. Fortunately the means were available, in the form of a variety of contemporary bone assemblages, to test Dart's idea, which even at its best was only a plausible explanation of what at the time justly appeared to be an anomalous situation. There was nothing in the evidence necessitating tool making, rather than merely permitting it as a possible explanation. In fact it is easy to view this hypothesis as growing from Dart's general assumptions about human behaviour. Also at that time tools and tool making behaviour were matters thought to demarcate humans from other animals. Dart had been criticized for his early opinion that australopithecines were hominids, and it would certainly have been convenient for him to be able to show that they partook of an essentially human behavior pattern, even in the absence of stone tools. I am not suggesting this was a conscious motive for the osteodontokeratic hypothesis, but it may have been an element in it.

This is perhaps the point to consider the relevance of modern hyaenas, and the bone assemblages they produce. One large accumulation occurs in Amboseli National Park, Kenya. Amboseli is a semi-arid grassland ecosystem, and whilst some of the area is occupied by the bare flats of dried up Lake Amboseli, there is a mosaic of woodland, grassland and swamp that supports a variety of large mammals (Western, 1973). One group of hyaenas (*Crocuta crocuta*) uses a natural trench in the calcrete duricrust of the lake flats as a den. It is about 20 m long by 4 m wide, and from this trench are openings into underground regions in the soft underlying sediment.

General accounts of the den are published (Hill, 1978, 1980c, 1983) and more detailed treatments are in preparation, but here I will summarize certain of its features that have a bearing on questions about early hominids.

Many bones occur in a high concentration on the surface at the den. Surface bone totalled over 2000 pieces larger than two centimetres long, with concentrations locally of up to $75/m^2$. About 600 bones could be taxonomically identified, and they come from at least 16 species and possibly 18. The most common is Wildebeest (*Connochaetes taurinus*), which is also the most common wild ungulate in the park. Other abundant species are Burchell's zebra (*Equus burchelli*), Grant's gazelle (*Gazella granti*), Thomson's gazelle (*Gazella thomsoni*) and Maasai cattle. For the most common of these species, the ranked order of abundance of bone fragments corresponds to the order of average annual abundance of those species in the living Amboseli community. The bones that are represented occur in anomalous proportions when compared with the numbers in a single skeleton. For example, out of a total of over 300 bones attributable to more than 17 individuals of wildebeest, there are only 21 vertebrae, of which only one is an axis, and there are no caudals and no sacra. There are only 5 pieces of femur. Most of the bones are broken, and the breakage shows certain consistencies in pattern. The same skeletal parts tend to be broken in similar and repeated ways. Amongst features of damage shown are bone flakes demonstrating characteristics shared by human artifacts (see Potts, 1982). The material also shows spatial patterning. Towards the main entrance are the majority of the skeletally and taxonomically distinguishable pieces, whereas at the other end of the den are concentrated many, more or less, unidentifiable bone fragments. There are no bones of any kind around the main entrance itself, and on the calcrete surrounding the trench are again many fragments, but also most of the articulated skeletal portions.

The den is used by a variable number of animals, but during the course of the investigation infants belonging to two females were there for much of the time, and congregations of up to 17 individuals gathered there for a short while in the evening and early morning. The den was used by the infants and occasionally by related juveniles. During the daytime the adults occupied another area some kilometres away. The den's main function is as a protection for the infants from other animals, including other hyaenas.

Looking now at the home base idea for sites such as FLK, first it is clear that a more social cooperative set of behaviour is being inferred from what are essentially similar data to those provided by Makapansgat. Secondly, it is equally clear that the criteria put forward as diagnosing this behaviour do not do so. The characteristics were a concentration much higher than general landsurface background densities, repeated patterns of breakage, anomalous skeletal proportions, many individuals of a range of species, and a patterned distribution. I showed above that all of these features are shared by the hyaena den assemblage, but hyaenas do not behave as early hominids are supposed to do, based on reasoning from this same evidence. The bones are not accumulated at a home base for example, as the den is not a home base as implied in the case of hominids. Adult hyaenas tend to congregate and rest several kilometres away. Nor does the presence of bones indicate food sharing in hyaenas. Food sharing is unknown amongst adult hyaenas, the infants are not fed with meat brought back to the den, and provisioning of juveniles is probably rare (Kruuk, 1972; Hill 1980b). There is not a clear division of hunting or territorial behaviour on the basis of sex.

Obviously the bone evidence does not successfully diagnose the human set of social behaviours drawn from it. The main general way in which hominid sites differ from hyaena sites is simply in the presence of stone tools. This is the only other material element which might be used in justification of these inferences, but neither the existence nor nature of the artifacts have been brought into the argument, other than to attempt to establish the human origin of the bone remains associated with them. It is not clear that a mere association of tools invites particular social inferences, although it is possible to suggest certain social consequences of the transport of stone tool material (Potts, 1982). However, the arguments under discussion here do not incorporate this.

The only remaining contributory element seems to be the notion that it was humans, rather than other agents, that accumulated the bone material, and as with Makapansgat, the main cause of the ideas and their elaboration comes from feelings about what humans do; it is this that tacitly justifies the accretion of the social narrative. It is fuelled by notions regarding the behaviour of contemporary hunter gatherers, and more general less explicit ideas about how humans behave and how they might be distinct from other beasts. But is has very little to do with the archaeological evidence, and particularly little to do with the nature or even the existence of the bone remains. I will treat this further in later sections after a short digression to indicate further difficulties with the material evidence, which are emphasized by essentially taphonomical data.

## 5. FURTHER PROBLEMS WITH THE MATERIAL

Perhaps I should stress that whether or not hyaena or other carnivore bone accumulations can be distinguished in detail from those produced by hominids is irrelevant to the objections I have raised to the inferences made from the nature of these early archaeological accumulations. But it is also true that the difficulty of separating these various bone collecting agents from one another does pose a supplementary set of problems to interpreting the hominid accumulations. At present I find it difficult to suggest ways in which the different kinds of assemblage can be distinguished reliably, but this situation may change with further work. As Potts points out in Chapter 6, the simple association with stone artifacts does count as significant contextual evidence. However, from the beginning, it has been obvious that these archaeological sites are affected by the behaviour of carnivores (Leakey M. D., 1971), and they are certainly mixed, even if not in the origin of bones then at least in terms of the origin of the modifications to which they have been subjected.

With these mixed assemblages it may ultimately be only possible to say that most of a particular accumulation was produced by hominids, entailing the recognition that the resolution of the data are limited in this way. But if the resolution of the data are limited like this, then the resolution of the questions that may be asked of them, and the answers propounded, must be constrained by those same limitations. It will be difficult to answer a question about a particular single bone in such a collection even though some of the interesting and basic questions may involve single bones in assemblages.

The example of FLK highlights some of these difficulties. Did hominids collect and modify the bones, or at least most of them? Potts goes into some detail over this question in Chapter 6. There is some direct evidence in the form of cut marks on the remains that were apparently made by stone tools (Potts and Shipman, 1981; Bunn, 1981). Even so the evidence of modification by carnivores show that it is to a certain degree a mixed assemblage. But accepting that hominids contributed largely to the assemblage, which species was responsible for it, or were both, or were neither? *Australopithecus boisei* was the first hominid to be found at the site, and before the discovery of *Homo habilis* at FLK and elsewhere, the evidence led to the conclusion that *Australopithecus* had made the artifacts and accumulated the bones (Leakey, L. S. B., 1959). It was in fact the relative excellence of the hominid cranium's preservation in comparison to the other remains that led to the comment from Leakey:

> . . . that this skull represents one of the hominids who occupied the living site; who made and used the tools and who ate the animals. There is no reason whatever, in this case, to believe that the skull represents the victim of a cannibalistic feast by some hypothetical more advanced type of man.

This is very reasonable. When the hypothetical more advanced type of man attained reality, the argument based on preservation apparently was less persuasive, and the larger brained, smaller toothed creature became the preferred candidate (Leakey *et al.*, 1964). If it was just one of them that collected the bones, then what is the relationship to it of the remains of the other at the site? Was it part of its food? Or is it part of the assemblage fortuitously, having perhaps been brought there by carnivores for some reason? This puzzle seems to illustrate that at present it is impossible to separate individual items of hominid food remains from other bone material. Neither are there any unequivocal procedures for determining what these scraps of bone mean in terms of diet or the behaviour involved in acquiring them. Were the animals hunted or scavenged? If scavenged, was it by opportunistic casual scavenging or by a more active process involving the expulsion of other carnivores from their kills? What proportion of the whole diet did the meat form? Do the bones represent just a small fraction of many meals, or a large fraction of a few? How many individual hominids were involved? These questions are not trivial, and the consequences of the different answers to them are quite different and significant for our beliefs regarding early hominid behaviour and the reliability that can be placed upon evidence and methods; and yet these important and elementary questions are very difficult to resolve.

## 6. METHODOLOGICAL PROBLEMS

Many of these questions could have been framed more formally as hypotheses to be "tested", but it is difficult at present to see what sort of information could possibly be obtainable that would serve to discriminate between them.

The same is probably the case with some aspects of the home base model and its elaborations. These are expressed in respectable Popperian terms (e.g. Isaac and Crader, 1982), yet it is not at all obvious that this is always appropriate.

This section adumbrates a few methodological factors that I see as problems in investigating early hominid behaviour. Whilst if true they would serve to explain at least partially some of the attitudes to the problems with material, their acceptance or rejection does not affect my earlier arguments based on the comparison of the nature of hominid sites with those of contemporary hyaenas.

It seems quite possible that sciences with an historical element such as palaeoanthropology are different in important ways from what might be termed "contemporary" sciences; those that deal with phenomena and subjects operating and observable in the modern world. In most aspects of such contemporary sciences, when an hypothesis is invented it has certain observable consequences; in principle data may be collected which might test it, and falsify or temporarily sustain the theory. This is not invariably the case with palaeoanthropological hypotheses.

Palaeobiology differs in that there is much about the past that will probably remain unknowable. Admittedly the areas which fall into this category should not be proscribed, but the past is almost certain to remain less accessible than the present. This obvious condition affects the status of hypotheses about the past. Despite such hypotheses not being operationally equivalent to those about the contemporary world, they are often tacitly regarded so. In fact they probably differ in status to a sufficient degree for the difference to be important. A contemporary hypothesis has the potential of being falsified by the collection of relevant data. In the historical case it may be equally easy to see what data needs to be collected to test an idea, but it may be information that is unobtainable.

This general lack of critical data seems to tend to a situation where mere congruence with available information is sufficient for the acceptance of a theory. The reasons for the acceptance of the idea may be simply that it was the first one to be proposed. The reasons for its proposition may involve very general notions which are separate from the material data, as is the case with many ideas about early hominid behaviour. The root of the ideas lies with the preference of the propounder, and various euphemisms are used to describe them, such as "model", "working hypothesis" or even sometimes "null hypothesis". Again, as in the example of the early hominids, this working hypothesis is mostly that which requires as little difference from the present day situation as is still tenable, or from what are regarded as contemporary analogies. It is at this point that notions of simplicity become involved. For example, Isaac (1978) states:

> . . . We know for a fact that somewhere along the line in the evolution of human behaviour two patterns became established: food sharing and a division of labor. If we include both patterns in our model of early hominid society, we will at least be parsimonious. . . .

In this case the idea is explicitly expressed, but its implicit assumption pervades the methodology involved in most work of this kind. By invoking simplicity a defence is formed for the theory, as it is one of the principles of a naive scientific methodology that the simplest hypothesis is the one to be provisionally accepted above all others. A situation most like some chosen aspect of the present day, and yet still consistent with the archaeological evidence is assumed to be the most simple or the most parsimonious interpretation. It seems that archaeology postulates similarities, although it might be more fruitful to discover differences, and assuming similarity between two or more conditions is wrongly thought to be a more simple position than assuming incongruities. Any element extraneous to this situation which is introduced into the argument, information about an unrelated species for example, is confused with the introduction of an additional parameter, and hence believed to violate principles of parsimony.

I believe this assumption about simplicity is false, for reasons I will discuss in the next section, and even if it is not then the analogical approach itself, uncritically applied, is unwise, for reasons that I will discuss here.

As the material evidence used to justify them in fact contributes little, ideas regarding early hominid behaviour involve an almost exclusive component of phylogenetic analogy; the hominids believed to have produced the sites which ostensibly provide data are no doubt more closely related phylogenetically to modern humans than to any other extant species. The "models" used in explanation derive largely from the supposed and actual way of life of various contemporary hunter-gatherer groups, or more often are based at a fairly superficial level on general ideas about their behaviour. At its simplest the structure of the argument seems to be, these modern technologically unsophisticated peoples have behaviours that tend towards the local accumulation of bone, and the social correlates of this are amongst those which demarcate present day humans from other primates and other animals in general. The Lower Pleistocene hominids are indeed hominids and not apes, and so these more extensive modern human attributes are uncritically applied to them also.

Now in an entirely contemporary situation it would be sensible, provisionally, to make this assumption; that for example a so far uninvestigated species of monkey might have certain behaviour in common with another species of the same genus. But here the lack of crucial data is less of a problem. The initial assumption can readily be tested by observations. Also, phylogenetic closeness is no imperative for similarity of behaviours. It may be part of an explanation for similar behaviours, but there are plenty of neontological examples to show that it is not a necessary condition. Phylogenetically close species may differ considerably in details of behaviour, and similar possibly inherited behaviour patterns may be directed toward radically different ends and functions.

Another point is that there is only one species of present day hominid to generalize from. Consequently it is intrinsically impossible to know whether the conclusions reached are generalities about *Homo sapiens* (or maybe even only a sub-group of *Homo sapiens*), or whether they relate to all members of the genus *Homo*, or to all hominids. Past species of hominid are characterized by morphologies different from the modern species, and this implies the potential of some differing behaviour patterns. In so far as inferences about past behaviour do not rely only upon conclusions drawn from similarities and differences in morphology, interpreted adaptively, then they have to be suspect, for the phylogenetic analogy made is monospecific.

In the historical case an additional factor is that it is the origin of a particular behaviour that is the actual subject of investigation. At some point in the past hominids presumably did not have a behaviour pattern that led them to collect bones in large quantities, and there is no real reason to imagine that this tendency arose instantaneously with the other social accretions that at the present day are seen to go with it. This may be a set of discrete behaviours arising separately at different times and not causally correlated. To assume the origin of social behaviours on this basis is to evade the possibility of detecting it.

## 7. PROBLEMS WITH THE NOTION OF SIMPLICITY

The assumption of simplicity can be attacked on more general grounds. There is admittedly a widespread and often justifiable tendency in science to aim for the simple interpretation. Ockham is presumably partly responsible for this, or at least he, or Ponce's more familiar version of his views (Thorburn, 1918), is largely used to justify it. Even though for a long time there were logical difficulties with this position, any suggestion that notions of simplicity, parsimony and economy of hypothesis should not be invariably adopted is either ridiculed or treated with the contempt accorded any gross heresy. Amongst those who have alluded to these difficulties in the general case is Quine (1963) who noted how very odd it was that the simpler of two theories is mostly not only regarded as the more desirable, but also as the more probable. In the context of archaeology Clarke (1968) has discussed some of these matters, and presented what he referred to as a "converse Ockham" approach.

Popper (1959) has put many of the tacit notions about simplicity on a firmer footing, by suggesting that the simplest of hypotheses, in the context of a particular problem, is also the most easily falsified, and hence in his terms the best to adopt initially. There remains the problem of what is meant by "simple". Jeffreys and Wrinch (1921) equated simplicity with a paucity of freely adjustable parameters, and this formulation was reached independently by Popper as the only definition that he believed could have any useful meaning, although he sees different consequences in it. For Popper it is the one which has the highest prior improbability, whereas Jeffreys and Wrinch believed it to be the most probable.

This fairly clear definition of what can usefully be meant by simplicity exposes the problems involved when it is considered explicitly or more often implicitly, in archaeology and palaeoanthropology. One set of problems comes from ignoring the definition of simplicity stipulated by Popper and by Jeffreys and Wrinch. In these cases, what a particular worker feels intuitively to be simple is defined as the simplest hypothesis, and often his intuition may be misleading. Another set of problems arises even when the definition by the smallest number of freely adjustable parameters it taken into account. Here the difficulty comes from the necessity of equating a complex, largely non-mathematical more verbal archaeological situation with a more strict mathematical definition. What, in the archaeological case, constitutes a parameter? A single theoretical system can often be fragmented into different numbers and kinds of parameter according to preference or point of view.

Yet another strategy involves reducing parameters by ignoring their existence. In the case of early hominid behaviour this is very easy to do because of the general paucity of information. The consistency of this meagre information with a model based on phylogenetic analogy is considered sufficient, and the action of factors known to exist but of which the details are unknown, receive much less attention or weight in arguments.

This discussion also demonstrates a tension which must exist between the aim to provide a highly falsifiable theory, and the wish to produce the most probable narrative. If Popper's view is correct, these two positions cannot coincide. A consequence of his arguments is that the theory that seems most probable in terms of what is already known is not the most simple, nor is it the most falsifiable. If the aim is to tell stories about the past that fit comfortably with what is known at present then this point need not be considered. But it is very important when attempting to invent bold imaginative scientific theories that might lead to further fresh ideas and discoveries.

## 8. CONCLUSIONS

Taphonomical investigations of modern bone assemblages make it possible better to evaluate evidence from archaeological and other fossil accumulations. The characteristics of a bone accumulation at a modern spotted hyaena den, for example, are shared by those at African lower Pleistocene archaeological sites. Consequently it is difficult to sustain the idea that these features are diagnostic of, or even evidence for, the proposed set of socio-ecological hominid behaviours that have been inferred from them. The behaviours include some which demarcate humans from other primates. Hyaenas, whilst producing the same bone material traces, do not behave in these ways.

Although it has been suggested that the bones at archaeological sites constitute the material evidence, in fact behavioural ideas about early hominids seem to rely almost exclusively upon crude analogy with certain technologically unsophisticated contemporary humans. Phylogenetic analogy has some merit, but it is limited, particularly when independent corroborating evidence is lacking. The hominids that produced these sites were morphologically different from modern humans. They belonged to a different species and possibly a different genus. It is difficult to know at what level any particular analogy applies; at the local population, species, genus or family. The particular social set of behaviours under consideration are not displayed by any modern ape, but they are unlikely to apply to the whole of the Hominidae, for there are fossil hominids which are not known to accumulate bones. This activity and the social behavior correlated with it in contemporary humans must have arisen at some stage during the evolution of the family. Phylogenetic analogy in this context is a dubious practice, for it permits the simple assertion of the condition that is the object of enquiry.

A more general notion of simplicity in science enters the argument in that what can be explained within an anthropocentric framework is apparently considered more simple than one which involves extraneous ideas and information from other species or conditions. The concepts of simplicity, falsifiability and probability are very confused in archaeological thought and practice, not least amongst those who claim some allegiance to a Popperian mode of investigation. This confusion plays a role in determining which of rival theories are preferred.

Less attention should probably be devoted to telling stories about the past based upon guesses that early hominids behaved as do selected modern human groups, supported by a mere congruence between the scanty material data and these scenarios. More scientific methods can be adopted. Some of this could focus upon the material evidence. The search for more detailed differences between archaeological assemblages and other bone accumulations should reveal the significance of new categories of observations that have not previously been considered as data. This further detailed taphonomical analysis of the bone material itself, as Potts will show in the following chapter, can produce innovative and plausible ideas. The social and behavioural consequences of gathering stone and making tools could also be explored further.

Information about hyaenas may also be incorporated in another, less adversely critical manner. As they produce similar sorts of bone assemblages in the course of dealing with what may in general be similar problems and ecological pressures on the African grasslands, then certain elements of their behavioural ecology may illuminate our ideas about early hominids. But I am not trying to suggest hyaenas, or even carnivores as a whole, simply as a replacement for the hunter-gatherer model. Rather they should be used as material and examples for understanding more basic principles of behavioural ecology that apply equally to early hominids as to any other animals. And of course this does not just apply to bone collecting. What are the consistent behavioural correlates, if any, of all known features of fossil hominids? Preferably these should be supported by functional ideas linking them to adaptation and evolution, and information from a much broader taxonomic base is available for this analysis, providing the scope for more subtle tests, than is provided by a restricted examination of contemporary humans. Maybe, in the evolution of ideas about early hominid behaviour, the assumption of adaptation will prove to be a more stable strategy than the simple adoption of naïve ethnographic comparisons.

## ACKNOWLEDGEMENTS

I thank Robert Foley for inviting me to write this chapter, and for his invitation to the Reading TAG meeting where I presented a preliminary version. I am grateful to David Harris for his criticism at that time. I thank Richard Potts for extensive discussion since then, and Misia Landau, Sally McBrearty and David Pilbeam for their comments on the manuscript. I am grateful to Glynn Isaac for helpful conversations, and for being allowed to see unpublished papers. The Government of the Republic of Kenya kindly gave permission for work on the hyaena den, which was facilitated by Joe Kioko and Bob Oguya, Wardens of Amboseli National Park. The hyaena research was financed by the Kenya Museum Society.

## REFERENCES

Behrensmeyer, A. K., Western, D. and Dechant Boaz, D. (1979). New perspectives in vertebrate paleoecology from a Recent bone assemblage. *Paleobiology* 5(1): 12–21.

Binford, L. R. (1981). *Bones: Ancient Men and Modern Myths*. New York: Academic Press.

Brain, C. K. (1967). Hottentot food remains and their bearing on the interpretation of fossil bone assemblages. *Sci. Pap. Namib Des. Res. Sta.* **32**: 1–11.

Brain, C. K. (1969). The contribution of Namib Desert Hottentots to an understanding of australopithecine bone accumulations. *Sci. Pap. Namib Des. Res. Sta.* **39**: 13–22.

Brain, C. K. (1980). Some criteria for the recognition of bone-collecting agencies in African caves. In: Behrensmeyer, A. K. and Hill, A. (eds) *Fossils in the Making*. Chicago: University of Chicago Press, pp. 107–130.

Brain, C. K. (1981). *The Hunters or the Hunted? An Introduction to African Cave Taphonomy*. Chicago, University of Chicago Press.

Buckland, W. (1822). Account of an assemblage of fossil teeth and bones. etc. *Phil. Trans. Roy. Soc.*, Lond. **112**: 171–237.

Bunn, H. T. (1981). Archaeological evidence for meat-eating by plio-pleistocene hominids from Koobi Fora and Olduvai Gorge. *Nature* **291**: 574–577.

Clarke, D. L. (1968). *Analytical Archaeology*. London: Methuen.

Dart, R. A. (1953). The predatory transition from ape to man. *Int. Anthrop. Ling. Rev.* **1**: 201–218.

Dart, R. A. (1954). The myth of the bone-accumulating hyaena. *American Anthropologist* **58**: 40–62.

Dart, R. A. (1957). The osteodontokeratic culture of *Australopithecus prometheus*. *Transvaal Museum: Pretoria. Memoir* No. 10.

Hay, R. L. (1976). *Geology of the Olduvai Gorge*. Berkeley: University of California Press.

Hill, A. (1975). *Taphonomy of Contemporary and Late Cenozoic East African Vertebrates*.

Ph.D. thesis. University of London.
Hill, A. (1976). On carnivore and weathering damage to bone. *Current Anthrop.* 17: 335–336.
Hill, A. (1978). Hyaenas, bones, and fossil man. *Kenya Past and Present.* **9**: 8–14.
Hill, A. (1979). Disarticulation and scattering of mammal skeletons. *Paleobiology* 5 (3): 261–274.
Hill, A. (1980a). Early post-mortem damage to the remains of some contemporary east African mammals. In: Behrensmeyer, A. K. and Hill A. (eds) *Fossils in the Making.* Chicago: University of Chicago Press. pp. 131–152.
Hill, A. (1980b). Hyaena provisioning of juvenile offspring at the den. *Mammalia* **44**: (4) 594–595.
Hill, A. (1980c). A modern hyaena den in Amboseli National Park, Kenya. *Proceedings of the 8th Panafrican Congress of Prehistory and Quaternary Studies*, Nairobi, pp. 137–138.
Hill, A. (1983). Hyaenas and early hominids. In: Clutton-Brock J. and Grigson C. (eds) *Animals in Archaeology.* B.A.R., Oxford. International Series **163:** 87–92.
Howell, F. C. (1978). Hominidae. In: Maglio, V. J. and Cooke, H. B. S. (eds) *Evolution of African Mammals.* Cambridge, Mass.: Harvard University Press.
Hughes, A. R. (1954). Hyaenas versus hominids as agents of bone accumulation. *Am. J. Phys. Anthrop.* (n.s.) **12**: 467–486.
Hughes, A. R. (1958). Some ancient and recent observations on Hyaenas. *Koedoe* **1**: 1–10.
Isaac, G. K. (1971). The diet of early man: aspects of archaeological evidence from lower and middle Pleistocene sites in Africa. *World Archaeol.* **2**: 278–298.
Isaac, G. K. (1976). The activities of early African hominids: a review of archaeological evidence from the time span two and a half to one million years ago. In: Isaac, G. L. and McCown E. R. (eds) *Human Origins: Louis Leakey and the East African Evidence.* San Francisco: Benjamin, pp. 483–514.
Isaac, G. L. (1978). The food sharing behavior of protohuman hominids. *Sci. Am.* **238**: 90–108.
Isaac, G. L. (1980). Casting the net wide: a review of archaeological evidence for early hominid land-use and ecological relations. In: Konigsson, L. (ed) *Current Argument on Early Man.* Oxford: Pergamon Press, pp. 226–251.
Isaac, G. L. (1983). Bones in contention: competing explanations for the juxtaposition of early Pleistocene artifacts and faunal remains. In: Clutton-Brock, J. and Grigson, C. (eds) *Animals and Archaeology.* Oxford. B.A.R. International Series *163* 3–19.
Isaac, G. L. (in press). Aspects of human evolution. In: Bendall, D. S. (ed) *Essays on Evolution: A Darwin Centenary Volume.* Cambridge: Cambridge University Press.
Isaac, G. L. and Crader, D. (1981). To what extent were early hominids carnivorous? An archaeological perspective. In: Harding, R. and Teleki, G. (eds) *Omnivorous Primates.* New York: Columbia University Press. pp. 37–103.
Jeffreys, H. and D. Wrinch (1921). On certain fundamental principles of scientific inquiry. *Phil. Mag.* **42**: 369–390.
Knox, R. (1822). Notice relevant to the habits of the hyaena in southern Africa. *Trans. Wernerian Nat. Hist. Soc.*, Edinburgh **4**: 383.
Kruuk, H. (1972). *The Spotted Hyaena.* Chicago: University of Chicago Press.
Leakey, L. S. B. (1959). A new fossil skull from Olduvai. *Nature* **184**: 491–493.
Leakey, L. S. B., Tobias, P. V. and Napier, J. R. (1964). A new species of the genus *Homo* from Olduvai Gorge. *Nature* **202**: 7–9.
Leakey, M. D. (1971). *Olduvai Gorge Vol. 3, Excavations in Beds I and II, 1960–1963.* Cambridge: Cambridge University Press.

Popper, K. R. (1959). *The Logic of Scientific Discovery*. London: Hutchinson.
Potts, R. B. (1982). *Lower Pleistocene Site Formation and Hominid Activities at Olduvai Gorge, Tanzania*. Ph.D. Thesis. Harvard University.
Potts, R. B. and Shipman, P. (1981). Cutmarks made by stone tools on bones from Olduvai Gorge, Tanzania. *Nature* **291**: 577–580.
Quine, W. (1963). On simple theories of a complex world. *Synthese* **15**: 103–106.
Sutcliffe, A. J. (1970). Spotted hyaena: crusher, gnawer, digester and collector of bones. *Nature* **227**: 1110–1113.
Thorburn, W. M. (1918). The myth of Occam's razor. *Mind* **27**: 345–353.
Walker, A. (1981). Diet and teeth: dietary hypotheses and human evolution *Phil. Trans. Roy. Soc.* Series B, Lond. **292**: 57–64.
Washburn, S. L. (1957). *Australopithecus*: the hunters or the hunted? *Am. Anthrop.* **59**: 612–614.
Western, D. D. (1973). *The Structure, Dynamics and Changes of the Amboseli Ecosystem*. Ph.D. thesis. University of Nairobi.

# 6 *Hominid Hunters? Problems of Identifying the Earliest Hunter/Gatherers*

*Richard Potts*

## 1. INTRODUCTION

The elaboration of meat-eating and hunting activities in hominids has occupied an important place in ideas about human evolution.[1] A wide diversity of ideas about the history of hominid "hunting" originate from studies of living social carnivores (e.g. Schaller and Lowther, 1969; King, 1975; Thompson, 1975, 1976); studies of non-human primates (e.g. Teleki, 1973, 1975; Strum, 1981; Zihlman and Tanner, 1979; Tanner, 1981); and studies of modern hunter/gatherers (e.g. Lee and Devore, 1968; Teleki, 1975, 1981). Certainly no consistent view about early hominid diet and foraging derives from these modern behaviour studies. Some reconstructions (especially those based on carnivorous mammals) emphasize the hunting and territorial characteristics of early hominids, whereas other models (especially ones based on non-human primates) stress omnivory and the particular importance of plant foods in the diet.

One problem with hypotheses about early hominid foraging which are based on contemporary behaviour research is that they are usually proposed as *answers* rather than as *questions* about early hominids. Carnivore, chimpanzee, modern hunter/gatherer, or other models of early hominid life rarely come with guidelines for testing and possibly falsifying these models for early hominids from a specific time and place. Behaviour studies often give rise to conflicting "theories" while information from the geological record remains the same for every viewpoint (Wilson, 1975, p 28; Isaac, 1981; Teleki, 1981, p 338). The geological record, which preserves fossil bones and artifacts, is the only source of information about specific early hominids. Thus, information derived only from the prehistoric context affords the opportunity to test various ideas proposed from modern studies.

HOMINID EVOLUTION
*ISBN 0-12-261920-X*

Excavated assemblages of associated stone artifacts and faunal remains have become one focus of information about early hominid diet and foraging. Usually, meat consumption is inferred by archaeologists from accumulations of bones connected with hominid activities (e.g. Isaac and Crader, 1981; Bunn, 1981). Where this dietary inference leaves off contemporary behaviour models pick up to provide more complex interpretations about early hominids. As a result, meat-eating is often equated with hunting (and occasional scavenging). Hunting, then, is understood as a new source of subsistence in hominid evolution, providing the basis for a hunter/gatherer way of life (Washburn and Lancaster, 1968; Lee, 1979, p 489–494). Finally, a "meat and plants" mode of subsistence is linked with the production of home bases around which food sharing and other human socioecological traits developed (e.g. Isaac, 1978). According to this interpretation, accumulation of bones by hominids signals home base activity (Isaac and Crader, 1981). Although this set of inferences dominates the thinking of palaeoanthropologists, these are not the only possible interpretations of the excavated remains (Binford, 1981; Isaac and Crader, 1981; Potts, 1982; Shipman, in press).

The purpose of this chapter is to dissect this chain of inferences regarding Lower Pleistocene hominid meat-eating and hunter-gatherer foraging. The main issue is whether the analytical methods and data used by archaeologists necessarily specify any particular set of inferences about early hominid diet and foraging. The methodological critique and ideas developed here arise largely from a personal study of excavated sites from Bed I Olduvai Gorge (Potts, 1982), an important source of information about Lower Pleistocene hominid subsistence (Leakey, 1971; Isaac, 1971; Isaac and Crader, 1981; Binford, 1981). Data obtained from this study will help to illustrate some problems with previous interpretations and some possibilities for more definitive inferences about early hominid meat-eating and foraging.

## 2. THEORETICAL CONSIDERATIONS: ECOLOGICAL INFLUENCES ON SOME EARLY HOMINIDS

### 2.1. Terminology and Conceptual Problems

Acquisition and consumption of animal protein have been noted in many primates. In turn, some primatologists have de-emphasized the adaptive impact that meat-eating and hunting had during hominid evolution (Harding, 1981, p 200; Teleki, 1975). However, this issue is muddled by the meaning of the terms "meat", "hunting", "predation" and "scavenging". It is necessary first to clarify the usage of these terms in this chapter.

"Hunting" is distinguished from "predation" by some researchers. Teleki (1975) sees the distinction as a phylogenetic one: "predation" refers to the activities of meat acquisition in nonhuman primates, whereas anthropologists typically use the term "hunting" for basically the same behaviour patterns (seizing, chasing and stalking prey) in humans. Isaac and Crader (1981, p 101) make more of a qualitative distinction: "hunting" is the active searching of relatively large, highly mobile prey; "predation" (which includes "hunting") also involves the acquisition/killing of small, relatively stationary prey (e.g. nestling birds). In contrast, Harding (1981, p 200) places the acquisition of all animal protein (including insects and small vertebrates) along a continuum with "hunting" behaviour. Harding's review (see also Gaulin and Konner, 1977) indicates that most primates do capture and eat insects and small vertebrates. Thus, it is no wonder that he sees "hunting" as a characteristic of many primates and as nothing special to hominid evolution. However, most discussion about the importance of hunting during hominid evolution has focused on large game, animals most notably hunted by the large mammalian carnivores. Thus, the confinement of the term "hunting" to macromammalian prey (greater than 10 kg in body weight) is appropriate for this discussion.

Similar confusion arises concerning the term "scavenging". Scavenging is claimed to be rare among chimpanzees (Teleki, 1975, 1981). However, McGrew (1979, p 458) noted that 75% of bushbucks eaten by Gombe chimpanzees were appropriated from baboons. Teleki (1981, p 330) describes this activity as "pirating" of fresh kills from baboons. This displacement of one animal by another at an animal carcass, however, matches what carnivore field observers call "scavenging" (e.g. Kruuk, 1972; Schaller, 1972). "Scavenging", then, involves the acquisition (sometimes by appropriation from other animals) of a complete or partial animal after it has died either from hunting by another species or from other causes. "Early scavenging" of a carcass (very soon after death) may permit access to the complete carcass, while "late scavenging" generally allows access to a smaller range of anatomical parts from the dead animal.

"Meat" is another term which is used in confusing, inconsistent ways. Definition of the term is partly linked to different usages of the word "hunting". "Meat" sometimes means all edible tissues obtained only from terrestrial vertebrates (e.g. Isaac and Crader, 1981, p 101). Alternatively, eggs and crustaceans have been included in a single dietary category with vertebrate tissues (Harding, 1981). In the realm of food science "meat" refers anatomically to food derived solely from muscle tissue (Eskin *et al.*, 1971). However, muscle, bone marrow and internal organs usually are treated by palaeoanthropologists as a single edible product from terrestrial vertebrates. Each of these tissues has different nutritional consequences.[2]

Furthermore, meat-eating, even when considering muscle consumption alone, is a more complex phenomenon than is suggested in discussions of early hominid behaviour. Upon animal death, blood flow stops and the muscle tissue rises in temperature for a period of time while increasing in pH (due to production of lactic acid via glycolosis). This latter process tends to denature proteins. It is not well understood how these processes interact with the digestive processes of various carnivores—nor of primates that eat mammalian muscle tissue. Nonetheless, these processes may mean different nutritional consequences for animals that consume meat immediately after death (e.g. by cheetahs after a kill) versus those that practice delayed consumption (e.g. by lions, which occasionally wait by kills before eating). Time plays a major factor in changing the digestibility and nutritive value of meat, not always for the detriment. Hence, the timing of hominid access to meat from a dead animal may hold significant digestive and nutritional consequences, though these are poorly defined at present.

Estimates of maximum dietary intake of meat by non-human higher primates are consistently less than minimum estimates for tropical hunter/gatherer humans. Estimates of meat intake (by weight) for Gombe chimpanzees (Teleki, 1981, p 327) are at least 10–20 times lower than occurs among the !Kung San and other hunter/gatherers (Lee and DeVore, 1968; Hayden, 1981). Furthermore, Hayden (1981, p 357) points out that meat yields more calories per unit weight than do most plant foods. Hence, the discrepancy between the caloric contributions of meat to the diets of human hunter/gatherers and of the most meat-eating non-human primates may be somewhat greater than is suggested by weight estimates. Since humans are deemed peculiar among primates in this dietary manner, it is valid to ask, "why did at least some hominids incorporate more meat in their diets?"

Before turning to this question, a final conceptual confusion concerns the importance of delayed meat consumption and meat sharing. A large but undocumented part of hunter/gatherer diets seems to derive from foods obtained during foraging trips but not taken back to camps (Hayden, 1981, pp 386–387; Mann, 1981, p 12). Therefore, a *strict* dichotomy should not be made between an individualistic feed-as-you-go strategy in nonhuman primates only as opposed to a food transport/sharing strategy practised by humans at all times. In addition, observations of meat distribution in chimpanzees have been used to de-emphasize the importance of food sharing in the evolution of hominid subsistence strategies (Teleki, 1975, p 1981). However, food sharing among modern hunter/gatherers often involves not only handing over of food but also recurrent transport of it to a defined social area, the base camp.[3] Descriptions of meat distribution among nonhuman primates do not fit this particular aspect of human food sharing (Isaac, 1978). In addition, although hunter/gatherer food sharing today involves transport of edible resources, transport alone need not always indicate food sharing (see Chapter 5, this volume). It is useful, therefore, to envision some independence between evidence for meat transport and evidence for meat sharing.

### 2.2. Why Eat More Meat?

Some explanations of increased meat-eating among hominids emphasize the nutritional and reproductive benefits accrued by hominids through individual or kin selection. For example, Isaac's (1976, 1978) model suggests that inclusive fitness was increased for those individuals who carried foods to other members of the social group at a home base. Since it comes in readily-transportable packages, meat is viewed as particularly important in facilitating a system of food transport to home bases (Isaac, pers. comm.). Furthermore, meat is a rich source of necessary amino acids (Yesner, 1980, p 95; Hayden, 1981, p 394). Thus, nutritional factors may have promoted the intake of meat in the diets of some hominids.

Such explanations, however, must be supplemented by information about the ecological conditions under which early hominid meat-eating occurred. For example, no answer to the question "why eat more meat?" would be complete without considering the ecological factors which permitted hominids to enter further into the realm of large carnivores than had been done by other primates. Here, an answer depends on one's outlook on species niches.

According to one view, within a heterogenous environment new ways of adapting (including dietary shifts within a clade) arise opportunistically. Not all resources within an environment are used. Thus, the adaptation of a species can vary and the evolution of a species is not highly limited by interspecific competition for resources (Stanley, 1981). Therefore, ecological innovations (e.g. new, viable uses of resources) are often available to new reproductive isolates without reference to particular environmental "pressures" which necessitate that evolutionary shift. From this point of view, one way evolutionary change in diet may come about is exemplified by dietary changes among Japanese macaques on the island of Koshima (albeit in experimental conditions in their natural habitat) (Kawai, 1965; Frisch, 1968; Kummer, 1971, p 117–124). These monkeys which had been mainly forest living, began to eat sweet potatoes and, later on, wheat placed by Japanese researchers on the beach. As the result of insightful behaviour by one female macaque, who figured out how to wash sweet potatoes and wheat grains for eating, the new dietary source became widely used. Consequently, these particular macaques began to occupy the beach zone, to enter the water, to swim, and even to utilize ocean resources. In this kind of scenario, then, an answer to "why some hominids began to eat more meat" may simply be phrased "because it was there", plus accepting the idea that all it took was one hominid or a group of hominids to make some insightful discovery about how to exploit that readily available resource. Strum (1981) makes a similar point about the elaboration of meat-eating and hunting among baboons at Kekopey (Gilgil), Kenya. Under the particular ecological and social conditions that have occurred at Kekopey, the innovative behaviour of one individual has incited a rise in meat-eating/hunting behaviours, which existed previously at a low level. Under certain, though unspecified, ecological conditions, one could imagine that a similar kind of opportunistic shift in the use of a food source could have led to a breakthrough within the hominid clade of long-term evolutionary importance.

An alternative viewpoint—that species within communities are highly interdependent, competitive and ecologically confined by biotic and abiotic factors—suggests that significant ecological/evolutionary changes result from adaptive "problems" or "pressures" posed by fluctuating environments rather than from incidental, opportunistic probing of niche space (see Chapter 4, for a further discussion of this issue). Since the main issue in this chapter concerns the increased consumption of meat acquired from large, terrestrial vertebrates, the potential amount of competition over such meat resources is an important factor to consider. In the Serengeti and elsewhere in East Africa a high degree of interference competition over large vertebrate carcasses has been documented. Such competition among carnivores (especially lions, hyaenas, cheetahs, leopards and wild dogs) leads to aggressive interactions over carcasses, sometimes a high degree of predation among carnivore species, and can significantly affect carnivore population sizes (Bertram, 1979, p 229; Kingdon, 1977; Potts, 1982, p 299).

Since carnivore-ungulate savanna communities seem to have developed in eastern and southern Africa by 4 m.y. ago (Barry, in press), the incorporation of meat form ungulates and other macrovertebrates into the hominid diet does not seem so simple as picking sweet potatos up off a beach. In the best documented cases the elaboration of meat-eating by nonhuman primates has occurred in areas where carnivores are rare, such that the primates are not subject to extensive predation or competition from carnivores (Teleki, 1975; Strum, 1981, p 283). It is clear, though, that very early hominids and later ones inhabited savanna-mosaic areas where a wide range of carnivores existed.[4] According to this view, then, an increase in meat-eating during hominid evolution involved some increased competitive ability among some hominids (relative to carnivores) in obtaining meat. This competitive ability, in turn, was selected because meat-eating served some function or set of functions critical to hominid survival and reproductive success. For example, either: (1) some hominids were faced with decreased availability of regular foods besides meat (due to climatic change or due to evolutionary diversification of monkeys, suids, or some other group which ate the same things hominids did); or (2) hominids that incorporated more meat in their diet had an increase in reproductive success, due to factors (considered previously) such as enhanced nutrition or enhanced transportability for social feeding. It is necessary to postulate that these benefits outweighed the costs of direct interaction and exploitative competition from carnivores.

At present it is difficult to choose between these two general ecological models. Although the latter view perhaps follows a more traditional concept of natural selection, the ecological conditions under which meat was acquired from large vertebrates in savanna-mosaics need not have involved high competitive costs with carnivores. As is discussed in the next section, a high percentage of African ungulate biomass today goes largely untouched by the large carnivorous mammals. Although predators are "closely packed" into their territories, carcasses and meat are available without recourse to predation and, therefore, without much competitive "pressure" from the large carnivores. Hence, it is difficult to dismiss *a priori* the idea that hominids entered opportunistically into a niche which, in part, involved exploitation of a readily available source of meat.

### 2.3. Lessons in Foraging for Meat

As remarked earlier, modern analogies do not provide the basis for final interpretations about early hominid behaviour. Yet field studies do indicate the ecological conditions which correspond to behaviours possibly found in early hominids. For example, the issue concerning hominid *hunting* for meat can be illuminated by field studies of carnivorous animals.

Studies of lions, hyaenas, and more general surveys of the large mammalian carnivores (Schaller, 1972; Kruuk, 1972; Bertram, 1979; Houston, 1979) indicate that the distinction between hunting and scavenging is not critical. Hunting and scavenging do not represent two mutually exclusive, species-specific modes of foraging. Rather, for a mammalian carnivore these two foraging activities represent *complementary* ways of acquiring meat. For a meat-eating species, while there is always more food available from living animals than from dead ones, the acquisition of meat by scavenging may be favoured in some areas more than in others (Schaller 1972; Kruuk 1972). Among the five large Serengeti carnivores (lion, hyaena, cheetah, leopard and wild dog), from 0% (for cheetah) to 33% (for the spotted hyaena) of the carcasses fed upon are obtained by scavenging (Bertram, 1979).

Moreover, in studies by Kruuk (1972, p 115) and Schaller (1972, p 435) the majority (60–70%) of carcasses scavenged by hyaenas and lions were taken from other carnivores; few scavenging events were of animals which died from non-predatory causes, such as disease or malnutrition. The act of carcass possession, especially when initiated by predation, seems to attract other carnivores. Yet, occasionally, scavengers may simply wait their turn for access to a carcass while another carnivore eats a portion of it.

The various carnivore species have adapted to interference competition in a variety of ways (Lamprecht, 1978; Schaller and Lowther, 1969). Cheetah, which fare poorly in interspecific encounters over carcasses, scavenge very little, if at all. These carnivores hunt silently and quickly, consume meat rapidly, and attract less attention during their kills by hunting individually or in very small groups. Leopards also tend to hunt individually. They acquire a carcass quickly and often consume it in a hidden spot (e.g. trees) where other carnivores cannot reach it. Hyaenas sometimes will take parts of carcasses away from the death site and hide them in water, where the meat usually is protected from other meat-eaters. In addition, hyaenas and wild dogs occasionally feed in large social groups. By this means, meat on the carcass is consumed quickly and efficiently; the carcass then is abandoned at the death site. Foraging in large groups also permits appropriation of carcasses from other carnivores and defense of carcasses from other carnivorous species (e.g. "mobbing behaviour" of hyaenas and wild dogs). Finally, the size and, in some cases, the sociality of lions enable efficient defense of carcasses from interspecific competitors. The ability to guard carcasses or to chase other carnivores away from their kills plays an important role in lion foraging behaviour.

Since most carcasses tend to be available to mammalian scavengers only after partial consumption by a different predator species, and since competition at carcasses can be great, Schaller and Lowther (1969) concluded that *no* large mammal could live just by scavenging. This conclusion was extended to early hominids:

> We believe that a hominid, relying upon scavenging alone for its meat, could under present or similar conditions exist only if a large part of its food supply consisted of vegetable matter, with meat being incidental and seasonal, such as is the case with the Bushman (Lee 1968). A carnivorous hominid would by necessity have to be both a scavenger and a hunter, using every available method to obtain meat, as the predators do today. (Schaller and Lowther, 1969, p 326).

In addition, although some carnivores transport bones to den areas (see Chapter 5, this volume), none seems to keep meat there. Noting that "an odorous collection of food remains would rapidly attract other carnivores," Schaller and Lowther (1969, p 335) point out that transport of meat to an area where young are kept would probably not be of survival value to modern carnivores or to early hominids. Such activity would demand defense of the transported meat/carcass *and* of the home base.

Most carnivore models of early hominid meat-eating result exclusively from generalizations about carnivorous mammals. However, studies of avian carnivores suggest that ecological conditions exist in which scavenging becomes a much more distinctive strategy of foraging than is observed among the Carnivora. In particular, some vultures scavenge all of their food by finding carcasses where mammalian carnivores are not present.

Information from the Serengeti best shows how some vultures are able to make use of large stores of meat by scavenging. The majority of food (estimated 70% by weight) available from ungulate carcasses is provided by migratory species. Furthermore, two-thirds of all deaths in migratory herds are from causes other than predation (Houston, 1979, pp 266–7). However, as noted earlier from Kruuk's and Schaller's research, the large mammalian carnivores in the Serengeti scavenge mostly from predator kills, and thus tend to interact with other large mammalian carnivores. At least in the Serengeti, there is a wealth of meat provided by migratory ungulates dying from non-predatory causes that is not tapped by the mammals. Houston (1974, 1979) has elucidated this aspect of ungulate mortality and meat supply. The migratory ungulates provide a readily available source of meat to those carnivorous animals that can locate it. In the case of the griffon vultures 100% of their diet is obtained by scavenging. Their foraging entirely depends on finding dead animals (Houston, 1974).

The amount of food that can be obtained by scavenging depends upon searching methods. Griffon vultures are able to follow migratory ungulate herds by highly efficient flight. Their gliding, airborne vantage permits sighting of carcasses and the ability to cover long distances in order to keep up with migratory herds. Griffons seem to focus exclusively on the large migratory herds, even during the breeding season when resident ungulates are closer to their nests (Houston, 1974). Although mammalian carnivores watch vultures to locate carcasses, the vast majority of the time [(84% of 64 cases observed by Houston (1979, pp 276–7)] mammalian carnivores do not appear at carcasses at which vultures arrive first.

In contrast, the large Carnivora have relatively fixed territories, which are related to stationary breeding areas. Thus, their foraging ranges are extremely limited compared with those of scavenging vultures. Although Serengeti hyaenas and lions locate their social groups partly in relation to the migration paths of herbivores, these migrants are within the foraging circle of these carnivores only a few months during the year (Pennycuick, 1979; Houston, 1979)[5]. At other times of the year, the large terrestrial carnivores depend on diverse resident ungulate species. Studies of resident ungulate mortality indicate that predation is responsible for most deaths and that all other resident animals dying from other causes are scavenged quickly (Kruuk, 1972; Melton and Melton, 1982). When the source of meat is confined to resident ungulates, scavenging evidently becomes a highly competitive matter among the mammalian carnivores. This is in contrast to situations in which migratory ungulates are the food source.

The point is that certain locomotor and socio-ecological strategies can favour relatively free access to a large meat supply and a completely scavenging mode of foraging, even though these activities do not characterize the large Carnivora. In the modern situation herbivore migrations involve few species with a huge number of individuals in the herds. For example, in the Serengeti migrations, wildebeest predominate, though Thomson's gazelle and zebra are also migratory (Maddock, 1979). A greater diversity of herbivore species (but with far fewer individuals in each) is formed by the resident ungulates. Thus, under ecological conditions similar to modern savanna-mosaics in East Africa, a mostly scavenging mode of foraging is conceivable in a terrestrial, nomadic mammal which could feed off of the large number of dead animals from a few migratory ungulate species.

## 3. ASSESSMENT OF EARLY HOMINID MEAT-EATING AND HUNTING

### 3.1. Taphonomic Developments

Early hominid diets and foraging can be studied only to the extent that data from the fossil record actually refer to hominid activities. Taphonomic, or site formation, studies have shown the possibilities for *multiple* interpretations of prehistoric patterns which were once deemed indicative only of hominid behaviour. For example, accumulations of faunal remains excavated by archaeologists may result (1) from hominids or (2) from other animals carrying bones; (3) from water aggregation of bones; or (4) from the cumulative death of animals at a particular spot over a period of time.

Increasingly, research on excavated remains is designed to involve "disproof" of those hypotheses which concern non-hominid agents of site formation (e.g. Binford, 1981, p 179; Potts, 1982). Any study of non-hominid hypotheses means that criteria must be established *a priori* for distinguishing among various non-hominid and hominid activities or processes.

Yet several recent analyses (Binford, 1981; Bunn, 1981; Potts, 1982; Potts and Shipman, 1981) have recognized that complex sets of geological and behavioural agents acted on the clusters of faunal remains excavated from Lower Pleistocene sediments in East Africa. Therefore, disproof of non-hominid hypotheses for these particular sites is not adequate. Ideally, as Binford (1981, pp 246–7) has stated, the analytic aim should be to discern the *degree* of contribution of *each* agent to the formation of an excavated assemblage. Yet no study to date has provided a means either (a) for identifying that particular subset of excavated remains which reflects only hominid activities, or (b) for deciding what percentage of an entire archaeological faunal assemblage is attributable to hominids versus other taphonomic agents.

### 3.2. Archaeological Indications of Meat-eating

Several lines of archaeological evidence are believed to show that early hominids ate "significant" amounts of meat. In a recent review article on excavated bone assemblages from the Lower Pleistocene, Isaac and Crader (1981) define criteria to assess the meat-eating habits of early hominids. These criteria are used to *test* the hypothesis "that tool-making hominids accumulated the bones in connection with their subsistence activities". However, it is admitted that even if excavated faunal and stone artifact assemblages do not exhibit these criteria, "it would still be legitimate to consider meat-eating as a possible component of Lower Pleistocene subsistence" (Isaac and Crader, 1981, p 47). Immediately this points out the difficulty of testing ideas about early hominid diet from the archaeological record. Because the amount (and adaptive significance) of meat in the early hominid diet is a complex ecological and evolutionary question, any diagnosis of this issue is bound to be challenged. Several recent studies, besides Isaac and Crader's, have focused on this problem (e.g. Binford, 1981; Bunn, 1981; Potts, 1982; Potts and Shipman, 1981; Shipman, 1983). What follows is a critique of some of the archaeological criteria proposed regarding hominid meat-eating.

#### (a) *Composition of the bone assemblage (by taxon, age, sex or body part)*

One of the test conditions for meat-eating which Isaac and Crader define is the presence of unusual features of bone assemblage composition which can only be explained by hominid activity. Binford's (1981) study also focuses on one aspect of assemblage composition (skeletal part abundance) to deduce the feeding characteristics of early hominids at Olduvai Gorge.

Faunal assemblage composition results from a complex set of taphonomic processes. The interaction among the following variables will determine the taxonomic, body part, and age/sex class representation in any faunal assemblage.

(1) *Availability*: those species, anatomical portions of carcasses, and age/sex classes which are most abundant in a certain habitat will have a greater chance of preservation in a fossil assemblage.

(2) *Selectivity*: selection (e.g. by carnivores) of favoured taxa, skeletal parts, or age/sex classes from those available will affect assemblage composition. For example, different skeletal regions possess different food values and, thus, may be chosen preferentially with regard to food utility (Binford, 1978).

(3) *Destruction*: diverse body parts, age and size classes, and taxa are

differentially affected by processes which decompose and destroy bones (Brain, 1969; Binford and Betram, 1977; Behrensmeyer *et al.*, 1979).

Availability and selectivity, of course, are effects of ecology and foraging behaviour. For this reason assemblage composition data have been used to identify particular modes of hunting or scavenging activity (e.g. Vrba, 1975; Klein, 1982; Binford, 1981). Thus, a particular combination of taxa, body parts, or age/sex classes in a fossil assemblage can be considered diagnostic of hominid activity *only if* one assumes that hominids foraged for animal bones in a particular manner. At present such assumptions are not valid because we are interested in the first place in finding out how hominids foraged. Assemblage composition data cannot be used for both purposes. The composition of a bone assemblage may reflect non-hominid foraging or non-behavioural accumulation processes (e.g. fluvial action), as well as hominid activity. Hence, evidence for early hominid activity must be kept somewhat independent from information about how taxa, bone parts and prey individuals were sampled or collected from the environment. It is necessary to show first that hominids were the main cause of a bone accumulation; then, the details of assemblage composition can be deemed indicative of particular hominid foraging activities.

### (b) *Hominid modifications of bones*

The identification of bone damages caused by hominids has been critical in many palaeoanthropological debates. For example, Dart's ideas about the Makapansgat bone assemblage and debates about human occupation of North America have drawn upon evidence from bone damages. Recently, the recognition of stone tool cutmarks has become a major focus in archaeological research on early hominid meat-eating. This section will summarize this initial phase of cutmark research.

The occurrence of stone tool cutmarks on animal bones from Olduvai and Koobi Fora establishes that hominids utilized animal tissues from at least some of the bones in faunal assemblages at these sites (Bunn, 1981; Potts and Shipman, 1981). To identify cutmarks on bones, methods were needed for distinguishing stone tool marks on fresh bone from carnivore and rodent tooth marks, from natural impressions made by veins and arteries, from metal tool marks (used in excavating and preparing fossils), and from other conceivable causes of bone surface scoring. The search for diagnostic methods has involved extensive observations of how bones (fresh, dry and fossilized) are modified by these various agents of bone surface damage. In part, this research has relied upon the application of scanning electron microscopy, or SEM, to areas of bone scoring (Potts and Shipman, 1981; Shipman, 1981).

Still, several key issues are not resolved concerning cutmarks on animal bones from Lower Pleistocene sites. First, it is not yet clear what percentage of fine, linear grooves on bones from Olduvai and Koobi Fora can be safely identified by eye (versus SEM identification). Bunn's study (1981) depended on macroscopic recognition of cutmarks with the aid of a strong light source. Although this research acknowledged potential confusion with other sources of bone damage, research by Shipman and Potts (Shipman, 1981; Potts and Shipman, 1981; Shipman and Rose, 1983) has emphasized that cutmark "mimics" confound attempts to identify macroscopically the causes of bone scoring. In particular, some carnivore tooth marks and metal preparation marks overlap in width with cutmarks (Shipman and Rose, 1983) and sometimes appear as sets of multiple grooves. These cases seem to require the magnification and resolution of SEM to make positive identifications. For some time stone tool cutmarks have been identified in archaeological research, especially on palaeo-Indians (e.g. Guilday *et al.*, 1962; Parmalee, 1965; Frison, 1974). Yet in these cases the possibility of multiple causes of fine, linear grooves on bone surfaces has not been deemed important to the extent that it has been recently in early hominid research. Furthermore, cutmarks from palaeo-Indian sites are usually identifiable as butchery marks, which involve the repeated occurrence of usually multiple grooves in areas of muscle or ligamentous attachment (Guilday *et al.*, 1962). For the Olduvai bones, however, a major problem is posed by a high frequency of fine surface marks which occur singly. For such specimens the chance of misidentifying a mark as a cutmark appears to be high (Potts, 1982). At present SEM examination of bone surface damages yields secure identification of causes. Yet for the time devoted to specimen preparation, this method only produces small samples of bone specimens with known causes of damage. Nevertheless, a study of cutmark frequencies and locations may well require SEM in situations where the most frequent type of bone surface damage is not stone tool cutmarks. This situation seems to occur at the Olduvai sites (Potts and Shipman, 1981).

A second issue concerns the inference of hominid meat-eating. Hominid meat-eating is an obvious way to explain the presence of stone tool cutmarks on animal bones. However, meat is not the only resource available from bones. Besides the extraction of edible products (which include muscle and bone marrow), acquisition of non-edible tissues also can produce stone tool cutmarks on bone. For instance, hides, ligaments and tendons (sinew), which contact bone in non-meaty areas of the skeleton, can be used in the production of carrying bags or of other equipment (Lee, 1979, pp 124, 129). Detailed data on cutmark locations should help to confirm or disconfirm whether human-like strategies of butchery and meat-procurement can be attributed to early hominids. Based on a very small sample ($n=20$) of ancient cutmarks from early sites at Olduvai, Potts and Shipman (1981; Potts, 1982) noted that half of these occurred on meat-bearing bones and half occurred on bones which do not bear meat, such as metapodials and phalanges. This pattern contrasts markedly with that of animal tooth marks; 80% of these marks were on meat-bearing bones. The principle suggestion from these data was that hominids and carnivores, in particular, utilized carcasses in different ways. However, this does not mean that hominid utilization of carcasses was like that of modern hunter/gatherers.

Wilson (1982) has suggested that tool *slicing* marks on non-meat bones from Olduvai reflect skinning operations which preceded the stripping of meat from the major limb bones. Several critical points are relevant in response to this suggestion. (1) Fifty per cent of the small sample of cutmarks observed on non-meat bones resulted from chopping or scraping actions with stone tools, not slicing (Potts and Shipman, 1981; Potts, 1982). (2) Although a much larger sample of clearly-determined cutmarks is needed from Olduvai, studies of human butchery marks give frequencies of cutmarks on non-meat bones which do not even begin to approach 50% of all identified cutmarks (e.g. see Parmalee, 1965). (3) Finally, no attempt has yet been made to distinguish cutmarks on non-meat bones which are a consequence of meat-procurement versus cutmarks which might result from primary utilization of non-meat tissues. Neither Wilson (1982) nor Potts and Shipman (1981) have identified criteria which uniquely signal either of these carcass exploitative goals.

In brief, the occurrence of stone tool cutmarks in meat-bearing areas undoubtedly reflects the handling of meat by early hominids. Yet adequate data have not yet been collected to test whether the chief focus of hominid bone modification was to acquire meat (muscle) or other animal tissues, or whether early hominid carcass utilization was similar to that of some recent hunter/gatherers.

Besides cutmarks, other types of bone modifications which reflect only hominid activity have been sought, though with less success. For example, spiral fractures of limb bone shafts are now known to result from stone tool fracture *and* non-hominid agents (evidence summarized by Binford, 1981; Potts, 1982, pp 210–213). The distinction between impact areas on bone made by carnivore teeth versus stone tools is another example. Sometimes, stone tool hammers leave broad, shallow indentations along the broken edges of limb bone shafts, whereas carnivore teeth leave smaller, biting indentations (Bunn, 1981). However, Binford (1981, p 157) reports that small, biting notches characterize the limb diaphyses broken for marrow by Eskimo; and Brain (1981, pp 141–142) illustrates well the wide amount of overlap in the morphology of carnivore and stone tool impact scars on bones.

It is apparent that bone damages produced by hominids and by non-hominid causes tend to overlap increasingly as more observations are made on bone modification. Humans seem "capable of producing modifications that can mimic most of the effects of any natural process" (Binford, 1981, p 180). For the present, then, the evidence for early hominid bone damage rests substantially on the sound recognition of cutmarks.

(c) *Dense aggregates of bones and artifacts*

Traditionally, an association between fauna and stone artifacts was sufficient to assume that the bone remains reflected hominid activity. Recent research, though, has tended to treat the concentration of faunal remains at Lower Pleistocene sites separately from the process of artifact accumulation (e.g. Isaac and Crader, 1981; Binford, 1981; Potts, 1982).

Inferences about collection of bones by hominids as a part of their subsistence activities have relied upon faunal remains which were more highly concentrated spatially than occurred on natural landscapes (Isaac and Crader, 1981; Potts, 1982). Yet it is important to realize that dense concentrations of bones cannot be assumed to be caused by hominids (see Chapter 5, this volume).

Binford (1981) has recently proposed that some of the faunal assemblages excavated at Olduvai Gorge were death sites of animals rather than aggregates due to hominid transport of bones. In contrast, Isaac and Crader (1981) see many of these same levels as evidence for bone accumulation by hominids in connection with meat-eating and food sharing activities. Based on skeletal part preservation for all sizes and species of animals combined, Binford stresses the similarity of the Olduvai faunal assemblages to a model of what animal kill sites look like. The suggestion, therefore, is that any hominid activities reflected by these bones at Olduvai occurred away from base camps. This inference is extremely interesting given the attention paid more and more to the activities of modern hunter/gatherers away from campsites, including their repeated occurrence in certain areas (e.g. near watering holes) away from the home base (Thomas, 1975; Binford, 1980; Foley, 1981; Hayden, 1981; A. Brooks, pers. comm.).

A recent personal study (Potts, 1982) of excavated remains from several levels in Bed I Olduvai examined the diverse non-hominid and hominid processes of bone accumulation inferred by Binford and by Isaac and Crader (among others). This Olduvai study exemplifies the kinds of taphonomic hypotheses pertinent to hominid meat-eating which can be tested against the archaeological record. Five of the Olduvai levels studied contained fauna and stone artifacts. A sixth level yielded only bone remains (Table 6.1).

None of the bone assemblages from these particular levels are primarily animal death sites or scatters adjacent to death sites. This interpretation conflicts with Binford's inference, e.g. for FLK "Zinj", DK level 3 and DK level 2 (Binford, 1981, pp 273, 281). These Olduvai assemblages do *not* show the following features which, together, characterize

Table 6.1. Information about the study sample of excavated levels from Bed I Olduvai.

| Level | Estimated excavation area ($m^2$) | Depth of level (cm) | # stone artifacts | Total weight artifacts/manuports (kg) | # macromammal bone specimens | MNI/taxa ratio | Axial/limb bone ratio |
|---|---|---|---|---|---|---|---|
| FLKN L/6 | 37 | 50 | 130 | 12·0 | 2256 | 1·7 | 0·58[b] |
| FLK "Zinj" | 290 | 9 | 2647 | 72·4 | >15 800 | 2·2 | 0·35 |
| FLKNN L/2 | 186 | 24 | 0 | 0 | 478[a] | 2·6 | 0·09 |
| FLKNN L/3 | 209 | 9 | 72 | 13·8 | 683[a] | 2·4 | 0·50 |
| DK L/2 | 345 | 68 | * | * | 1832 | 1·6 | 0·14 |
| DK L/3 | 345 | 9 | 1163* | 93·9 | 1366 | 1·6 | 0·15 |

* For artifact data levels 2 and 3 at site DK are combined.

[a] Does not include numerous bone fragments sieved from FLKNN levels 2 and 3 combined.

[b] Includes an almost complete elephant carcass; ratio = 0·28, without the elephant skeleton.

animal death sites (including multiple kill sites), as established from extensive studies of bones on modern African landscapes (e.g. Hill, 1975; Behrensmeyer *et al.*, 1979; Behrensmeyer and Boaz, 1980).

(i) *Areal concentration of bones and carcass mixing.* Any carcass starts out as a cluster of articulated bones. Yet animal death sites in modern African savannas with high biomass are often isolated from one another (Hill, 1975; Behrensmeyer and Boaz, 1980). As a result, the average areal concentration of bones tends to be low over modern landscapes in East Africa. Bones are often found clumped, but such clusters are small and mostly belong to a single carcass. Thus, bones from different carcasses tend not to mix or overlap to a great degree over the landscape, except where some form of transport (by water or by animal) is involved (Hill, 1975, 1979, pers. comm.).

The particular Olduvai assemblages examined here all show a *combination* of high bone concentration and high carcass mixing which is uncharacteristic of most kill/death sites in African habitats today. For Olduvai the spatial density of specimens, based on the minimum number of identifiable skeletal elements from macromammal species alone, were 27–2320 times greater than occurs for all bones and bone fragments in high biomass savanna areas in East Africa today (Potts, 1982). These particular Olduvai bone aggregates are also more concentrated spatially than occurs generally in excavated levels within the lake margin zone, where all known Bed I sites occur. Furthermore, the assemblage with the sparsest concentration of remains (FLKNN level 2) contained a minimum number of 23 individuals identified from 324 macromammal specimens. The densest concentration (FLK "Zinj") contained at least 36 different individuals identified from 614 specimens. In each case the bone specimens showed no recognizable spatial sorting according to taxon. This is indicative of an extremely high degree of carcass mixing.

(ii) *Taxonomic/ecological diversity of the fauna.* Multiple carcasses can occur at a death site usually in two main situations: (a) drought conditions where many animals may die together from starvation, and (b) multiple kill sites. First, drought does not account for the Olduvai faunal accumulations. The Olduvai levels exhibit neither geological nor palaeontological evidence (Shipman, 1975) of drought-produced assemblages. Secondly, multiple kills tend to occur either as one predation event (e.g. Kruuk, 1972, p 195); or over a period of time at an ambush site, as noted for some carnivores in India (Schaller, 1967; A. K. Behrensmeyer, pers. comm.). Kills at such sites, however, involve one *preferred* species of animal. This largely results from the repeated presence of a particular prey species at that site, or from the occurrence of a single species in a herd. Hence, bone accumulations at multiple kill sites can be expected to have a low taxonomic diversity relative to the number of prey individuals.

In contrast, high taxonomic diversity is characteristic of each Olduvai faunal accumulation. Various species of bovids, suids, equids and carnivores are represented in each assemblage. Remains of hippos, rhinos, giraffids, proboscideans and primates also occur in some levels. Faunal body sizes in every bone assemblage range at least from small bovids and carnivores (12–72 kg) to large bovids (320–820 kg). Table 6.1 gives the ratio of MNI (minimum number of individuals) to the number of macromammal taxa recognized in each Olduvai assemblage. Averaged over the six Bed I assemblages, the most numerous species is represented by about 6 MNI out of a total of over 32 MNI. Thus, no single species predominates. Besides the fact that many species are represented, there is a wide ecological diversity of animals in each assemblage. The faunas include browsers, grazers and mixed feeders. Animals ranging from the forest living *Cercocebus* to open grassland bovids and equids are sometimes found in the same assemblage. These are not combinations of species which are likely to be preyed upon in areas the size of the excavated levels in Bed I Olduvai (see Table 6.1). The point is that the taxonomic and ecological diversity within the Olduvai bone accumulations exceeds expectations (based on modern descriptions) of multiple kill sites. However, these faunal assemblage characteristics do not necessarily rule out transport of bones by animals or by other agents (Brain, 1981; see Chapter 5, this volume).

(iii) *Skeletal element representation*. Studies of animal carcasses (death sites) in modern East African environments indicate that limb bones of the skeleton tend to be dispersed early after death, while axial elements (vertebrae and ribs) tend to stay longer at the original death site (Hill, 1979; Hill and Behrensmeyer, in press; Kruuk, 1972, p 126; Shipman and Phillips-Conroy, 1977). Thus, axial components rather than limb bones can be expected to accumulate in an attritional assemblage of kills on a stable land surface. Data on land surface assemblages in East Africa bear out this prediction (Hill, 1975; Behrensmeyer and Boaz, 1980).

This characteristic of animal death sites again is not met by any of the Olduvai assemblages. The latter show frequencies of limb elements relative to axial elements which far exceed what occurs in even complete skeletons (Table 6.1). One problem is that axial bones tend to be preferentially destroyed over limb bones under conditions of bone attrition. Yet in documented cases of extreme bone attrition acting on *complete* skeletons axial bones still outnumber limb bones (Brain, 1969; Binford and Bertram, 1977; Potts, 1982). This is not the case at Olduvai. Since differential preservation does not account for the high relative abundance of limb bones, these elements must have been numerous compared with axials in the original bone accumulations. This inference is not in accord with a kill/death site interpretation for these ancient faunal assemblages.

For the Olduvai assemblages, then, all of these general descriptive characteristics (spatial, taxonomic and skeletal) indicate that skeletal parts were transported from animal death sites (or from the landscape at large) to well delimited areas of accumulation.[6]

(d) *Hominid accumulation of bones*

Recognizing a particular agent of bone *modification* does not automatically specify the primary agent of bone accumulation. Bones may become modified in diagnostic ways either before or after they have been transported to an accumulation area. The presence of cutmarks *per se* does not mean that the spatial association between *all* bones and stone artifacts at a site occurred through hominid behaviour. Although cutmarks occurring on some bones from a dense concentration does indicate hominid handling of bones, so does the occurrence of cutmarks on bones scattered over an ancient land surface. Since cutmarked bones dispersed over the landscape can become clustered by non-hominid processes, other taphonomic indications must be examined to assess the contribution of processes, such as water action and carnivore activity, to early archaeological bone accumulations.

To continue with the example from Bed I Olduvai, each of the six levels represents an area relatively undisturbed by hydraulic action. Essentially, these levels do not show the combination of sedimentary, palaeontological and lithic patterning which is diagnostic of extensive water action (Potts, 1982). Although water transport of bones and artifacts can be largely rejected as a major process of site formation, the possible activity of carnivores greatly confounds any secure connection between bone concentration and hominid activity. Each of the Olduvai assemblages associated with stone artifacts contains bones with carnivore tooth damages *and* definite cutmarks (Potts and Shipman, 1981; Potts, 1982).

In each Olduvai level the bones show wide variation in their surface features. This variation relates directly to the length of bone exposure to climatic weathering. Stages and rates of weathering established by Behrensmeyer (1978) for various East African habitats were applied to the most complete limb specimens from each assemblage. The range of bone weathering stages within each level reflect at least a 4–6 year period of bone exposure and accumulation for each assemblage (Potts, 1982). Interestingly, the sets of cutmarked and carnivore-damaged bones in each artifact-associated level exhibit the same range of long-term weathering stages. This means that if the bones were cut or gnawed before they became dry and started to weather, then these bones were modified by hominids and carnivores over the whole period of site formation. This assumption of fresh bone modification is probably correct because any usable soft tissues on or inside bones are present only before the bones (in most cases) have passed through one weathering stage. In addition, the bone damages which characterize each assemblage (e.g. opening up of marrow-rich limb elements, fragmentation of bones and differential attrition of skeletal parts) reflect the activity of both hominids and carnivores directly at the sites of bone accumulation. Thus, the cutmarks and carnivore tooth marks evidently were made over a series of visits by both agents to the sites (Potts, 1982).

The picture which emerges for each of these Olduvai bone assemblages is that bones were transported away from spots where animals died and were accumulated by some behavioural agent(s) at these sites. Both hominids and carnivores were responsible for modifying bones.

Although the mere presence of stone tools is not sufficient to favour a hominid activity interpretation, it is important to ascertain whether the processes of stone artifact deposition necessarily help resolve the problem of bone accumulation. Taphonomic interpretations are usually based on contextual evidence (e.g. sedimentary context of bones). In five out of the six Olduvai levels bone and stone artifact concentrations do represent contextual evidence for each other. Thus, the questions of artifact accumulation and of bone accumulation parallel one another.

Each of the artifact-bearing levels contained exogenous stone material weighing from 12 kg (for 116 pieces in FLKN level 6) to 72 kg (for 2647 pieces in FLK). Spatially, these stone artifacts are more highly concentrated than occurs generally in excavated levels within the Bed I lake margin zone, or than occurs over known landscape assemblages from Late Stone Age time in East Africa (Potts, 1982; cf. Foley, 1981).

Since the bone accumulations are neither primarily animal death sites nor scatters around such sites, the major problem with a carnivore accumulation hypothesis is that it does not account for the mass of artifacts also concentrated in these confined areas. At least in modern hyaena dens with bone accumulations, little is left in the way of marrow, meat or other soft tissues to expect any benefit for hominids from transporting large quantities of stones to such den areas. The chief problem with the hypothesis that *only* hominids accumulated the bones is that it ignores possible contributions from carnivores and from background bones scattered over the landscape by various processes.

Several informative differences among the Olduvai faunal assemblages can be observed when the bones associated with stone artifacts are compared with the set of bones from FLKNN level 2, which contained no artifacts. Most of the differences between the artifact- and non-artifact-associated assemblages are in features of assemblage composition and bone damage which characterize at least some hyaena bone accumulations, especially one intensively studied from Amboseli, Kenya (Hill, 1981, this volume; Potts, 1982; Hill and Potts, in prep.). The bone concentration from FLKNN level 2 exhibits those features to a much greater degree than do the artifact-associated assemblages.[7] Although it is likely that a carnivore was primarily responsible for the bones aggregated in FLKNN level 2 (Potts, 1982), this inference does not automatically imply that the artifact-bearing levels were loci of hominid activity only.

Methods have not yet been developed to distinguish subsets of bones in an accumulation which have had such different taphonomic histories. Therefore, cautious use of the current methods of inference makes it necessary to identify the sources of bone accumulation only according to a *greater* or *lesser* degree of hominid activity directly at the site. By virtue of the stone artifacts and the cutmarks on bones, it is clear that hominid activity is exhibited to a much greater degree overall in the artifact-bearing levels than at FLKNN level 2. This inference is based upon (1) the apparent production of cutmarks on bones over the entire period of accumulation in the levels containing artifacts and (2) the primacy of carnivore concentration of bones at FLKNN level 2 (Potts, 1982). Faunal remains in the artifact levels also reflect a greater degree of transport by hominids than occurs in the assemblage at FLKNN level 2. Although such relativistic statements appear "weaker" than previous conclusions about early archaeological sites, they reflect better the web of taphonomic factors involved in the formation of each of these sites. As it turns out, general comparisons among assemblages showing greater and lesser hominid participation are informative about the diet and palaeoecology of early hominids. Two examples are presented below. One concerns the acquisition of bone marrow from limb bones, and the other concerns hominid foraging for animal carcasses.

### 3.3. Bone Marrow Processing

It has been suggested recently that hominid marrow consumption rather than meat-eating *per se* is indicated by the bone assemblages at Olduvai (Binford, 1981). This idea results from analytical methods applied to assemblage composition data. The inference is that the Olduvai hominids had access only to skeletal parts of low food utility. Hominids obtained these parts by scrounging from the carcasses and carcass scatters abandoned by carnivores at prey death sites (Binford, 1981, pp 282, 294). Although this latter inference appears to be incorrect, there are differences in bone preservation among the Olduvai assemblages that relate to distinctions (observed in modern contexts) between the marrow acquisition activities of tool-using humans and of carnivores.

Tool-using humans tend to acquire marrow by cracking and destroying limb bone diaphyses while leaving the articular ends intact. In contrast, carnivores tend to attack articular ends rather than diaphyses; they thereby gain access first to marrow in trabecular bone and possibly later to marrow in the diaphysis ("cylinder") after it has been structurally weakened by the absence of both articular ends (Bonnichsen, 1973, 1979). Similar, extensive observations on modern wolf and human bone assemblages (Binford, 1981) help to define criteria for distinguishing marrow processing by tool use from typical carnivore strategies of marrow acquisition. In short, high frequencies of articular ends with more than one-third of the diaphysis attached ("shafts" and "shanks") are characteristic of animal-gnawed assemblages. These pieces result from the absence of diaphysis destruction when one of the articular ends is left intact. Articular ends with one-third or less of the shaft attached ("isolated ends") are predominant in assemblages of limb bones cracked by humans for marrow. The latter bone damage pattern results from a tendency of humans to acquire marrow by diaphysis destruction (Binford, 1981, pp 172–173).

Table 6.2 provides data on the frequency of isolated articular ends in the modern hyaena assemblage from Amboseli (based on preliminary data) (Potts, 1982). Isolated ends form 42% of the identified ungulate limb bone pieces in this sample. This table also presents data for Olduvai assemblages. FLKNN level 2 is the only one which possesses isolated ends at a frequency *less* than 50% (the figure is 33% for this assemblage). Each of the other levels, judged to have had greater degrees of hominid influence, contain ungulate bone assemblages with *greater* than 50% isolated epiphyses. The lowest frequency of isolated ends is 59% at DK level 2.

Thus, on this relative scale, artifact-associated bone assemblages from Olduvai exhibit preservation patterns typical of marrow processing through

Table 6.2. Percentage of ungulate long bone specimens (possessing at least one articular end and part of the diaphysis) which show "isolated ends" versus more complete diaphyses ("shafts" and "shanks").

| Assemblage | # Long bone specimens | "Isolated" articular ends[a] (%) |
|---|---|---|
| Amboseli hyena | 74 | 42 |
| FLKNN L/2 | 43 | 33 |
| Artifact-associated: | | |
| FLKN L/6 | 32 | 62 |
| FLK "Zinj" | 97 | 82 |
| FLKNN L/3 | 33 | 73 |
| DK L/2 | 95 | 59 |
| DK L/3 | 73 | 73 |

[a] See text for definition of "isolated ends".

the diaphysis. It is probable that the hominid contribution to these assemblages partly involved processing and consumption of diaphysis marrow. This kind of marrow utilization is not well represented in at least one accumulation made primarily by carnivores at Olduvai.

## 3.4. Foraging for Animal Tissues

A wide variety of evidence from the fossil record has been used to assess early hominid hunting and scavenging. For example, Vrba (1975) suggested that the range of body sizes and the percentage of juvenile individuals in a faunal assemblage indicate whether the assemblage was collected by hunters or scavengers. A low frequency of juveniles and a wide range of body weights reflect scavenging, whereas hunting results in access to a high percentage of juveniles and to a relatively narrow range of prey body sizes (Vrba, 1975; but see Vrba, 1980). These criteria are based upon modern carnivore predation (Vrba, 1976, p 65).

One problem with this method is that the percentage of juveniles in any bone assemblage tends to be decreased by processes of differential bone attrition (e.g. Behrensmeyer *et al.*, 1979). In each Olduvai assemblage, for example, the relative frequency of juvenile bovids (MNI) is within the range of juvenile representation in modern animal populations (Potts, 1982). Yet patterns of preferential bone destruction are evident in each faunal assemblage. Therefore, it is difficult to say whether the original juvenile frequencies in these accumulations were high or, alternatively, merely matched the frequencies in the living bovid populations at Olduvai. A further, critical problem is that this modern carnivore model need not apply any better to early hominids than would a reconstruction based on some modern hunter/gatherers known to *hunt* a wide size range of animals.

Tool use also has been invoked in the hunting versus scavenging question. Noting a wide size range of animals from Olduvai sites, Isaac and Crader (1981, p 86) suggest that hominids both hunted and scavenged, and that scavenging was probably more important than hunting in acquiring medium- and large-sized animals. The reason they give for this latter inference is the simplicity of early hominid technology.

Strictly speaking, analogies to modern predators or assumptions about what early hominids were capable of doing with their stone tools are not adequate for deciding the hunting/scavenging issue. Analysis of this foraging problem has required a set generalizations derived from ecological and taphonomic observations which is then applied to the fossil record.

Earlier, I reviewed a range of ecological conditions under which carnivorous animals hunt and/or scavenge. Although a combination of hunting with scavenging would make the most sense for a carnivorous mammal, this generalization applies to carnivores but not necessarily omnivores. Furthermore, this review pointed out that conditions do exist under which a carnivore can be a total scavenger. This case is exemplified by the griffon vultures. These scavengers succeed in foraging by following migratory herds composed of a few ungulate species, which provide large quantities of dead meat for scavengers. We can reconcile the archaeological record with such a scavenging strategy by envisioning that migratory hominids made sites along the annual migration circuit of ungulates.

There are at least two palaeoecological inferences about the early hominids at Olduvai which pertain to this "total scavenger" hypothesis. The first inference is based on rather circumstantial evidence. As noted in "lessons in foraging for meat" (see pp. 135–8), a total scavenger strategy demands a nomadic life, unless (as in vultures) a highly efficient, rapid form of movement allows one to return every day hundreds of kilometers to a base area. In contrast, the palaeogeographic locations of hominid traces at Olduvai suggests an areal confinement of hominid activity by ecological factors. All sites in Bed I are located in the lake margin sedimentary facies. No stone artifacts or other evidence of hominids are known from sediments outside of the lake margin before Middle Bed II, though such sediments are exposed in Bed I. It could be argued that this initial confinement of site production to the lake margin means that free movement by hominids with migratory ungulate herds was prohibited. On the other hand, it is possible that hominids may have followed migratory ungulates only through such lake margin zones in the Olduvai vicinity.

A second inference relevant to scavenger strategies concerns the taxonomic diversity of Olduvai faunal assemblages. Each bone accumulation in Bed I Olduvai preserves a wide diversity of ungulate species. In no assemblage is there a numerical dominance of only one or a few species. Furthermore, each faunal assemblage associated with artifacts shows higher taxonomic and ecological diversity (per number of identified bones) than does the probable mammalian carnivore accumulation from FLKNN level 2. The diversity of fauna preserved in each Olduvai assemblage matches more closely the diversity found among resident ungulates than among just migratory species. Moreover, each assemblage includes species (e.g. browsing animals, monkeys) which are not known to migrate as part of large annual migrations in East Africa. At least a partial dependence on resident animals by the bone collector(s) at Olduvai is suggested. It is in just such a situation that the generalization about carnivorous mammals as hunters *and* scavengers seems to hold best. Mortality in resident ungulate populations mostly results from hunting. Residents which die from other causes are quickly scavenged. Thus, resident ungulates provide an inadequate food source for a total scavenger. One implication is that Olduvai hominids which utilized the carcasses of resident ungulates participated in the competitive milieu associated with those carcasses (see pp. 137–8).

A possible problem with this chain of inference is the assumption that the ancient ungulate ecosystems were similar to those today. We do not really know if migratory species existed in Bed I Olduvai times, or whether they were few in number and present in large herds. Yet the occurrence of savannas (and, thus, seasonal climates) and of ungulates with at least skeletal adaptations similar to modern species (Gentry and Gentry, 1978) may validate these assumptions. Carnivora similar to those today existed at ancient Olduvai, and they undoubtedly had the same breeding constraints as pertain to all mammalian carnivores today. Such constraints mean that the carnivores must have depended on resident ungulates throughout part of the year. Therefore, it is likely that a diversity of resident ungulates was maintained in the ancient savannas as it is today.

While these ecological arguments are suggestive, other faunal patterns can be identified which are diagnostic of ways hominids acquired carcasses. Observations of disarticulation of ungulate carcasses in African environments (Hill, 1979; Hill and Behrensmeyer, in press) have suggested a method to evaluate the relative timing of hominid access to carcasses (Potts, 1983). In brief, a relatively high frequency of forelimb elements in a bone assemblage collected by an animal indicates relatively early access to carcasses in its environment. This deduction results from observations that forelimbs tend to be the first skeletal unit to disappear from an animal death site. Hindlimbs tend to disarticulate and are removed somewhat later in the sequence. Finally, axial bones typically remain at the death site. Early access to a carcass may occur either via hunting or early scavenging. In contrast, scrounging, or late scavenging, would tend to yield assemblages composed of a high proportion of axial bones. For example, accumulations made by porcupines, which scrounge the environment for dry, disarticulated bones, are dominated by vertebrae (e.g. see Brain, 1980, 1981). The method used to assess relative bone frequencies from different skeletal units is discussed fully elsewhere (Potts, 1983).

When applied to the early Olduvai assemblages, the results of this method are as follows:

(1) In the probable carnivore accumulation at FLKNN level 2, a clear pattern of early access to all carcasses is indicated by a consistently very high representation of forelimb bones across all carcass size classes.

(2) None of the levels exhibiting a greater degree of hominid activity shows a consistent pattern of early access to carcasses as noted at FLKNN level 2. For some body size categories of prey animals, forelimb elements dominate; in other categories, hindlimbs are relatively abundant. Therefore, the faunal assemblages with greater influence from hominid activity contain parts of

carcasses obtained sometimes early and sometimes later in the sequence of carcass disarticulation. In no assemblage associated with artifacts, though, is a scrounging style of bone collecting indicated.

In addition, it is worth noting that the frequency of meat-bearing limb bones relative to non-meat limb bones is about the same in the artifact-associated levels as in that from FLKNN level 2. The latter assemblage falls within the range of variation shown by the other Olduvai assemblages on a ratio of non-meat bones to meat-bearing bones of the limb (Table 6.3). While different processes or strategies of foraging may be reflected by the artifact- versus non-artifact-associated bone assemblages, these data indicate that the degree of meat-bearing bone transport (and/or relative destruction) was similar in both kinds of assemblage.

Table 6.3. Frequencies of macromammal meat and non-meat limb bones from 6 levels in Olduvai Bed I. These frequencies are based on the minimum number of bones for each skeletal element. A ratio is given between the numbers of these two kinds of limb bone.

| Level | Non-meat bearing limb bones[a] | Meat-bearing Limb bones[b] | Ratio (non-meat/meat) |
|---|---|---|---|
| FLKNN L/2 | 92 | 61 | 1·5 |
| Artifact-associated: | | | |
| FLKN L/6 | 76 | 46 | 1·7 |
| FLK "Zinj" | 121 | 88 | 1·4 |
| FLKNN L/3 | 93 | 36 | 2·6 |
| DK L/2 | 181 | 96 | 1·9 |
| DK L/3 | 123 | 62 | 2·0 |
| Living macromammals (approx.) | 48 | 12 | 4·0 |

Mean ratio for artifact-associated assemblages = 1·9.
[a] Metapodials, carpals, tarsals, and phalanges; [b] Scapula, humerus, radius, ulna, femur and tibia.

In summary, then, a variety of skeletal and palaeoecological evidence indicates that Olduvai hominids were not "total scavengers". Rather, their bone collection activities were generally similar to those found in some present-day mammalian carnivores. A combination of hunting and scavenging probably occurred. Collections of bones transported by hominids reflect a combination of early and later access to carcasses which resulted from opportunistic foraging.

### 3.5. Dependency on Meat-eating?

From the preceding discussions several observations can be made about early hominid meat-eating and its significance. I will phrase these points briefly in response to three questions.

(a) *Did early hominids eat meat?*

The early archaeological evidence pertains to only those Lower Pleistocene hominids that produced sites. The following inferences together point toward the conclusion that some early hominids occasionally ate meat:

(a) occasional transport of animal bones by hominids to areas where stone artifacts and raw materials had also been taken;

(b) the occasional acquisition of bones which tend to be available at a carcass only early in the natural process of skeletal disarticulation;

(c) frequencies of meat-bearing limb bones in the artifact-associated assemblages which were similar to that in the probable carnivore accumulation at Olduvai;

(d) the documented presence of stone tool cutmarks in meat-bearing areas.

(b) *How significant was meat in the diets of Lower Pleistocene hominids?*

No method currently exists to calculate the rate of meat consumption or the percentage of meat in the diet. Minimum estimates of this sort depend on knowing the amount of meat on the bones, the rate of bone accumulation at sites, what percentage of the meat on bones was consumed by hominids, and the number of hominids that fed on the meat.

Estimation of the amount of meat on the bones would be biased by inferences about which particular animals were obtained by hunting and which by scavenging. Ideas about which bones were transported for marrow or for muscle tissues also would greatly influence this calculation. At first glance it seems reasonable to assume that hominid acquisition of bones from large ungulates is proof of extensive meat-eating. This kind of bone evidence may even point to feeding strategies devoted to the collection of more meat than can be eaten by the foraging individuals—and, thus food sharing. However, adjusting even very slightly for size-related preservation biases, the relative abundances of animals ranging from small (e.g. antilopine bovids) to very large (e.g. *Elephas*, *Giraffa*) in the Olduvai assemblages are well within the range of body size frequencies observed today in savanna mosaics (Potts, 1982). Hence, there is no evidence for an accumulation of bones from large animals beyond that expected from an opportunistic searching of the environment. Furthermore, there is no evidence that whole or nearly whole carcasses were transported to the sites of bone concentration at Olduvai.

In the Olduvai assemblages the overall quantity of bones is indeed impressive. Yet, based on surface weathering features, the bones evidently were not collected all in one season or in a single year. Unlike the relatively short-term use of particular camps by modern hunter/gatherers (e.g. see Yellen, 1976), bone accumulations at Olduvai sites were probably made over a period of several years; bones were modified by hominids (tool cutmarks) over that period. Yet the calibration of bone weathering patterns is not fine enough to discern seasonal fluctuations in bone accumulation rates or even to estimate minimum annual rates of bone transport to these sites.

The presence of cutmarks or of other damages possibly made by hominids does not necessarily imply that meat-eating was as significant to early hominids as it is to modern tropical hunter/gatherers. Attraction of carnivores to the Olduvai assemblages suggests that hominids did not fully utilize all food attached to bones brought to the sites. Moreover, the assumption that edible muscle tissue was the only animal material attractive to hominids should be treated cautiously. Cutmark location patterns may help to determine hominid use of meat versus non-meat tissues. However, meat may be obtained from a bone without producing cutmarks. Alternatively, obtaining meat may lead to destruction of the bony part to which meat was connected (which, thus, eliminates any chance for recognizing a cutmark). At least this latter process leaves observable traces and, therefore, can help to assess cutmark data about the skeletal locations of hominid tool use.

Most archaeological discussions of hominid meat-eating (this one is no exception) examine only the macromammal remains aggregated on sites. Dietary interpretations of the archaeological record have yet to consider fully other sources of animal protein. More significantly, as yet there are no archaeological data about plant foods in the diets of Lower Pleistocene hominids. The only detailed functional study of early stone artifacts has documented microscopic wear traces from both plant tissues and meat (Keeley and Toth, 1981). This finding, though, is not evidence of what plant foods were consumed by hominids nor of the quantities involved. A recent scanning electron microscope study of tooth microwear indicates that some Lower Pleistocene hominids had diets broadly similar to those of frugivorous higher primates (Walker, 1981). At present it is not known whether the hominids that accumulated bones at Olduvai had this same pattern of tooth wear. From an archaeological perspective, the inability to estimate the percentage of meat in the early hominid diet (or to state the adaptive significance of meat-eating) is exacerbated by an ignorance about plant food consumption by early hominids.

(c) *What was the adaptive significance of site production by hominids?*

From this survey of the early archaeological record at Olduvai, adaptive implications can be drawn only from inferences about the processes of faunal and stone artifact concentration. It is evident that some Lower Pleistocene hominids repeatedly transported bones to areas where they also carried tools and stone materials for tool production. Generally, large accumulations of stone artifacts and bones have been viewed as the remnants of home bases; presumably, food was brought for sharing by group members that foraged independently. This interpretation has been applied to a number of Lower Pleistocene sites (Isaac, 1976; Isaac, 1978; Lee, 1979, p 493; Isaac and Crader, 1981), such as those examined in this chapter. One concept of the early hominid home base, as a safe area where hominid social activity was centered, was defined even before African Lower Pleistocene sites had been described (Washburn and DeVore, 1961, p 101). The transport of food is traditionally viewed as a fundamental feature of the hunting/gathering adaptation. The home base, or campsite, is the focus of this activity and also the place where humans sleep, engage in some tool production and food preparation tasks, consume food, and meet with other social group members. Any particular campsite may be temporary. Yet wherever the camp is located, it is nonetheless a predictable base of social activity and food collection to the group members.

However, the evolution of hunter/gatherer home bases must be viewed in a broad palaeoecological context. By foraging for animal tissues hominids entered into the feeding domain of carnivores. A migratory scavenging adaptation or a scrounging style of scavenging might have greatly reduced competition with the cats, hyaenas and other carnivorous mammals. Yet the evidence from the Olduvai faunal assemblages does not indicate either of these two foraging strategies. The faunal assemblages associated with stone artifacts do suggest occasional early access to animal carcasses. This foraging pattern is identified largely from the accumulation of limb bones, which led to the production of bone assemblages generally similar to that probably produced by a carnivore at FLKNN level 2. These inferences imply that hominids were open to both indirect and direct competition with mammalian carnivores. The transporting of bones by hominids to particular places may have occurred in response to high amounts of carnivore competition at carcasses. On the other hand, the activities of hominids may have been constrained by competition with carnivores at the sites themselves, to which bones were transported.

The numerous bone breakages and tooth marks made by carnivores suggest that hominid activities in these areas of dense bone concentration at Olduvai, in fact, were limited by carnivore activity. As noted earlier, both the carnivore damages and stone tool cutmarks apparently were made over the entire period of bone collection at the Olduvai sites. Furthermore, the concentration of carnivore tooth marks on meat- and marrow-bearing bones and the occasional presence of complete, undamaged meat- and marrow-bearing limb bones[8] in the Olduvai assemblages suggests reasons for carnivore attraction to these sites.

In the light of these taphonomic inferences it is extremely difficult to apply to the Olduvai sites the simplified "home base" interpretation which was summarized above. The maintenance by modern hunter/gatherers of open campsites which are undisturbed by wild carnivores during human occupation has a history of carnivore avoidance of humans behind it. While little may be done today to protect their living sites (Yellen and Lee, 1976, p 37), hunter/gatherers potentially have fire and domesticated carnivores (dogs) to keep the wild carnivores away from the camps. However, the *origin* of home bases is a different adaptive problem from the *maintenance* of home bases by hunter/gatherers today. The site production system inferred from Bed I Olduvai involved small hominids that partially impinged upon the carnivore community in East Africa over 1·5 million years ago. These hominids transported bones and accumulated them in well defined areas, and apparently lived without the benefits of controlled use of fire or of domesticated dogs. In addition, the Olduvai bone concentrations show no evidence of extensive meat or bone processing (e.g. comminution of all marrow-bearing bones) which would have eliminated remains attractive to carnivores.

Thus, there appears to have been an absence of certain behaviours at these particular sites which would have permitted their uses as areas where social group members (including those with infants) consistently returned to sleep, to feed leisurely, and to carry out other human home base activities. These relatively large concentrations of stone artifacts and bones do indicate that hominids transported resources dispersed over their environment. But there is currently no diagnostic evidence which requires that they lived at these sites as modern hunter/gatherers live at campsites. One alternative to the home base interpretation is that these accumulations at Olduvai were useful caches of stone artifacts to which bones, in particular, and possibly other food sources were taken for processing (Potts, 1982). Stone tool/raw material caches possibly were visited very briefly as food was found by hominids (individually or collectively) and then were revisited over the long term. They may have been set up either (1) close to other important, more predictable resources (e.g. water or plant foods), or (2) throughout the foraging range to minimize the distance, time and energy of bone/meat transport. Carcass parts were transported away from death sites to areas containing stone materials which were adequate for processing a portion of the meat/bones obtained during the forage. In contrast to the home base idea, the evidence for carnivore attraction to these sites, specifically to available meat-bearing bones, suggests that a *minimizing* of time at these sites would have been adaptive for hominids. Although such "adaptationist" arguments do not stand as evidence, the site characteristics and inferences made here do not lead to the view that these particular Olduvai sites were hunter/gatherer home bases in a modern human sense. The evolution of such home bases required that the foci of hominid social activity and of resource transport were at the same places. All that can be inferred currently about the early sites at Olduvai is that they were areas to which resources were carried. These may have been the *antecedents* to true home bases of hunter/gatherers.

## 4. CONCLUSIONS

This chapter has focused on ideas about meat-eating, foraging and hunting/gathering among Lower Pleistocene hominids. The Olduvai example shows well how difficult it is to assess ecologically and evolutionarily complex issues such as these. Interpretations from the archaeological record about early hominid diet and foraging must involve a complicated set of inferences, with one inference based on the preceding ones. An essential point is that simple taphonomic observations from archaeological sites do not yield definitive answers to questions about hominid diet and foraging. For example, cutmarks on bones and the inference that hominids ate meat do not necessarily add up to a socioecology of hominids centered at home bases.

In particular, this analysis has been based on several pairs of concepts which need to be disassociated from one another:

### 4.1. Bones and Meat-eating

Meat is generally assumed to be the chief reason for early hominid interest in bones. While this assumption may be correct, the possibility was raised that other carcass resources were also used by hominids. The archaeological record specifies very little about what hominids actually consumed. In addition, while information is available on nutritive values of foods, nothing is known about the digestibility of meat in higher primates. This latter factor bears upon the theoretical value we place on meat protein to those early hominids that ate meat.

### 4.2. Meat and Hunting

The inference that hominids occasionally ate meat does not automatically imply that hominids consistently killed prey to obtain meat. Ecological conditions which would have favoured scavenging or hunting were set out early in this chapter. From an ecological view a combination of hunting and scavenging, which permits opportunistic access to food, would make the most sense for a mammalian carnivore. While early hominids are not viewed here as carnivores, the few test data on Olduvai palaeoecology and bone assemblage composition do indicate that hominids had both "early" and "relatively later" access to carcasses in foraging areas where carnivores also lived.

### 4.3. Cause of Bone Modification and Cause of Bone Accumulation

Recognizing the causes of bone damage does not always yield a clear interpretation about the agent(s) responsible for the accumulation of those bones. Stone tool cutmarks occur in the Olduvai assemblages, but so do damages caused by carnivore teeth. In the Olduvai example inferences about processes of bone accumulation had to be based on other data. The concentration of stone artifacts is important contextual information for understanding the causes of bone accumulation and the sequence of hominid/carnivore bone modifications. This assertion is not a matter of ascribing bone accumulations to hominid activity simply *because* they are associated with stone artifacts. Rather, the process of stone artifact transport and concentration represents an issue which parallels the problem of bone accumulation. At five of the six Olduvai levels examined, dense concentrations of stone artifacts occurred, including apparently unmodified chunks of exogenous raw material (manuports). It is clear that hominids transported these stones to the same delimited areas as the bones (some with cutmarks). This activity appears inconsistent with the idea that the bones were *primarily* collected as food by carnivores. This latter inference is not based on spatial associations alone; rather it is an argument based on the ecology and foraging of known carnivore bone collectors.

### 4.4. The Place of Bone Accumulation and the Place of Social Activity

The areas to which hominids carried artifacts and bones are usually viewed as the primary places of hominid food consumption and social activity. At the early Olduvai sites hominids were not consistent in protecting excess meat and marrow from carnivores. Modern-like home base activities of hunter/gatherers would have corresponded with these centers of bone accumulation when such excesses were controlled. Schaller and Lowther (1969, pp 378–9) suggested that protection of excess meat would have occurred by hominids sharing the food. While hominids sometimes handled parts of large ungulates and possibly large portions of muscle tissue, they also may have abandoned considerable portions on these sites of bone accumulation. Meaty and marrow-rich skeletal portions were part of what they abandoned, and these were certainly enough to attract carnivores. This idea should not be deemed inappropriate behaviour for a hominid. After all, even large carnivorous mammals at least occasionally leave major portions of carcasses uneaten. That is the basis for sequences of scavenging well known among African carnivores. It is likely that early hominids were the first primates to engage in, and to contribute carcass material for, such sequences.

Identification of these four sets of issues is critical to the chain of inferences examined in this chapter. For Bed I Olduvai at least, this chain of inferences points to an interpretation of archaeological sites which is neither "kill/death site" nor "home base". None of these sites examined here comfortably falls within either of these two categories of site supposedly produced by human hunters. The recognition of unique hominid behaviours which preceded a human hunting/gathering pattern may be possible. This is likely to arise only from further detailed studies of archaeological materials from a taphonomic and palaeoecological viewpoint.

## NOTES

1. Several summaries of the development of ideas regarding hominid meat eating and hunting are available: for example, Harding and Teleki (1981, pp 1–9); Binford (1981); King (1975).
2. Although the food science literature is vague on this topic, it is clear that muscle provides protein with all of the essential amino acids; marrow is largely fat; and internal organs provide protein and various concentrations of vitamins (e.g. vitamin A is provided by the liver in high amounts and by the kidneys, lungs and adrenals in a lower quantity) (Scott, 1968). In addition, blood contributes proteins and trace elements.
3. This does not mean that all food transported back to camps by hunter/gatherers necessarily is shared by an individual or beyond the nuclear family level (cf. Gould, 1981). As Hayden (1981, pp 386–387) notes, the actual extent and nature of food sharing in hunter/gatherers has rarely been studied.
4. Diverse carnivore assemblages are known from these early hominid sites: Laetoli (Barry, in press); Hadar (Johanson *et al.*, 1982); Koobi Fora (Leakey, 1976); Omo (Howell and Petter, 1976); and Olduvai (Leakey, 1971).
5. The influence of relatively fixed territories on foraging is supported by the lappet-faced vulture. Like the mammalian carnivores, this non-griffon species has fixed territories and has access to migratory herbivores for only a short period of time during the year. It is less frequently seen scavenging from ungulate carcasses than the griffon vultures and may occasionally kill small animal prey (Houston, 1979, p 274).

6. Among the Olduvai levels examined here FLKNN level 3 is one suggested by Binford (1981, pp 274–5, 280–1) to resemble most closely a second-order, transported assemblage (primarily of disarticulated limb elements obtained near animal kills). I would agree that this particular assemblage is transported. However, Binford's inference is surprising in light of the criteria given above. The macromammal assemblage from FLKNN level 3 actually contained the *most* axial elements relative to limb bones of all Bed I assemblages examined here (see Table 6.1). Furthermore, this level, unlike the others, preserved a nearly complete skeleton of the medium-sized bovid *Kobus sigmoidalis*. This one aspect of the assemblage represents the closest characteristic to a death site shown in any of the six levels, except for an elephant carcass in FLKN level 6. Nonetheless, overall, the accumulation at FLKNN level 3 shows the characteristics given here of a transported bone assemblage (Potts, 1982).
7. For example, in contrast to the other faunal accumulations, FLKNN level 2 shows: (1) clear priority of bone damages by carnivores over the one identified case of hominid modification (a cutmark); (2) the highest frequency of complete limb bones; and (3) the highest frequency of special carnivore bone modifications, matching those observed at some modern hyaena dens.
8. In the five bone assemblages with stone artifacts, the average representation of *complete* macromammal long bones is about 9% of the total number of long bone specimens identified to family level ($n$=558). In contrast, only 1·2% of the total number of macromammal long bone pieces ($n$=162) from 16 !Kung San campsites were complete (Yellen, 1977, pp 146–236).

## ACKNOWLEDGEMENTS

I wish to thank K. Wolf for help in preparing the manuscript, and R. Foley, A. Hill, E. Messer and K. Wolf for valuable discussion of this chapter. Data collection was carried out with the permission of M. D. Leakey, and the governments of Kenya and Tanzania. Funding was provided by NSF (BNS 7819174) and by the Boise Fund, University of Oxford.

## REFERENCES

Barry, J. (in press). The large carnivores from the Laetoli region of Tanzania. In: *Laetoli Monograph*. Berkeley: University California Press.

Behrensmeyer, A. K. (1978). Taphonomic and ecologic information from bone weathering. *Paleobiology* 2: 150–162.

Behrensmeyer, A. K. and Boaz, D. D. (1980). The recent bones of Amboseli Park, Kenya, in relation to East African palaeoecology. In: A. K. Behrensmeyer and A. Hill (eds), *Fossils in the Making*. Chicago: University of Chicago Press, pp. 72–92.

Behrensmeyer, A. K., Western, D. and Boaz, D. D. (1979). New perspectives in vertebrate palaeoecology from a Recent bone assemblage. *Paleobiology* 5: 12–21.

Bertram, B. (1979). Serengeti predators and their social systems. In: A. Sinclair and M. Norton-Griffiths (eds.), *Serengeti*. Chicago: University Chicago Press, pp. 221–248.

Binford, L. R. (1978). *Nunamiut Ethno-archaeology*. New York: Academic Press.

Binford, L. R. (1980). Willow smoke and dogs' tails: hunter-gatherer settlement systems and archaeological site formation. *Amer. Antiquity* **45**: 4–20.

Binford, L. R. (1981). *Bones: Ancient Men and Modern Myths*. New York: Academic Press.

Binford, L. R. and Bertram, J. B. (1977). Bone frequencies—and attritional processes. In: L. R. Binford (ed.), *For Theory Building in Archaeology*. New York: Academic Press, pp. 77–153.

Bonnichsen, R. (1973). Some operational aspects of human and animal bone alteration. In: B. M. Gilbert (ed.), *Mammalian Osteo-Archaeology: North America*. Missouri Arch. Society, pp. 9–24.

Bonnichsen, R. (1979). *Pleistocene Bone Technology in the Beringian Refugium*, Archaeological Survey Canada, Paper No. 89.

Brain, C. K. (1969). The contribution of Namib Desert Hottentots to an understanding of Australopithecine bone accumulations. *Sci. Pap. Namib Desert Res. Station*, No. 39.

Brain, C. K. (1980). Some criteria for the recognition of bone-collecting agencies in African caves. In: A. K. Behrensmeyer and A. Hill (eds), *Fossils in the Making*. Chicago: University of Chicago Press, pp. 107–130.

Brain, C. K. (1981). *The Hunters or the Hunted*. Chicago: University of Chicago Press.

Bunn, H. T. (1981). Archaeological evidence for meat-eating by Plio-Pleistocene hominids from Koobi Fora and Olduvai Gorge. *Nature* **291**: 575–577.

Eskin, N. Henderson, H. and Townsend, R. (1971). *Biochemistry of Foods*. New York: Academic Press.

Foley, R. (1981). *Off-site Archaeology and Human Adaptation in Eastern Africa*, BAR Intern. Series 97.

Frisch, J. E. (1968). Individual behavior and intertroop variability in Japanese macaques. In: P.C. Jay (ed.), *Primates*. New York: Holt, Rinehart and Winston, pp. 243–252.

Frison, G. C. (1974). *The Casper Site*. New York: Academic Press.

Gaulin, S. and Konner, M. (1977). On the natural diet of primates, including humans. In: R. Wurtman and J. Wurtman (eds), *Nutrition and the Brain*, vol. 1. New York: Raven Press.

Gentry, A. W. and Gentry, A. (1978). *Fossil Bovidae (Mammalia) of Olduvai Gorge*. London: British Museum.

Gould, R. A. (1981). Comparative ecology of food-sharing in Australia and Northwest California. In: R. Harding and G. Teleki (eds), *Omnivorous Primates*. New York: Columbia University Press, pp. 422–454.

Guilday, J. E., Parmalee, P. W. and Tanner, D. P. (1962). Aboriginal butchering techniques at the Eschelman Site (36 La 12), Lancaster County, Pa.. *Penna. Arch.* **32**: 59–83.

Harding, R. (1981). An Order of omnivores: nonhuman Primate diets in the wild. In: R. Harding and G. Teleki (eds), *Omnivorous Primates*. New York: Columbia University Press, pp. 191–214.

Harding, R. and Teleki, G. (eds) (1981). *Omnivorous Primates*. New York: Columbia University Press.

Hayden, B. (1981). Subsistence and ecological adaptations of modern hunter/ Gatherers. In: R. Harding and T. Teleki (eds), *Omnivorous Primates*. New York: Columbia University Press, pp. 344–421.
Hill, A. (1975). *Taphonomy of Contemporary and Late Cenozoic East African Vertebrates*. Ph.D. dissertation, University of London.
Hill, A. (1979). Disarticulation and scattering of mammal skeletons. *Paleobiology* 5: 261–274.
Hill, A. (1981). A modern hyaena den in Amboseli National Park, Kenya. *Proc. Pan African Cong. Preh. Quat. Stud.: Nairobi*, pp. 137–138.
Hill, A. P. and Behrensmeyer, A. K. (in press). Disarticulation patterns of some modern East African mammals.
Houston, D. C. (1974). Food searching behaviour in griffon vultures. *E. Afr. Wildl. J.* **12**: 63–77.
Houston, D. C. (1979). The adaptations of scavengers. In A. Sinclair and M. Norton-Griffiths (eds), *Serengeti*. Chicago: University Chicago Press, pp. 263–286.
Howell, F. C. and Petter, G. (1976). Carnivora from Omo Group Formations, Southern Ethiopia. In: Y. Coppens *et al.* (eds), *Earliest Man and Environments in the Lake Rudolf Basin*. Chicago: University Chicago Press, pp. 314–331.
Isaac, G. (1971). The diet of early man: aspects of archaeological evidence from Lower and Middle Pleistocene sites in Africa. *World Arch.* **2**: 278–299.
Isaac, G. (1976). The activities of early African hominids. In: G. Isaac and E. McCown (eds), *Human Origins*. Menlo Park: Benjamin, pp. 483–514.
Isaac, G. (1978). The food-sharing behavior of protohuman hominids. *Sci. Amer.*, April.
Isaac, G. (1981). Emergence of human behaviour patterns. *Phil. Trans. R. Soc. Lond.* **292**: 177–188.
Isaac, G. and Crader, D. (1981). To what extent were early hominids carnivorous? an archaeological perspective. In: R. Harding and T. Teleki (eds), *Omnivorous Primates*. New York: Columbia University Press, pp. 37–103.
Johanson, D. C., Taieb, M. and Coppens, Y. (1982). Pliocene hominids from the Hadar Formation, Ethiopia (1973–1977). *Amer. J. Phys. Anthrop.* **57**: 373–402.
Kawai, M. (1965). Newly acquired precultural behavior of the natural troop of Japanese monkeys on Koshima Islet. *Primates* **6**: 1–30.
Keeley, L. H. and Toth, N. (1981). Microwear polishes on early stone tools from Koobi Fora, Kenya. *Nature* **293**: 464–465.
King, G. E. (1975). Socioterritorial units Among carnivores and early hominids. *J. Anthrop. Res.* **31**: 69–87.
Kingdon, J. (1977). *East African Mammals*, Vol. III, Part A. London: Academic Press.
Klein, R. G. (1982). Age (mortality) profiles as a means of distinguishing hunted species from scavenged ones in stone age archaeological sites. *Palaeobiology* **8**: 151–158.
Kruuk, H. (1972). *The Spotted Hyena*. Chicago: University of Chicago Press.
Kummer, H. (1971). *Primate Societies*. Chicago: Aldine-Atherton.
Lamprecht, J. (1978). The relationship between food competition and foraging group size in some larger carnivores. *Z. Tierpsychol.* **46**: 337–343.
Leakey, M. (1971). *Olduvai Gorge*, Vol. III. Cambridge: Cambridge University Press.
Leakey, M. G. (1976). Carnivora of the East Rudolf succession. In: Y. Coppens *et al.* (eds), *Earliest Man and Environments in the Lake Rudolf basin*. Chicago: University of Chicago Press, pp. 302–313.
Lee, R. B. (1979). *The !Kung San*. Cambridge: Cambridge University Press.
Lee, R. B. and DeVore, I. (eds) (1968). *Man the Hunter*. Chicago: Aldine.

Maddock, L. (1979). The 'migration' and grazing succession. In: A. Sinclair and M. Norton-Griffiths (eds), *Serengeti*. Chicago: University of Chicago Press, pp. 104–129.

Mann, A. (1981). Diet and human evolution. In: R. Harding and G. Teleki (eds), *Omnivorous Primates*. New York: Columbia University Press, pp. 10–36.

McGrew, W. C. (1979). Evolutionary implications of sex differences in chimpanzee predation and tool use. In: D. A. Hamburg and E. R. McCown (eds), *The Great Apes*. Menlo Park: Benjamin/Cummings, pp. 441–464.

Melton, D. A. and Melton, C. L. (1982). Condition and mortality of waterbuck (*Kobus ellipsiprymnus*) in the Umfolozi Game Reserve. *Afr. J. Ecol.* **20**: 89–103.

Parmalee, P. W. (1965). The food economy of Archaic and Woodland peoples at Tick Creek Cave Site, Missouri. *Missouri Arch.* **27,** no. 1.

Pennycuick, C. J. (1979). Energy costs of locomotion and the concept of 'foraging radius'. In: A. Sinclair and M. Norton-Griffiths (eds), *Serengeti*. Chicago: Chicago University Press, pp. 164–184.

Potts, R. (1982). *Lower Pleistocene Site Formation and Hominid Activities at Olduvai Gorge, Tanzania*. Ph.D. dissertation, Harvard University.

Potts, R. (1983). Acquisition of faunal resources by early hominids at olduvai Gorge, Tanzania. In: Clutton-Brock, J. and C. Grigson (eds) Animals and Archeology 1: Hunters and their Prey. Oxford BAR (Inter. Services) 163: pp. 51–62.

Potts, R. and Shipman, P. (1981). Cutmarks made by stone tools on bones from Olduvai Gorge, Tanzania. *Nature* **291**: 577–580.

Schaller, G. B. (1967). *The Deer and the Tiger*. Chicago: University of Chicago Press.

Schaller, G. B. (1972). *The Serengeti Lion*. Chicago: University of Chicago Press.

Schaller, G. B. and Lowther, G. R. (1969). The relevance of carnivore behavior to the study of early hominids. *Southwest. J. Anthrop.* **25**: 307–341.

Scott, P. P. (1968). The special features of nutrition of cats, with observations on wild Felidae nutrition in the London Zoo. *Symp. Zool. Soc. London* **21**: 21–36.

Shipman, P. (1975). Implication of drought for vertebrate fossil assemblages. *Nature* **257**: 667–668.

Shipman, P. (1981). Application of Scanning Electron Microscopy to taphonomic problems. *Ann. N.Y. Acad. Sci.* **276**: 357–385.

Shipman, P. (1983). Early hominid lifestyle: hunting and gathering or foraging and scavenging. In: Clutton-Brock, J. and C. Grigson (eds) Animals and Archaeology 1: Hunters and their Prey. Oxford BAR (Inter. Services) 163: pp. 51–62.

Shipman, P. and Phillips-Conroy, J. (1977). Hominid tool-making versus carnivore scavenging. *Amer. J. Phys. Anthro.* **46**: 77–87.

Shipman, P. and Rose, J. (1983). Early hominid hunting, butchering, and carcass-processing behaviors: approaches to the fossil record. *J. Anthropolog. Anthropol.* **2**: 57–98.

Stanley, S. M. (1981). *The New Evolutionary Timetable*. New York: Basic Books.

Strum, S. C. (1981). Processes and products of change: baboon predatory behavior at Gilgil, Kenya. In: R. Harding and G. Teleki (eds), *Omnivorous Primates*. New York: Columbia University Press, pp. 255–302.

Tanner, N. M. (1981). *On Becoming Human*. Cambridge: Cambridge University Press.

Thomas, D. H. (1975). Non-site sampling: Up the creek without a site?. In: J. B. Mueller (ed.), *Sampling in Archaeology*. Tuscon: University of Arizona Press, pp. 61–81.

Thompson, P. R. (1975). A cross-species analysis of carnivore, primate, and hominid behavior. *J. Human Evol.* **4**: 113–124.

Thompson, P. R. (1976). A behavior model for *Australopithcus africanus*. *J. Human Evol.* **5**: 547–558.
Teleki, G. (1973). *The Predatory Behavior of Wild Chimpanzees*. Lewisburge: Bucknell University Press.
Teleki, G. (1975). Primate subsistence patterns: collector-predators and gatherer-Hunters. *J. Human Evol.* **4**: 125–184.
Teleki, G. (1981). The omnivorous diet and eclectic feeding habits of chimpanzees in Gombe National Park, Tanzania. In: R. Harding and G. Teleki (eds.), *Omnivorous Primates*. New York: Columbia University Press, pp. 303–343.
Vrba, E. S. (1975). Some evidence of chronology and palaeoecology of Sterkfontein, Swartkrans, and Kromdraai from the fossil Bovidae. *Nature* **254**: 301–304.
Vrba. E. S. (1976). *The Fossil Bovidae of Sterkfontein, Swartkrans and Kromdraai*. Transvaal Mus. Mem. 21, Pretoria.
Vrba, E. S. (1980). The significance of bovid remains as indicators of environment and predation patterns. In: A. K. Behrensmeyer and A. Hill (eds), *Fossils in the Making*. Chicago: University of Chicago Press, pp. 247–271.
Walker, A. (1981). Dietary hypotheses and human evolution. *Phil. Trans. Roy. Soc. London* **292**: 56–64.
Washburn, S. and DeVore, I. (1961). Social behavior of baboons and early man. In: S. Washburn (ed), *Social Life of Early Man*. Chicago: Aldine, pp. 91–105.
Washburn, S. and Lancaster, C. S. (1968). The evolution of hunting. In: R. B. Lee and I. DeVore (eds), *Man the Hunter*. Chicago: Aldine, pp. 293–303.
Wilson, E. O. (1975). *Sociobiology*. Cambridge: Harvard University Press.
Wilson, M. C. (1982). Cut marks and early hominids: evidence for skinning. *Nature* **298**: 303.
Yellen, J. E. (1976). Settlement patterns of the !Kung: an archaeological perspective. In: R. Lee and I. DeVore (eds), *Kalahari Hunter-Gatherers*. Cambridge: Harvard University Press, pp. 48–72.
Yellen, J. E. (1977). *Archaeological Approaches to the Present*. New York: Academic Press.
Yellen, J. E. and Lee, R. B. (1976). The Dobe-/Du/da environment. In: R. Lee and I. DeVore (eds), *Kalahari Hunter-Gatherers*, Cambridge: Harvard University Press, pp. 28–46.
Yesner, D. R. (1980). Nutrition and cultural evolution: patterns in prehistory. In: N. Jerome, R. Kandel, and G. Pelto (eds), *Nutritional Anthropology*. Pleasantville: Redgrave, pp. 85–115.
Zihlman, A. and Tanner, N. (1979). Gathering and the hominid adaptation. In: L. Tiger and H. M. Fowler (eds), *Female Hierarchies*. Chicago: Beresford Book Service.

# 7 *Mental Abilities of Early Man: A Look at Some Hard Evidence*

*John A. J. Gowlett*

## 1. INTRODUCTION: THE VITAL PLACE OF HUMAN MENTAL EVOLUTION

We human beings have a very powerful desire to set ourselves in context, to see ourselves in perspective, both as individuals and as a species. This desire stems from the scope and power of the human mind, and thus one of the great challenges for science must be to investigate how that mind has evolved.

This is no new preoccupation. Charles Darwin found it worthwhile to devote two whole chapters of *The Descent of Man* to the subject of human mental faculties (Darwin, 1874).

The central problem can be framed as follows. There is a gap in the present day between us and our nearest living relatives, the apes, a substantial gap labelled by genus and species. This gap is largest, and most difficult to comprehend, in terms of mind, but as Darwin noted, "there is no fundamental difference between man and the higher mammals in their mental faculties" (Darwin, 1874, p 99). On the other hand, "Of the high importance of the intellectual faculties there can be no doubt, for man mainly owes to them his predominant position in the world". This fundamental importance of the intellect is too often minimized in studies of human evolution which rely very much on drawing parallels with animal social behaviour, but again both the importance of the issue, and the need to relate man to the animal world carefully and without prejudice, have long been appreciated:

> It would be no less wrong than absurd to deny the existence of this chasm, but it is at least equally wrong and absurd to exaggerate its magnitude, and resting on the admitted fact of its existence, to refuse to enquire whether it is wide or narrow. (Huxley, 1863, p 104)

The dilemma, then, is how to relate human beings to the apes, and other

HOMINID EVOLUTION
ISBN 0-12-261920-X

intelligent mammals, without on the one hand putting us on such a pedestal that the relationships cannot be traced, and without on the other hand, simply explaining away human qualities as quickly as possible through a process of reduction which avoids all awkward questions. I believe that these issues are faced squarely by most primate studies of present-day evidence, for example, of chimpanzee behaviour, but the problem of bridging the gap must remain, because of the lack of living intermediate forms. In studies based on the evidence of the past, where the gap must have been bridged, by the process of evolutionary change, we can lose sight of the issues all too easily.

In starting this paper, I have emphasized the views of pioneer workers of evolution, partly because mental evolution had an importance to them often since overlooked; partly because unencumbered by all the specific data of recent years, they could clearly distinguish between the wood and the trees; and partly to emphasize that good ideas about human evolution are not entirely linked with the recent spate of fieldwork.

This paper is intended to provide some specific examples which show that mental evolution can profitably be investigated in the past, at least in the period for which cultural evidence survives. As no other animal, not even the chimpanzee, makes environmental modifications (tools) which have much chance of surviving in the record of the past, this may seem to be chiefly an essay upon human distinctiveness. But, the activities to be described can be related to those of other intelligent mammals, if we accept that "operational chains" (described below) provide a clear link between the advanced behaviour of man, and some other primates such as the chimpanzee. First, the context of the approach must be sketched in more fully.

At almost all times during the last hundred years serious consideration has been given to the issues of mental evolution (e.g. Drummond, 1901; M'Dougall, 1925; Schmidt, 1936; Huxley, 1941, 1951; Rensch, 1959; Hallowell, 1960; Holloway, 1969; Wynn, 1979, 1981; Oakley, 1981). Nevertheless, it has always been difficult to maintain focus on mental evolution as one of the most important questions, if not the most important question of evolution, since there are several powerful distracting factors. Mental faculties in the present are insufficiently understood, so their study in the past must be more difficult. At times the view has held weight that, since we cannot prove even the existence of minds other than our own, we should not use the word at all. Fortunately, scientists such as Sir Charles Sherrington and Sir Julian Huxley were never daunted by this, and I think it has remained widely accepted that in some way or other the mind can be studied (Sherrington, 1940; Huxley, 1941, 1951; Chomsky, 1976). Some still prefer to use only the term "brain", but here I am using "mind" simply to mean the brain in action, with all its store of experiences. When J. Z. Young (1978) talks of "programs of the brain", it is perhaps only a different shade of meaning to say "workings of the mind". My fundamental point is that for the past two million years—in man, and in man alone—we have considerable access to the output of "programs of the brain", and these workings of the mind cast much light on its evolution.

The prospects for future work are exciting, but progress depends upon a genuine interest in analysing and explaining past activities and abilities, rather than explaining them away so as to tell a plausible and appealing story.

## 2. PROGRESS OF RECENT YEARS

To set the evidence about mental abilities in perspective, it is necessary to look briefly at the nature of the major advances which have been achieved in studies of human evolution during the last quarter of a century.

A great concentration of effort, in several related disciplines, has been placed in the period which centres on two million years ago (providing a bias in favour of studies of that particular period). Potassium argon dating provides a firm chronological framework for the work (e.g. Evernden and Curtis, 1965; Dalrymple, 1972). As a result, we know certainly that tool-making goes back for at least two million years; we know that at this time there was more than one hominid species existing; we know that a lineage which can be termed *Homo* appears as much as two million years ago, and that the species *Homo erectus* emerges 1·6 or 1·7 million years ago (Wood, 1978; see Chapter 3, this volume).

Although there was for a long time no clear record extending back beyond the stone tools, much progress has now been made. The knowledge that human bipedalism goes back at least 3·5 million years (Johanson and White, 1979; Leakey, 1981) and probably more than four million years (see also Tuttle, 1981; J. D. Clark, reported in *Time* magazine, June 1982), shows clearly that hominids had found a successful bodily adaptation, before an avenue of development opened up which favoured the rapid evolution of the human (as opposed to hominid) brain.

At this very exciting time, it is perhaps frustrating that the archaeological record does not go back far enough to allow those earlier times to be interpreted at all fully in behavioural terms. We can see that it is very much more difficult to make behavioural interpretations when cultural evidence is absent. But this very fact emphasizes the difference of the earlier and later records. At least, the new discoveries now relieve the temptation to pick out in the archaeology the whole course of human evolution, when in fact only the latest, and most human, parts of the story are there. There is, however, still a widespread tendency to *assume* the primitiveness of the earlier archaeological record, rather than to test it for complexity.

## 3. HYPOTHESES OF RECENT YEARS AND THEIR PROBLEMS

A feeling has often been expressed that human evolution cannot be accounted for in general terms of natural selection, as something that just happened, but that there must be specific influences or mechanisms to account for the differentiation from the apes. They, in spite of being very different from one another, hang together as a group, both structurally and behaviourally, whereas human beings stand apart on both counts. We, it is felt, are the ones who became different, and there must be an obvious explanation.

The traditional major archaeological model is, however, essentially descriptive. It is that of technological progress: the stone tools which are the chief index in the Stone Age become more complex through time. This model, which has ancient roots (see Piggott, 1960), demands the existence of culture to get its starting point. Oakley's definition of "Man the Toolmaker" sets the beginning of humanity at the time of the first tools, in an arrangement which is arbitrary, but still useful (Oakley, 1951). Yet the preadaptation of early hominids, both physically and above all mentally, to a way of life where the transmission of experience accumulated by past generations, and the execution of complex actions, such as are shown in tool-making, are important factors which cannot be treated directly from the technological model.

Further insight comes from the view expressed well by Julian Huxley that cultural development represents a fundamental change of evolutionary level, from ordinary biological evolution to psychosocial evolution, in which change can happen much more rapidly, and in which, ideally, it can be guided by the species concerned. In this sense, culture, as a concept, embodies not just material objects, but all the abstracted rule systems by which human beings operate, and which are handed down from individual to individual. It has become widely accepted that in such a system, biological evolution and cultural evolution affect one another in a positive feedback relationship, thus providing both change and its cause (cf. Huxley, 1951; Dobzhanksy, 1963; Tobias, 1981).

This view has never been effectively challenged, but it can be argued that it accounts for a process that is going on, rather than for the start of it. It does not satisfy the desire for a single perceptible factor, an "initial kick" which starts off human evolution. Consequently hypothesis after hypothesis has emerged in which such a kick has been found, and in some of them its momentum dominates the whole story.

The "hunting hypothesis" stems from one such search for mechanism. Popular quite recently, it has its origins almost 80 years ago. In Carveth Read's words, "The adoption of the hunting life . . . is the essential variation upon which everything else depends" (Read, 1920, p 1). It was in the social context of the pack that he saw this happening, and he spoke with conviction too: "Such a variation *must* have occurred at some time, since Man is everywhere more or less carnivorous". Two of Read's assumptions, however, are less certain: that the earliest known men were hunters; and that weapons are among the earliest known artifacts. Then too, there is the whole phenomenon of systematic gathering of plant foods, which seems just as essential in the human adaptation—can this be left out of the hunting model?

Although the hunting hypothesis has a propensity to stress human aggressive tendencies, this idea is taken to extremes in the "killer ape" interpretation popularized by Ardrey (1961, 1976). This is just one in a series of extreme presentations, which stress one characteristic, or analogy, at the expense of all others (see Chapter 5, this volume, for further discussion). Territorial animals, Naked Apes and suchlike, they succeed one another in an apparently endless progression.

A hypothesis which must be treated much more seriously is that of foodsharing. This was seen by Le Gros Clark (1967) as an important social step in the process of hominization, perhaps one of the preconditions for the development of culture. Foodsharing, in one form or another, is known to take place in other primates, and its importance has been succinctly reviewed by McGrew (1979). Far more emphasis has been laid upon the foodsharing hypothesis by Isaac, who has adopted it as the major mechanism for determining the course of human evolution, not in the times before material culture, but in the early Pleistocene when stone tools were already being made (Isaac, 1978).

This approach implicitly rejects the notion of "man the toolmaker", reducing, in effect, the value of cultural evidence, and labelling the toolmakers themselves as "protohuman". It seems that the evidence of material culture must be treated in this way in order to elevate foodsharing to the status which the hypothesis demands.

Foodsharing must be of importance in the sense that Le Gros Clark suggested, but as a postulated major mechanism in the Lower Pleistocene, if accepted uncritically, it has the potential to be greatly misleading. Earlier this century, the status of the Neanderthals was undermined by Boule, who *expected* to see a primitive, brutish creature which would fit in with current ideas on human evolution (Boule, 1911–13). As is now well known, he greatly overemphasized the primitive characters, with lasting ill effect. We must now make quite sure that a similar reduction does not happen to the hominids of the Lower Pleistocene, unless indeed it is the correct interpretation. A stimulating debate on this issue can only do good for a subject which has been overladen with oversimplified single-shot popular theories, in comparison with which the food-sharing hypothesis deserves serious, but not uncritical, consideration.

## 4. AN ALTERNATIVE APPROACH: MENTAL ABILITIES

It is possible that in the complexity of the real world, human evolution can best be accounted for by an intimate combination of both "kicks" (hard or gentle) and "feedback". Another approach thus relies not so much on looking for single mechanistic "kicks", as on analysing the whole pattern of human abilities. We might do well to be careful of any simple mechanisms, since "the significance of adaptation can only be understood in relation to the total biology of the species" (Huxley 1941, p 449). Culture is now an integral part of that biology, and the mental abilities which it entails need to be tied in. In the earlier phases of human evolution, a whole range of attributes of change must be considered, for example those listed by Holloway (1972), which include both physiological and psychosocial factors.

Although this virtually rules out simple explanations, if the major direction of human evolution has been cultural—leading towards complete dependence upon experience accumulated and organized by past generations—and the adaptation is long-term and complex, the mind must have developed in the long-term to regulate this. Mind does not exist in a vacuum, but in Sherrington's view was initially concerned with motor action. The more effectively a mind can provide an internal view of the external world, and map out possible courses of action, the more effectively the creature can respond. Craik (1943) postulated that the workings of the human mind could be parallelled by analogy with mechanical devices, and that in effect the human mind operated by simulating the processes of the outside world. He recognized that human beings were able to do this much better than any other species, to much higher levels of prediction. Indeed, if there is to be a single line of human success, we can suggest that it is based on the mind, but we must recognize that the mind itself depends upon all the other adaptations.

This fits in on a wider scale, for Rensch (1959) has characterized all the major advances of evolution as "brain victories", and in that sense human evolution is but a specific example of a general trend (see Chapter 4, this volume).

Mind, then, reflects what must be done in the real world: but to have a useful study, it must also be recognizable in the real world. In human evolution, a foundation has been provided by studies of early hominid neurology made by Holloway (1972, 1974). His studies of cranial endocasts indicate a basically human-like brain organization both in early *Homo* and in australopithecines, in spite of the much smaller brain size. Holloway, however, points out the limitations of anatomical evidence [though further progress in such studies has been made in recent years (Holloway and De La Coste-Lareymondie, 1982) ], and it is fortunate that the archaeological evidence has much potential to provide specific information, a line of enquiry which Holloway himself has recommended (Holloway, 1969).

Archaeological documentation of mental abilities depends upon psychological studies of behaviour, such as the work of Miller *et al.* (1960), who developed the idea that algorithms could be used to describe behavioural processes. They suggested that the fundamental unit in behaviour was not, as had been widely believed, the reflex, but a feedback loop performing a congruity test (the TOTE unit). Clearly a model based on such loops has all the rigidity of a standard computer algorithm, but we can hope that artificial intelligence routines will continue to develop so as to give a much better fit with human behaviour (cf. Sacerdoti, 1977). In the meantime, where enough archaeological information is available, it is possible to offer less formal flow charts of the processes involved in early human activities, illustrating some, but not all, of the decisions, predictions and selections involved.

The examples which follow are chiefly those of technology, for this is what is preserved: technology yields the hardest evidence. There is no reasonable doubt, however, that a whole range of comparable activities, involving comparable qualities of mind, must be implied. Miller *et al.* (1960) drew attention to links between planning ability and language; similarly Holloway (1969) drew attention to the similarity of semantic structure underlying language and stone tool manufacture. There will always be difficulty in bringing together these very separate strands of evidence, but we can be sure that human abilities, if present at all, were not limited to one area of application.

## 5. CASE STUDIES

### 5.1. Navigation in Space and Time

Human beings possess remarkable abilities of orientation in relation to the external environment. Not only are we aware of the present, but we are aware of space around us, on the scale of a house, a city, or perhaps the whole globe, and we are aware of time, on a personal scale of days, weeks and years in past and future, of millenia on a historical or archaeological scale.

It is easy so to take these abilities for granted as not to question their evolution. The complex machinery which underlies them must have an evolutionary history which can be studied, perhaps in much the sense that Chomsky (1976) urged the study of the mind as an organ.

As in so many characteristics, other animals give us a base line for comparison. Studies of chimpanzees show their abilities to find their way in an environment, and to locate and relocate objects. This is far from surprising, in our nearest living relative.

We can nevertheless pick out enhanced features in human beings: (1) the absolute range of space and time embraced by our mental activity; (2) the ability to assemble chains of activity through space and time; (3) the ability to simulate alternative complex strategies; (4) the use of a formalized frame of time.

Such abilities can be documented in the past, provided that there is a cultural record. Stone artifacts and cobbles, which have been characterized as "Stone Age visiting cards" (see Isaac, 1981a) provide an ideal basis for such documentation, revealing not only that somebody was present, but sometimes where they came from and went to, and the time frame in which they did it.

Time and again on Old Stone Age sites, we find that stone tools have been carried in from areas where suitable raw materials were available, to areas where they were destined for use. Often the rocks used are so characteristic that they can be assigned to a certain source area beyond doubt.

Four examples should suffice. First, the well known early sites at Olduvai Gorge (Tanzania). Here early human beings occupied flat areas by the side of a lake, in an area where the main stone sources were volcanic mountains at least a few kilometres away, and "inselbergs" or isolated individual hills which rise from the plains. For example, it is known that dark grey lava came from Lemagrut Mountain, green phonolite from Sadiman, and green quartzite probably from the Naibor Soit inselberg. Leakey and Hay conclude that artifacts in some cases were transported 8–12 miles to the living sites (Leakey, 1971; see also Foley, this volume).

Similar compelling evidence comes from the Acheulean site of Olorgesailie in Kenya (Isaac, 1977), where the presence of small quantities of non-local rocks is interpreted as attesting "either to a territorial range that includes such outcrops or to trading contacts with adjoining groups" (Isaac, 1977). Obsidian, used for at least three of the hundreds of hand-axes used at Olorgesailie, occurs at nearest 16 km from the sites. Quartz and quartzite, which occur 40 km or more away to the east and west, at nearest, were very rare on the sites, but occur persistently in small pieces.

The Acheulean site of Kilombe, probably over 700 000 years old, provides supporting evidence (Gowlett, 1978, 1980, 1982). Like most major Acheulean sites, it is situated close to one raw material source, but other materials are found. Local phonolite makes up over 90% of the finds, but trachytes and olivine basalt, from the range 5–15 km, make up the remainder. Again, one obsidian biface was found, this time at least 60 km from any known source.

The final raw material example comes from Gönnersdorf, a Magdalenian site in Germany, little more than 10 000 years old. Thousands of stone tools were found at Gönnersdorf, but only about 60% of them consist of a quartzite from the Middle Rhine area. The others are of flint, mostly of Northern flint which occurs 150 km away at closest, the remainder of Maas flint, found about 75 km away (Bosinski, 1969).

We can infer that throughout prehistoric times raw materials for stone tools were selected for their particular properties, that the individual outcrops were known, and that transport of pieces was regular and deliberate.

This provides us with invaluable information about the geographic knowledge—and hence mental abilities—of early human beings. Above all, it shows us that the effective timescales of their planning or foresight, ran at least to a day or more, and that they had detailed knowledge of "mental maps" of areas at least 20 km across. Could these findings be explained along the lines that the rocks "just happened" to be encountered in the course of foraging, and were then transported? Not only the continuity of evidence, from early past to present day, argues against this, but also the precise knowledge demonstrated by the rocks selected. Recently, McGrew *et al.* (1979) have observed the care with which chimpanzees select stalks for fishing for termites, occasionally moving perhaps 20 metres to commence the task. On what grounds would we say that hominids who are *known* to have used more elaborate tools, and transported them much further, did so with less planning foresight?

## 5.2. Stone Tool Making—Analysis of a Skill

Other animals use tools, and on occasion make them, but human beings are tool makers *par excellence*. If we wished to enhance the distinction from other animals, we could talk of "regular and habitual tool-making", but even this does not suffice as a rigid distinction, since it is increasingly plain that human beings of the present day are distinguished above all by the *complexity* of their tool-making.

As we are separated from our near relatives the apes (who are naturally somewhat like us) through being a species with a specific history, there is actually no need to force out definitions which state bluntly what we can do and they cannot. Thus tool modification by other primates does not erode the significance of human tool-making; it serves, if anything, to highlight how much further we have gone in that direction than any other species.

The human mind allows human beings to sustain much longer chains of related actions than any other animals. Just as the human world-view encompasses more space and time, so plans and actions can be projected further ahead.

The great majority of human actions leave no substantial trace, so we are particularly fortunate that the enduring stone tools of the Old Stone Age provide remarkable records of past decision-making processes. An Acheulean hand-axe may have been made through a process of detaching hundreds of flakes, and thus preserves, almost literally, a blow-by-blow account of the manufacture—and indirectly of the processes of mind necessary for that manufacture.

The essence of this process is that the tool-making is carried out by removing individual flakes of stone, that these are removed in sequence—an operational chain—and that each individual step must be subordinate to the ultimate goal. Indeed, after each individual blow, there must be a check on whether the tool is still compatible with its ultimate intended form.

It is no wonder then that the importance of early tool-making, and its close connection with all later technology, have been appreciated for many years. "It may seem a far cry from the first generalised stone implement to the latest highly specialised aeroplane," wrote O. G. S. Crawford (1921, p 6), "but once the first step is taken the rest is comparatively easy". Later, Julian Huxley (1941) expressed a very similar view: "There is no essential difference between man's conscious use of a chipped flint as an implement and his design of the

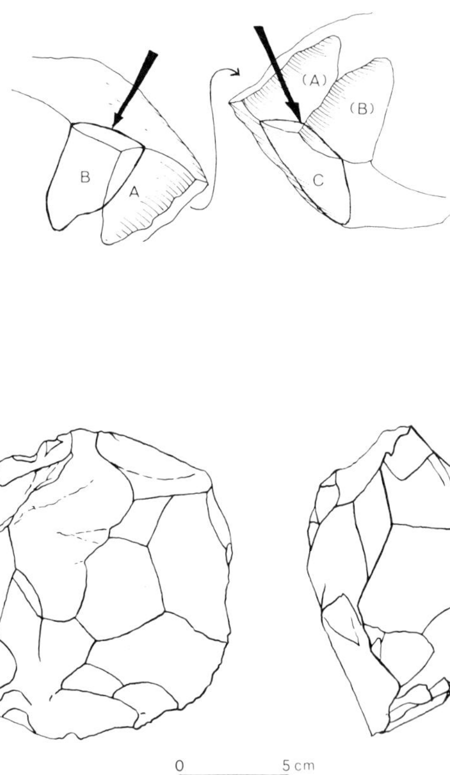


Fig. 7.1. Above: a simplified illustration of the process involved in bifacial working. Below: a large Oldowan discoid from Chesowanja, in plan and side views. Bifacial technique is as extensively used as in Acheulean bifaces, but there are no signs of the imposed long axis.

most elaborate machine. . . ."

Early stone industries, from Africa in particular, provide concrete examples of early mental abilities, recorded indelibly in the ancient sequences of flaking.

If we consider artifacts from the lower levels of Olduvai Gorge, which offer us the largest series of early tools, we find not just simple flakes struck by a single blow, but core-tools or cores which have been shaped through the removal of a whole series of flakes. We cannot always tell whether the prime object was to produce a tool of a set shape, or whether to continue the flaking process economically so as to yield the maximum number of flakes. We can be certain, however, that either way, the task required very considerable abilities: of manipulation, concentration, planning and of "visual imaging".

One of the best examples is provided by the tools termed "discoids" (Fig. 7.1). It can be seen that these have been worked from both sides, yielding or preserving a "bifacial" edge.

The striking of the individual flake requires manual dexterity, and hand-eye co-ordination, as well as an appreciation of the fracture properties of stone. More than this, it also requires an ability to "see" where that flake will come off [cf. comments on visual imaging in Galton (1907) ]. A modern flint knapper simulates the process internally by "visual imaging"—I have seen the American flint knapper Bruce Bradley demonstrate this by outlining the intended flake in chalk on the flint block, then proceeding to strike it. We might argue that early flint knappers proceeded without this ability, but we would then be in difficulty to explain the end product.

The routine, or operational chain (cf. Roche, 1980), is not merely a sequence of strikes, for then it would be without purpose. Even if the only goal is to produce a series of flakes, each must be taken off so as to give the maximum opportunities for continued working. In fact, the flaking process often had another end, that of producing shaped core-tools, such as Oldowan choppers or Acheulean hand-axes. Each of these processes requires a "routine" of stored knowledge, then unfolded through time.

It has become unfashionable to talk of "mental templates", but this expression is of great value for referring to sets of linked abstractions, held in the mind, which serve as a pattern for activities in the outside world (Deetz, 1967). That Oldowan tools were designed to have a set form can be disputed, so we might leave templates of form until the Acheulean (see below); it is much more difficult to dispute that there was a routine for the actual process of manufacture: *a procedural template* (Fig. 7.2).

I would acknowledge that the word "template" sounds too hard and formal, but that comes very close to the point which I am trying to make: those tools can only be made when the controlling mind is able to "hold to" its template, while yet being flexible enough to adjust to the manufacture of the individual tool.

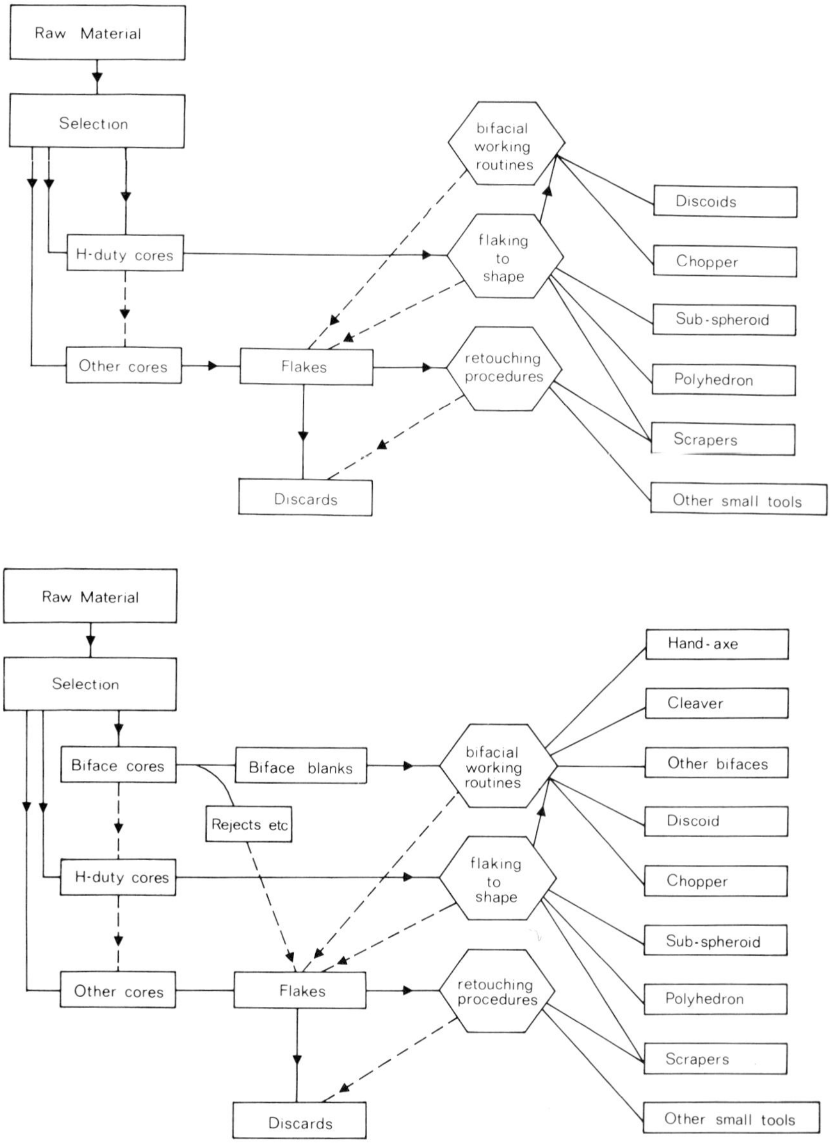


Fig. 7.2. Above: the process by which an Oldowan industry is made—an approximate procedural template. Below: the equivalent procedural template for an Acheulean industry, showing elaborations over the Oldowan which may have arisen gradually.

Additionally, we have to consider the social context and the specific context of the tool-making. According to Hallowell (1960), "Tool-making as observed in *Homo sapiens* is a skilled act—learned in a social context where speech exists, and usually performed with reference to a purposeful act at some *future* time" (p 324). Whether or not we can concede language to the earliest tool-makers, the whole process of tool-making must fit in with other activities: various needs have been felt, various choices of material and location made, before the process has even begun, and equally, the anticipated use begins only after the manufacture has ended.

There is an approach which seeks to minimize the importance of stone tools, and in particular to minimize the abilities which are required for their manufacture. This is no more or less than an explaining-away procedure.

As a performance it has two main themes: (1) that tool-making abilities are not of special significance, since once under tutelage an ape made stone tools; (2) that form in the tools is not determined by set plans, "formal design content", but that flaking is "opportunistic".

Those points are worth considering briefly:

(1) Manufacture of flakes by an orang utan (Wright, 1972). A situation can be created where the separate steps of an operational chain are made plain to an ape. The striking of one single flake in a "primed" setting speaks for the abilities of the orang-utan, but does nothing to undermine our appreciation of the abilities required for making *complex* tools.

(2) Lack of formal design content. A flake can only be struck successfully when the right material is available, the right hammer, the right force and the right striking angle. Oldowan cores and core tools show that these preconditions were brought together time and again. Do we really suppose that this was in some sense "accidental", or that the right conjunction was rediscovered by trial and error each time a tool was made?

It could be suggested that the tools do not really demonstrate much, because they could have been made in no other way. In other words, there are only a few simple ways of working stone, so hominids practising these needed no special abilities. This view would demonstrate a failure to analyse and recognize fully the mental faculties which are required to produce the end products which we find. Following the same approach to its conclusion, we could say that there are only a few ways of making aeroplanes that will fly, hence human beings are not very clever to have achieved this.

The conclusions drawn here also contrast strongly with those of Wynn (1979, 1981), who attempted to use the concepts of Piaget's "operational intelligence" to interpret early stone tools. In relation to the Acheulean, he argues for abilities as developed as those of modern man, but for the preceding Oldowan, he suggests that the stone artifacts show no mental abilities not present in the gorilla. A diagram of the procedural steps necessary for making Oldowan and Acheulean industries shows that both involve much longer operational chains than ever employed by the gorilla, and also how the Oldowan could have served as a basis for the later emergence of the Acheulean. The common basis of mental abilities required for making these two traditions is much the most apparent feature.

Undoubtedly the simplest stone tools are simple by our standards, but equally, there is no doubt that they were made in an impressive context of knowledge of the properties of stone: selection, visualizing, conceptualizing of form in three dimensions, mastery of multi-step routines, and transfer of that knowledge to other individuals (cf. Fig. 7.3).

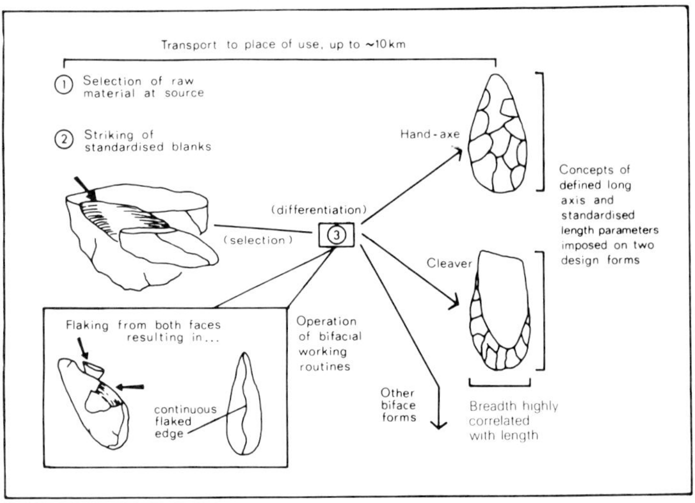


Fig. 7.3. The complex nature of biface manufacture in the Acheulean, about a million years ago, is summarized in this diagram, which attempts to show various aspects of the process.

These requirements cannot be wished away. It cannot be emphasized too strongly that they were the common property of early human beings as much as two million years ago, and that this is hard knowledge, not speculation.

### 5.3. Other Skills: Fire Control and Animal Butchery

In the next few years we are likely to gain much more specific information about other early human skills in addition to stone tool-making. We are not likely to learn much about wood-working, for reasons of preservation, though it was almost certainly practised (the oldest preserved wooden tools are 200–

300 000 years old). Nor are we likely to learn much about actual hunting procedures, though suggestive information is sometimes recovered (e.g. the BK site at Olduvai where a herd of *Pelorovis* was apparently hacked down while bogged in mud (Leakey, 1971); or the site with giant baboon remains at Olorgesailie (Isaac, 1977; Shipman *et al.*, 1981). The two skills which are most likely to leave tangible remains are fire-making and butchery of animals.

Fire is sometimes regarded as a key sign of humanity, a characteristic which is present or absent, and the presence of which must represent an advance in mental abilities. Dogmatic assumptions are however, unsafe. Other intelligent animals certainly make use of fire, for example predators which position themselves to pounce upon other animals fleeing from the flames of a bushfire. It is unreasonable to think that even early tool-using hominids would not take advantage of such fire. As Darwin (1874, p 126) said, "the nature of fire would have been known in the many volcanic regions where lava occasionally flows through forests". Why should we suppose otherwise?

Fire use is certainly less well-documented than stone-tool use, for the evidence is less likely to be preserved. We can hope to find hearth arrangements, burnt fuel from the fire, and perhaps baked earth from the surrounds of the fire. Often hearth stones were absent, or later moved; wood burns to fine ash and is easily removed by wind and rain; charcoal flecks can be dispersed by worms; soil becomes baked like brick, but only in a small proportion of fires.

Most of the record of fire use comes from caves, where chances of preservation are highest. As preserved caves mostly belong within the last half million years, so does most of the fire record. Thus cave sites such as L'Escale in France, the Cave of the Hearths in South Africa, and Choukoutien in China provide much evidence of the use of fire, sometimes in the form of hearths with ash and charcoal, sometimes also with charred bone.

Such evidence for caves goes back for about 0·5 million years, but as large numbers of more recent open air sites preserve no evident traces of fire, we may suspect that in earlier times apparent absence of signs does not add up to absence of the phenomenon.

Recently, signs have been gathering that fire has been used for more than a million years. Some of the sites with the most suggestive evidence have been found in East Africa, others in China. In general, the evidence is not conclusive, since of course archaeologists do not find burning fires, but only the flimsy traces mentioned already. Nevertheless the evidence from individual sites, and the pattern, amount to more than can be dismissed easily.

Chesowanja, near Lake Baringo in Kenya, is probably the site from which the strongest case can be built (Harris *et al.*, 1981; Gowlett *et al.*, 1981). Here, in sediments dated to over 1·4 ± 0·07 million years old, a hearth-like concentration of stones is associated with lumps of burnt clay. Many stone tools, including the smallest waste fragments, are found in association with the other finds, but an absence of larger blocks in the area surrounding the "probable hearth" suggests that an area may have been specially cleared.

On a later site, where the evidence would accord with expectation, the Chesowanja evidence would probably be regarded as unequivocal evidence for controlled fire, since there are numerous comparable examples of hearth arrangements in similar form. As natural bush fires burn fairly regularly in Africa, they also can be invoked as an explanation. Further detailed studies at Chesowanja are likely to resolve the matter beyond reasonable doubt.

Other possible early fire sites, which must be taken equally seriously, are Yuanmou in China (Li *et al.*, 1977), FxJj20 at Karari at East Turkana (Isaac and Harris, 1978), and Gadeb in Ethiopia (Barbetti *et al.*, 1980). All these sites are likely to be more than a million years old.

Whatever we conclude from the individual site, the early use of fire is a reasonable hypothesis, supported by suggestive evidence from numbers of sites where burnt material and artifacts are found together.

The processes of fire using and fire making closely parallel those of stone tool-making, in structure through time, and in the steps of the routine. The need for a fire must be foreseen, the raw materials must be selected and collected, the problem of kindling, or acquiring, fire must be solved. It did not necessarily require a leap in mental abilities to master fire, over and above those shown in early stone tool-making. The hypothesis that fire was in use must, therefore, be a reasonable one to test, for its use would be quite compatible with all the other evidence which we have.

Animal bones on archaeological sites can tell us what was eaten, but also about methods of preparation, for food or other purposes.

Recently a controversy—much overplayed—has arisen about whether the animal bones found on early occupation sites are actually linked with those sites (Isaac, 1981b; Binford, 1981; see Chapter 6, this volume). On occasion, animal bones unrelated to the human occupation may by chance accumulate within the site. For some curious reason, the explanation which might possibly hold in a minority of cases, has been converted into the *force majeure* explaining all of them.

Yet, as there are large numbers of later Palaeolithic sites in which stone tools and bones are unequivocally associated directly, there has never been much real doubt that the same was true for the great majority of early sites which have been found in good sedimentary contexts.

A site such as Mwanganda's Village (Clark and Haynes, 1970) provides an example of articulated animal bones associated with clusters of stone tools. In these circumstances there are scant grounds for doubting the association with human activity, and there are many other examples throughout the duration of the Palaeolithic.

Careful studies of cut-marks on the bones tend to confirm this view (Bunn, 1981; Potts and Shipman, 1981). Cut-marks made by stone tools, not animal claws or teeth, are there where they would be expected to be.

Although hunting rather than scavenging is not always easily established (see Chapter 6, this volume), there is a good case for saying that carcases were being butchered in a way that amounted to a skill, probably with separately appreciated goals of getting meat, tendons and skin for different purposes. Such butchery would certainly be most easily and effectively carried out as a routine, which would closely parallel—and indeed sometimes be linked in with—the operational chains involved in stone working.

### 5.4. Foundations for the Origins of Mathematics and Art

We have seen that early human beings, over a million years ago, had minds which could handle extents of time and space, much as we can, and construct long chains of activity through them, using set routines, but able to rewrite these flexibly in detail.

If this is the basic pattern of operation of the human mind, and if it was established so early, we can ask how and when the developments were made which led towards all the striking phenomena of the last thirty or forty thousand years—art, religion, mathematics, a proliferation of abstract, symbolic activities and highly developed economic patterns.

Stone tools again provide the most precise insights into the workings of the mind, and two examples are given here which show that the foundations for later developments extend far, far beyond the last 100 000 years.

For a very long period, from about 1·4 Myr to about 0·15 Myr ago, one particular class of stone tools was made to a common pattern. These are the "hand-axes" of the Acheulean. This long stability is generally interpreted as "primitiveness", but the hand-axe is a logical end-step of stone-working.

The hand-axes from Kilombe in Kenya are particularly suitable if we wish to study the expression of form, since they occur in large numbers from a single horizon (Gowlett, 1978). Although we do not know what they were used for, or why there were so many of them, we can learn about the human mind from them. The Kilombe site is believed, from palaeomagnetic evidence and other grounds, to be over 700 000 years old, making it almost certain that a form of *Homo erectus* was responsible for the tool-making.

Standardization and form both tell us a good deal. First, the very high correlations between basic dimensions (length, breadth and thickness) are remarkably similar for different parts of the site (Table 7.1). Correlations between length and breadth are the highest, apparently reflecting both the primary design target, and the greater technical difficulty of

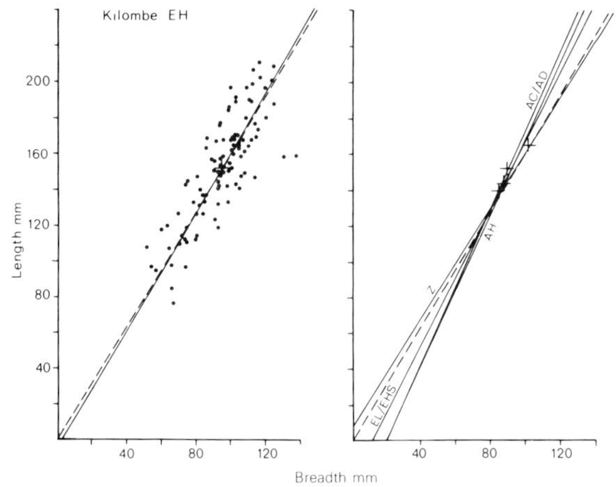


Fig. 7.4. A developed sense of geometric proportions is demonstrated by the Kilombe bifaces. The plots are of length and breadth. Left: data points from EH excavation bifaces. The dashed line provides a perfect ratio through the mean value; the solid line is that fitted to the data. Right: comparable fitted lines for other Kilombe samples. The crosses represent sample means for the two dimensions.

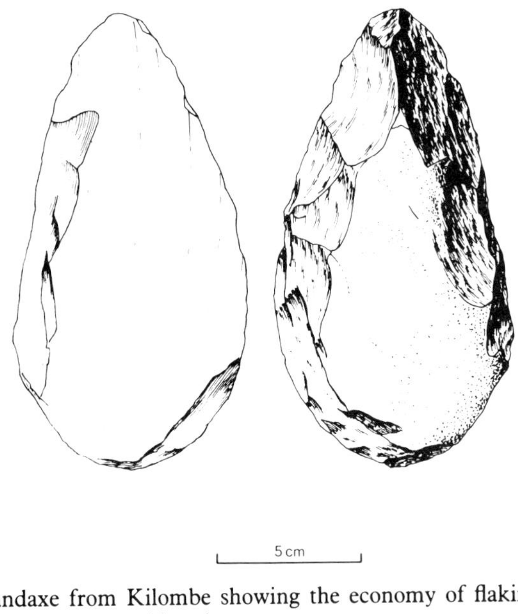


Fig. 7.5. A handaxe from Kilombe showing the economy of flaking by which the desired form was achieved.

controlling thickness. The consistency of the correlations shows a high degree of standardization, and must imply a well-defined mental image of the desired end-product. This could evidently be transmitted successfully, for it is most unlikely that all the Kilombe hand-axes were produced by the same hand.

The standardization is high enough that, with a view to studying proportions, lines can be fitted to the length and breadth data. The hand-axes were generally made in the range 8–24 cm long. Very few are smaller than 8 cm long or 5 cm broad. Nevertheless, when lines are fitted to the data points, and extended, they pass either through or very close to the origin of the two axes (Fig. 7.4).

Fairly obvious as this is, it provides us with a great deal of information. It shows us that *Homo erectus* of 700 000 years ago had a geometrically accurate sense of proportion, and could impose this on stone in the external world. In effect, without pen, paper or ruler, mathematical transformations were being performed. The artifacts which we see, and can measure, were present in the mind. In appreciating this, we need not be the less impressed that the designs could actually be expressed in stone with such accuracy (Fig. 7.5). In effect, we are not seeing one ability, but a related complex of abilities.

The early human beings at Kilombe were not knowingly practising

Table 7.1. Biface ratios, correlations and basic measurements from different parts of the main Kilombe site. Dimensions in millimetres.

| Sample | Specimens | Breadth/Length Ratio | | | Thickness/Breadth Ratio | | | Correlations | |
|---|---|---|---|---|---|---|---|---|---|
| | | Mean | | s.d. | Mean | | s.d. | B & L | T & B |
| EH | 105 | 0·62 | ± | 0·08 | 0·47 | ± | 0·12 | 0·83 | 0·46 |
| AH | 27 | 0·62 | ± | 0·09 | 0·45 | ± | 0·11 | 0·93 | 0·54 |
| AC/AD | 115 | 0·60 | ± | 0·07 | 0·48 | ± | 0·09 | 0·80 | 0·48 |
| EL/EHS | 61 | 0·61 | ± | 0·07 | 0·45 | ± | 0·09 | 0·86 | 0·51 |
| Z | 16 | 0·61 | ± | 0·05 | 0·55 | ± | 0·12 | 0·85 | 0·58 |

| Sample | Specimens | Length | | | Breadth | | | Thickness | | |
|---|---|---|---|---|---|---|---|---|---|---|
| | | Mean | | s.d. | Mean | | s.d. | Mean | | s.d. |
| EH | 105 | 152 | ± | 30 | 94 | ± | 18 | 43 | ± | 10 |
| AH | 27 | 144 | ± | 43 | 87 | ± | 20 | 38 | ± | 8 |
| AC/AD | 115 | 152 | ± | 31 | 89 | ± | 14 | 42 | ± | 9 |
| EL/EHS | 61 | 140 | ± | 27 | 84 | ± | 14 | 37 | ± | 9 |
| Z | 16 | 165 | ± | 25 | 101 | ± | 16 | 56 | ± | 15 |

mathematics or art, but in their activities the fundamental preconditions for these were expressed.

A second example demonstrates that the Acheulean is not the period of stasis which is sometimes supposed. Late Acheulean sites in the Kapthurin Formation near Lake Baringo in Kenya preserve numbers of hand-axes beautifully made by a specific technique (Leakey *et al.*, 1969; Gowlett, 1980; new work reported in Van Noten, 1982). The sites were dated to about 230 000 years by potassium argon (Tallon, 1978).

One side of the bifaces is carefully shaped on the core, and then the final form is created, in one fell swoop, by detaching a single giant flake which constitutes the biface. This technique, sometimes known as "Victoria West" or "proto-Levallois", represents a logical refinement of the process of striking biface blanks seen at much earlier sites such as Kilombe. Instead of three or four flakes being taken off to shape the blank (Fig. 7.6), there are now sometimes 40 or more.

The Kapthurin industry shows consummate skill, and includes a series of

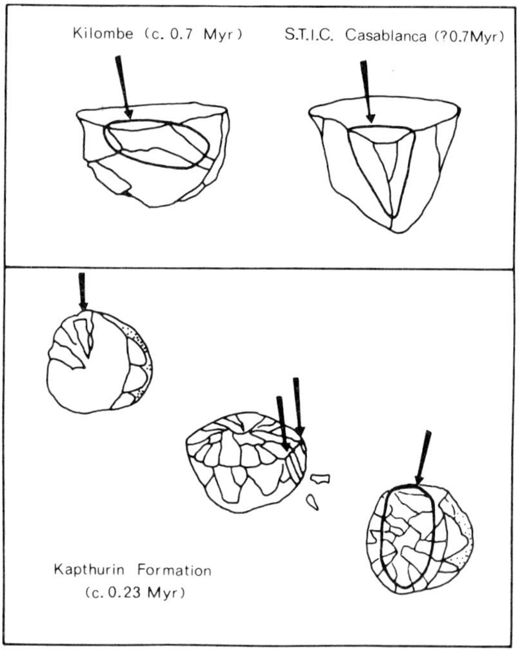


Fig. 7.6. Continuity and progress in the Acheulean. Early methods of striking blanks for bifaces (above) are contrasted with the elaborate procedure found later at Kapturin (below). 15 centimetres is about the average length for a biface blank.

technical advances which are usually assumed to belong to the last 50 000 years. Nevertheless, there is a continuity of ideas going back to the early Acheulean. This evidence is consistent with the gradual transition from *Homo erectus* to *Homo sapiens* which can be traced back from human remains, but throughout the Acheulean, it is now apparent that material culture—and hence probably other aspects of culture—reach high levels of complexity earlier than was supposed.

The problem-solving abilities shown in the manufacture of stone tools, by techniques which cannot be surpassed today, would—if projected onto other aspects of life—probably entail complex economic strategies, and levels of social interaction. There is not yet the evidence to make such a projection, but the elaborate nature of the surviving evidence cannot be ignored on that count.

## 5.5. The Last 100 000 years

We can see that mankind has approached the last 100 000 years armed with a battery of mental skills, many of which have been assumed to have developed much more recently.

Does this mean that mental abilities reached their present levels long ago, and that cultural content, as opposed to cultural capacity, has simply taken time to accumulate? That is one view which may be taken (Wynn, 1979), but however advanced early human beings were, it is unlikely that there have been no important developments in the last 100 000 years. Only, however, if we have accurately appreciated the earlier abilities, can we discern what "extra" elements have appeared more recently. That is why it is so critically important not to downgrade Lower and Middle Pleistocene man as a matter of convenience, in the way that Boule downgraded the Neanderthals.

To deal successfully with the last 100 000 years, when cultural change is relatively fast, archaeologists need dating methods of much higher resolution than are available for earlier periods. Such a chronology is only now becoming a possibility, since most radiocarbon dating has been limited to the past 30 000 years. Radiocarbon accelerator dating (Hedges, 1979, 1981) now offers hope of bridging the critical gap, and uranium series and thermoluminescence dating are also increasingly valuable.

So far, we have few clues as to the rates of cultural change, especially from 100 000 to 50 000 years ago, and little knowledge as to what were the key developments. But in the whole period we do see: (1) the appearance of anatomically modern man; (2) the development of finely shaped bone tools; (3) the development of new stone technologies; (4) the appearance of representational art; (5) the appearance of formal burials with grave goods; (6) finally, in the last 10 000 years, a whole host of economic, social and technological developments.

Poorly documented as the origins of all these are, we can at least trace continuity in the development of mental faculties from the archaeological record. Representational art, in a painting on a cave wall, or in a finely shaped bone object, requires essentially a development of the abilities shown in stone working, in the proportions of the Kilombe bifaces, and in the technology of Kapthurin.

The artist must analyse form in the outside world, say, in the form of an animal, then he must store this in the head, break it down, re-work it, and then be able to recreate that form in the outside world. The abilities required for this are highly impressive, but very probably they are closely related to many other abilities required in everyday life—recognizing other individuals, recognizing roles in society, predicting the movements of game, and interpreting the weather. Few of us become artists, but all possess a common stock of interrelated abstracting abilities, which may seem esoteric, but which relate to the real world.

To trace these interrelationships further in the last 100 000 years must surely be a major archaeological objective.

## 6. CONCLUSIONS

Human evolution is an unrestricted science, in the sense intended by Pantin (1968): there are a great many variables to account for, and few of them can be charted accurately. Scientific method, especially the application of particular techniques, is of great assistance in the documentation of the activities of early man. We can never afford to forget, however, that there is a human element, even millions of years ago, which cannot be reduced out of existence.

As human beings are distinguished so much by their minds, the study of those minds must be a legitimate objective of evolutionary studies. It is true that we must beware of projecting ourselves, our own minds, onto early human beings, but the danger of passing to the other extreme, of failing to accord to early man abilities which are there, and can be demonstrated to be there, is even greater. As far as possible, the true gradient of human mental evolution needs to be established, and downgrading is no better than upgrading as a substitute for this.

The human mind reflects the human way of life: complexity reflects complexity. Complex routines and practices observable in the Lower Pleistocene are best accounted for by the progress of natural selection itself, rather than by single shot mechanisms, such as the hunting hypothesis or the foodsharing hypothesis. In a demanding environment, there is every incentive to develop more foresight, better planning abilities. The archaeological record enables us to demonstrate that the foundations for these had been laid at least two million years ago.

## ACKNOWLEDGEMENTS

This is a revised version of a paper presented at the British Association for the Advancement of Science in 1982. I am grateful to R. A. Foley and G. N. Bailey for reading through the original paper and offering helpful suggestions. I thank Evelyn Hendy for her help in editing and typing.

## REFERENCES

Ardrey, R. (1961). *African Genesis*. London: Collins.
Ardrey, R. (1976). *The Hunting Hypothesis*. London: Collins.
Barbetti, M., Clark, J. D., Williams, F. M. and Williams, M. A. J. (1980). Palaeomagnetism and the search for very ancient fireplaces in Africa. *L'Anthropologie* **18**, 2–3, 299–304.
Binford, L. R. (1981). *Bones: Ancient Men and Modern Myths*. New York: Academic Press.
Bosinksi, G. (1969). Der Magdalenien-Fundplatz Feldkirchen-Gönnersdorf, Kr. Neuwied, Grabung 1968. *Germania* **47**: 1–52.
Boule, M. (1911–1913). L'Homme fossile de La Chapelle-aux-Saints. *Annls de Paleont.* **6**, **7** & **8**.
Bunn, H. T. (1981). Archaeological evidence for meat-eating by Plio-Pleistocene hominids from Koobi Forra and Olduvai Gorge. *Nature* **291**: 574–577.
Chomsky, N. (1976). *Reflections on Language. London: Temple Smith.*
*Clark, J. D. and Haynes, C. V. (1970). An elephant butchery site at Mwanganda's village, Karonga, Malawi, and its relevance for Palaeolithic archaeology. World Archaeology* 1(3): 390–411.
Craik, K. J. W. (1943). *The Nature of Explanation*. Cambridge: Cambridge University Press.
Crawford, O. G. S. (1921). *Man and his Past*. Oxford: Oxford University Press.
Dalrymple, G. B. (1972). Potassium-argon dating of geomagnetic reversals and North American glaciations. In: W. W. Bishop and J. A. Miller (eds) *Calibration of Hominoid Evolution*. Edinburgh: Scottish Academic Press. pp. 107–134.
Darwin, C. (1874). *The Descent of Man* (2nd edn) London: John Murray.
Deetz, J. (1967) *Invitation to Archaeology*. Garden City: New York, Natural History Press.
Dobzhansky, T. (1963). Cultural direction of human evolution—a summation. *Human Biology* **35**: 311–316.
Drummond, H. (1901). *The Ascent of Man*. London: Hodder and Stoughton.
Evernden, J. F. and Curtis, G. H. (1965). Potassium-argon dating of late Cenozoic rocks in East Africa and Italy. *Current Anthropology* **6**: 343–385.
Galton, F. (1907). *Inquiries into Human Faculty and its Development*. London, Dent: Everyman Edition (1st edn 1883).

Gowlett, J. A. J. (1978). Kilombe—an Acheulian site complex in Kenya. In: W. W. Bishop (ed.) *Geological Background to Fossil Man*. Edinburgh: Scottish Academic Press. pp. 337–360.

Gowlett, J. A. J. (1980). Acheulean sites in the central Rift Valley, Kenya. In: Leakey, R. E. and Ogot, B. A. (eds) *Proceedings of the 8th Panafrican Congress of Prehistory and Quaternary Studies, Nairobi, 1977*. Nairobi, TILLMIAP, pp. 213–217.

Gowlett, J. A. J. (1982). Procedure and form in a Lower Palaeolithic industry: stoneworking at Kilombe, Kenya. *Studia Praehistorica Belgica* 2 [Cahen, D. (ed.) Tailler! Pour quoi faire? Prehistoire et technologie lithique II], pp. 101–110.

Gowlett, J. A. J., Harris, J. W. K., Walton, D. and Wood, B. A. (1981). Early archaeological sites, hominid remains, and traces of fire from Chesowanja, Kenya. *Nature* **294**: 125–129.

Hallowell, A. I. (1960). Self, society and culture in phylogenetic perspective. In: Tax, S. (ed.) *Evolution after Darwin*, Vol II. Chicago: University of Chicago Press, pp. 309–372.

Harris, J. W. K., Gowlett, J. A. J., Walton, D. A. and Wood, B. A. (1981). Palaeoanthropological studies at Chesowanja. In: *Las industrias mas Antiguas*, preprint, *Union Internacional de ciencias prehistoricas y protohistoricas, X Congreso*, Mexico, D.F., 1981.

Hedges, R. E. M. (1979). Radioisotope clocks in archaeology. *Nature* **281**: 19–24.

Hedges, R. E. M. (1981). Radiocarbon dating with an accelerator: review and preview. *Archaeometry* **23**: 1, 3–18.

Holloway, R. L. (1969). Culture: a *human* domain. *Current Anthropology* **10** (4): 395–412.

Holloway, R. L. (1972). Australopithecine endocasts, brain evolution in the hominoidea, and a model of human evolution. In: Tuttle, R. (ed.) *The Functional and Evolutionary Biology of Primates*. Chicago: Aldine, pp. 185–203.

Holloway, R. L. (1974). The casts of fossil hominid brains. *Sci. Am.* **231** (1): 106–115.

Holloway, R. L. and De La Coste-Lareymondie, A. M. (1982). Brain endocast asymmetry in pongids and hominids: some preliminary findings on the palaeontology of cerebral dominance. *Am. J. phys. Anthrop.* **58**: 101–110.

Huxley, J. (1941). *The Uniqueness of Man*. London: Chatto and Windus, pp. 1–34.

Huxley, J. (1951). New bottles for new wine: ideology and scientific knowledge. *Huxley Memorial Lecture for 1950, Jl. R. anthrop. Inst.* **43**: 7–23.

Huxley, T. H. (1863). *Evidence as to Man's Place in Nature*. London: John Murray.

Isaac, G. Ll. (1977). *Olorgesailie: Archaeological Studies of a Middle Pleistocene Lake Basin in Kenya*. Chicago: University of Chicago Press.

Isaac, G. Ll. (1978). The food-sharing behaviour of protohuman hominids. *Sci. Am.* **238**: 90–108.

Isaac, G. Ll. (1981a). Stone Age visiting cards: approaches to the study of early land-use patterns. In: Hodder, I. Isaac, G. and Hammond, N. (eds) *Pattern of the Past: Studies in Honour of David Clarke*. Cambridge: Cambridge University Press, pp. 131–155.

Isaac, G. Ll. (1981b). Emergence of human behaviour patterns. *Phil Trans. R. Soc. Lond. B* 292: 177–188.

Isaac, G. Ll. and Harris, J. W. K. (1978). Archaeology. In: Leakey, M. G. and Leakey, R. E. (eds) *Koobi Fora Research Project Vol I: The fossil hominids and an introduction to their context*. Oxford: Clarendon Press, pp. 64–85.

Johanson, D. C. and White, T. D. (1979). A systematic assessment of early African hominids. *Science, N.Y.* **202**: 321–330.

Leakey, Margaret, Tobias, P. V., Martyn, J. E. and Leakey, R. E. F. (1969). An

Acheulian industry with prepared core technique and the discovery of a contemporary hominid at Lake Baringo, Kenya. *Proc. Prehist. Soc.* **35**: 48–76.

Leakey, M. D. (1971). *Olduvai Gorge*, Vol. 3. Excavations in Beds I and II, 1960–1963. Cambridge: Cambridge University Press.

Leakey, M. D. (1981). Tracks and tools. *Phil. Trans. R. Soc. Lond. B* **292**: 95–102.

Le Gros Clark, W. E. (1967). Human food habits as determining the basic patterns of economic and social life. In: Kuhnau, J. (ed.) *Proceedings of the 7th International Congress of Nutrition*. Braunschweig, Viewig & Sohm. Vol 4, 18–24.

Li, P., Chien, F., Ma, M. Pu, C. Hsing, L. and Chu, S. (1977). Preliminary study on the age of Yuanmou Man by palaeomagnetic technique, *Scientia Sinica* **20**: 645–664.

M'Dougall, W. (1925). Mental evolution. In: Bower, F. O. *et al. Evolution in the Light of Modern Knowledge*. London: Blackie, pp. 321–354.

McGrew, W. C. (1979). Evolutionary implications of sex differences in chimpanzee predation and tool use. In: Hamburg, D. A. and McCown, E. R. (eds) *The Great Apes*. Menlo Park, California: Benjamin. pp. 441–463.

McGrew, W. C., Tutin, C. E. G. and Baldwin, P. J. (1979). Chimpanzees, tools, and termites: cross-cultural comparisons of Senegal, Tanzania, and Rio Muni. *Man* **14**: 185–214.

Miller, G. A., Galanter, E. and Pribram, K. H. (1960). *Plans and the Structure of Behaviour*. New York: Holt, Rinehart and Winston.

Oakley, K. P. (1951). A definition of man. *Science News* **20**, 69–81. Reprinted in: Montagu, M. F.A. (ed.) *Culture and the Evolution of Man*. New York: Oxford University Press, 3–12.

Oakley, K. P. (1981). Emergence of higher thought 3·0–0·2 Ma B.P. *Phil. Trans. R. Soc. Lond.* B **292**: 205–211.

Pantin, C. F. A. (1968). *The Relations between the Sciences*. Cambridge: Cambridge University Press.

Piggott, S. (1960). Prehistory and evolutionary theory. In: Tax, S. (ed.) *Evolution after Darwin, the University of Chicago Centennial, Vol II, The Evolution of Man*. Chicago: University of Chicago Press, pp. 85–97.

Potts, R. and Shipman, P. (1981). Cutmarks made by stone tools on bones from Olduvai Gorge, Tanzania. *Nature* **291**: 577–580.

Read, C. (1920). *The Origin of Man*. Cambridge: Cambridge University Press.

Rensch, B. (1959). *Evolution above the Species Level*. London: Methuen.

Roche, H. (1980). *Premiers Outils Taillés d'Afrique*. Paris: Société d'Ethnographie.

Sacerdoti, E. D. (1977). *A Structure for Plans and Behavior*. New York: Elsevier.

Schmidt, R. R. (1936). *The Dawn of the Human Mind*. London: Sidgewick and Jackson.

Sherrington, C. (1940). *Man on his Nature*. Cambridge: Cambridge University Press.

Shipman, P., Bosler, W. and Davis, K. L. (1981). Butchering of giant geladas at an Acheulian site. *Current Anthropology* **22** (3): 257–264.

Tallon, P. W. J. (1978). Geological setting of the hominid fossils and Acheulian artefacts from the Kapthurin Formation, Baringo District, Kenya. In: Bishop, W. W. (ed.) *Geological Background to Fossil Man*. Edinburgh: Scottish Academic Press, pp. 361–374.

Tobias, P. V. (1981). The emergence of man in Africa and beyond. *Phil. Trans. R. Soc. Lond. B* **292**: 43–56.

Tuttle, R. H. (1981). Evolution of hominid bipedalism and prehensile capabilities. *Phil. Trans. R. Soc. Lond. B* **292**: 89–94.

Van Noten, F. (1982). Excavations in the Kapthurin Formation. *Nyame Akuma* **20**: 17–19.
Wood, B. A. 1978. In: Chivers, D. J. and Joysey, J. A. (eds), *Recent Advances in Primatology*. London: Academic Press, pp. 349–372.
Wright, R. V. S. (1972). Imitative learning of a flaked stone technology—the case of an orang utan. *Mankind* **8**: 296–306.
Wynn, T. (1979). The intelligence of later Acheulean hominids. *Man* **14**: 371–391.
Wynn, T. (1981). The intelligence of Oldowan hominids. *J. Hum. Evol.* **10**: 529–541.
Young, J. Z. (1978). *Programs of the Brain*. Oxford: Oxford University Press.

# 8 *Hominids and Fellow Travellers: Human Migration into High Latitudes as Part of a Large Mammal Community*

*Alan Turner*

## 1. INTRODUCTION

Previous chapters have emphasized the tropical origin of the hominids. This chapter examines their dispersal into the temperate zones of Eurasia and the eventual colonization of the New World and Australia.

It is clear that Plio-Pleistocene hominids were full members of the tropical large-mammal community (see Chapter 4, this volume). Vrba (in press) has stressed the value of studying non-hominid (specifically bovid) members of that community as a guide to the understanding of ecological differences between hominid species and for insight into the determinants of speciation, extinction and evolutionary rhythm. Both Hill and Potts (see Chapters 6 and 7, this volume) have pointed to the ecological similarities between hominids and the large predators, similarities which lead on the one hand to comparative models of hominid predation strategy and on the other to confusions in the interpretation of bone-accumulating agencies. Numerous authors have now stressed the fact that both hominid and non-hominid components are involved in many instances of fossilized behaviour patterns recorded in vertebrate assemblages (Binford, 1981; Brain, 1981; Turner, 1981b, 1981d; Vrba, 1980 and see Gamble, Chapter 10, this volume).

The present-day distribution of man is the end result of dispersal from this tropical origin point during the course of the Pleistocene. But the entire Cenozoic has been characterized by mammalian dispersals and adaptive radiations (Kurtén, 1968, 1971, 1980; Maglio, 1978; Eisenberg, 1981). Conventional discussions of the patterns of hominid dispersal, and associated assessments of evolutionary relationships, tend to consider only the evidence of archaeology and physical anthropology in arriving at interpretations. While

HOMINID EVOLUTION
ISBN 0-12-261920-X

possibly correct, many of the explanations put forward on this basis offer little scope for evaluation save in terms of plausibility. By considering man as a special case we leave ourselves with little option other than the acceptance of special interpretations. Although Cenozoic dispersals may not have involved wholesale movements of past large-mammal communities, it is none the less probable that an understanding of the dispersal pattern of any one species will be enhanced, and the basis of our conclusions strengthened, if we consider it as a part of a larger process. This aspect of hominid membership of a palaeo-community has received relatively little attention in the literature, but clearly we cannot simply assume that hominids were immune to general determinants of mammalian dispersal.

Table 8.1. Partial glacial–interglacial sequences in Britain, northern Europe and North America, with oxygen isotope stages, time scale and probable correlations (broken lines). (Sources: Shackleton, 1975; West, 1977; Heusser and Shackleton, 1979; Kurtén and Anderson 1980).

| | Oxygen isotope stages | Age (Myr) | | British sequence | N. European sequence | N. American sequence |
|---|---|---|---|---|---|---|
| | 1 | | | | | |
| | | 0·1 | Upper | (G) Devensian | (G) Weichselian | (G) Wisconsinan |
| | 3 | | | (I) Ipswichian | (I) Eemian | (I) Sangamonian |
| | 5 | | | (G) Wolstonian | (G) Saalian | (G) Illinoian |
| | | 0·2 | | (I) Hoxnian | (I) Holsteinian | ? (I) Yarmouthian |
| | 7 | | | (G) Anglian | (G) Elsterian | (G) Kansan |
| | | 0·3 | | (I) Cromerian | (I) Cromerian | (I) Aftonian |
| Pleistocene | 9 | | Middle | (G) Beestonian | — | (G) Nebraskan |
| | 11 | 0·4 | | | | |
| | 13 | 0·5 | | | | |
| | 15 | 0·6 | | | | |
| | 17 | 0·7 | | | | |
| | 19 | | Lower ↓ 2·0 Myr | | | |

Table 8.1 gives a timescale for the events to be discussed, and requires a note of caution (see also Chapter 2, this volume). Correlation between glacial and interglacial sequences from different parts of the world, and the assignment of absolute dates, is notoriously difficult. Terrestrial sequences are an expression of the evidence for successive events in a given area. Sequences shown in Table 8.1 are those most widely accepted for each area (West, 1977; Kurtén and Anderson, 1980), but one should not assume automatic correlation between each numerically equivalent stage. The only

well-accepted correlations and dates are those for elements of the last interglacial–glacial cycle, with dates as shown (Heusser and Shackleton, 1979), and the age of the Cromerian.

## 2. PLEISTOCENE ENVIRONMENTS AND LARGE MAMMAL DISTRIBUTION

It is now clear that the Pleistocene period was one not only of massive climatic shifts but also of frequent such changes (see Chapter 2, this volume). The effect of these shifts on the large-mammal fauna may be clearly seen in the temperate zones, both in terms of the increase in numbers of cold-adapted species over time and in the stratigraphic alternation of cold and warmth-tolerant forms (Kurtén, 1968; Maglio, 1975; Stuart, 1974). Vrba (in press) has pointed to similar cyclical effects observable in the African Plio-Pleistocene bovid fauna which may be correlated with gross environmental variations.

Such climatic variations not only affected the distributions of mammals within parts of the world but also made possible significant exchanges between these areas. Major changes in sea-level accompanied the climatic shifts, and one important result was the periodic appearance of land bridges. North America was at times in contact with Eurasia via the Bering land bridge, while Sumatra, Java and Borneo were connected to Asia. Britain became a continuation of the northwestern tip of Europe. In the Aegean, mammal dispersals were clearly affected by changes in the geography of the region (Dermitzakis and Sondaar, 1978). Only Australia and Antarctica among the continental land masses remained isolated.

Interchanges of mammal species between Africa and Eurasia were a feature of the entire Cenozoic (Maglio, 1978). During the latter part of the Pleistocene, however, such intercontinental movements became less marked, although Africa north of the Sahara Desert barrier has continued to exchange species with Eurasia to the present day (Cooke, 1972). For the hominids, Pleistocene dispersal may be put into two broad categories, as was proposed by Campbell (1972). First, movement out of the tropics into the Eurasian temperate zones and, secondly, the colonization of regions subjected to arctic conditions. Of course such a two-stage classification cannot encompass the complexities of Pleistocene large-mammal dispersal as a whole, but it retains advantages for the purpose of the present discussion and will therefore be employed. In the framework of the discussion presented by Campbell these two categories represented increasing abilities to exploit harsh environments, but they may equally well reflect increasing requirements to do so.

## 3. COLONIZATION OF THE TEMPERATE ZONES OF EURASIA

As Campbell (1972, p 46) emphasized, extensive seasonal temperature fluctuation is a significant feature of temperate biomes. Although the average temperature may be moderate, winter can bring several months of snow and frost. Mammals living in these areas must either possess physical adaptations to such cold conditions or exhibit behavioural repertoires which enable them to minimize their effects.

Isaac (1975, 1978) pointed out that it is difficult to produce clear evidence of hominid dispersal into Eurasian temperate zones prior to the Lower–Middle Pleistocene boundary at approximately 700 000 B.P. Interestingly, a recent report on an early site, Isernia La Pineta in central Italy, gives a date almost precisely on the boundary (Coltorti *et al.*, 1982). Bordes and Thibault (1977) argued for a "well-established" human population in Europe by that time, going on to suggest that material from several sites in southern France and southern Spain could indicate hominid presence perhaps as far back as 1·9 Myr. Most of these instances offer dubious support for hominid activity in the area at such an early date, however. Even if the assigned dates for the deposits are realistic, the associations of lithic and other materials tend to be tenuous. Furthermore, claims for hominid presence and activity often rest on interpretations of animal-bone frequencies and damage patterns which could equally well have resulted from large-predator bone consumption, and such arguments must clearly be treated circumspectly. Both Brain (1981) and Binford (1981) have recently produced useful discussions of this latter problem.

Repenning and Fejfar (1982) have recently suggested that the site of 'Ubeidiya in Israel, where considerable numbers of implements have been found in apparent association with an extensive mammalian fauna, may be dated close to 2·0 Myr. This suggested age exceeds previous proposals for a date just prior to 700 000 (Bar-Yosef, 1975), and if correct would lend some weight to claims for hominid appearance in temperate Eurasia well within the Early Pleistocene. However, I feel that several aspects of the argument for an early date of hominid presence at the site are less than compelling.

The assignment of a date close to 2·0 Myr rests on a consideration of archaic elements in the mammal fauna. This fauna, previously thought to correlate with "Cromerian" assemblages, is argued by Repenning and Fejfar to be of late Villafranchian (Villanyian) age on the basis of recent refinements in European micromammalian biochronology. But it is by no means clear that correlation with European sequences can place such restrictively early dates on the 'Ubeidiya assemblage as Repenning and Fejfar claim. Moreover, the mixed aspect of the fauna should be reiterated. What appear to be later-occurring species are present in the assemblage, and this mixing is even apparent in the case of the scanty hominid remains. Repenning and Fejfar manage to convey the impression that the hominid specimens are indeed of *Homo erectus* in their discussion of possible invasion of Africa by this species, whereas Tobias (1966, p 133) was categorical in his view that the material could only be assigned to *Homo* sp. Also, the investigation of the relative ages of the hominid and other mammal bones undertaken by Molleson and Oakley (1966) seems to have been entirely overlooked in the discussion by Repenning and Fejfar. This investigation found a "marked disparity" in the relative ages, with the hominid material being considerably the younger. Quite how such results affect the relative chronologies of the stone tools and the mammalian remains is unclear, but evidently some problems with the entire assemblage from the site remain to be considered. And finally, but perhaps most significantly, the discussion of Quaternary stratigraphy by Horowitz (1979) makes it quite clear that the 'Ubeidiya Formation, part of the Jordan Group, is underlain by what is termed the Cover Basalt which is dated to 2·0–1·7 Myr. Repenning and Fejfar (1982, Fig. 1) show the 'Ubeidiya Formation as contemporaneous with the formation of the Cover Basalt, which cannot be correct. As both Bar-Yosef (1975, p 583) and Horowitz (1979, p 142) stress, the base of the 'Ubeidiya Formation has not been exposed, so that the timespan between deposition of the Cover Basalt and the lowest part of the 'Ubeidiya sequence is hard to determine. The bone and stone-tool assemblages were mostly recovered from what are considered to have been beach deposits, and seem likely to reflect a series of depositional events which may include reworking.

It would thus seem that claims for hominid presence in Europe before the Middle Pleistocene are not free from dispute. If hominids were present, then their numbers were clearly not great, and significant dispersal into temperate Eurasia by early man would seem to have been a Middle Pleistocene phenomenon.

While an African origin for man is well-established, there are indications that hominids may also have been present in the tropical areas of southern Asia prior to 700 000 B.P. (Howells, 1980; Isaac, 1978). If so, this would argue for an earlier hominid dispersal confined to the tropical or semi-tropical regions. If early hominid presence in southern Asia is likely, then was hominid appearance in temperate Eurasia the result of dispersals from Africa, tropical Asia or both? On balance an African source is perhaps most likely, for two major reasons. First, as Howells (1980) points out, while distinctions between African and European biface industries on the one hand and East Asian chopping tools on the other are less absolute than previously claimed, the case can perhaps still be made with sufficient force to indicate separation. Secondly, and perhaps more importantly, Stringer (1980, 1981; Stringer *et al.*, 1979) suggests that European Middle Pleistocene hominids show more affinities with African rather than with Asian material, on the basis of finds from Petralona in Greece and the African specimens from Bodo and Broken Hill. The extent to which any southern Asian groups may have contributed to the overall occupation of Eurasia remains unclear at this stage, however.

If we look beyond the pattern of hominid dispersal, we see that the Lower-Middle Pleistocene boundary in Eurasia was marked by a major change in mammalian faunas (Maglio, 1975) which was particularly evident in western parts of the area. Some species, such as reindeer, *Rangifer tarandus*, and woolly rhinoceros, *Coelodonta antiquitatis*, appear to have originated in the east and spread westwards during this period. The wolf, *Canis lupus*, is first recorded in the Early Pleistocene of the Middle East, appears in Europe in the early Middle Pleistocene and reaches northern Asia by the latter part of that period. Other species, such as the hippopotamus, *Hippopotamus amphibius*, the forest elephant, *Elephant antiquus*, and, in particular, the lion, *Panthera leo*, the leopard, *Panthera pardus* and the spotted hyaena, *Crocuta crocuta*, seem to have migrated northwards from Africa. Migrations by these carnivores occur in the same direction as the major dispersal of the hominids and within the same general timespan (Table 8.2). This observation, when considered in conjunction with the similarities in ecology between potentially carnivorous hominids and the large predators, may be significant. Kurtén (1968; Kurtén and Anderson, 1980) emphasizes the enormous size of the first European lions, a feature suggesting successful predation on large game, and this development is mirrored in the case of both the leopard and the spotted hyaena. Both size and movement in parallel argue for good conditions for predators moving into the area at that time, and for advantages in the behaviour and abilities of these animals in comparison with the then existing carnivores of the region. *Crocuta* appears to replace two other large hyaenids, *Hyaena perrieri* and *Hyaena brevirostris*, throughout Eurasia, while the lion and leopard between them became the dominant felids in the place of the machairodont cats although the tiger, *Panthera tigris*, is present in Asia from Middle Pleistocene times onwards. Man therefore dispersed into temperate Eurasia at a time when other large predators, originating in a similar area of the world, were also finding conditions there to their advantage. While we can assess the suitability of the then indigenous herbivore fauna for potential human exploitation and make a judgement on the likely advantages which accrued to the hominids from their dispersal, the fact that other large predators seem to have moved in parallel highlights the wider nature of such advantages. A detailed study of Middle Pleistocene large-predator biogeography should reveal a number of interesting features of the migratory patterns of these species and open a series of profitable perspectives on the palaeoecology of the large-mammal community. Such perspectives would remove the dependence on material culture-orientated interpretations of hominid dispersal which considerations of early man in isolation from his prey and competitors so often seems to involve.

Of course the biological communities in which man appeared in Europe

Table 8.2. Earliest appearances of man and some large predators in various parts of the world, expressed in millions of years B.P. Ages for Britain in parenthesis refer to time of major occupation.

| Species | Africa | Temperate Eurasia | Britain | N. America | Australia |
|---|---|---|---|---|---|
| *Homo* sp. | *ca.* 2·0 | *ca.* 0·7 | *ca.* 0·5 (0·1) | *ca.* 0·03 | *ca.* 0·05 |
| *Panthera leo* | > 2·0 | *ca.* 0·7 | *ca.* 0·7 | > 0·125 | — |
| *Panthera pardus* | > 2·0 | *ca.* 0·7 | *ca.* 0·15 | — | — |
| *Crocuta crocuta* | > 2·0 | *ca.* 0·7 | *ca.* 0·7 (0·15) | — | — |
| *Canis lupus* | — | *ca.* 0·7 | *ca.* 0·7 | *ca.* 0·5 | — |

have no precise modern counterparts (see Chapter 9, this volume). Any assessment of the palaeoecology of, and interactions between, members of these communities must be based on principles derived from studies of modern populations. A further difficulty is raised by uncertainties about the suitability of a strictly uniformitarian stance in such reconstructions. Hoffman (1979) has discussed these and other difficulties of community palaeoecology, and points out that such problems may be minimized by concentrating on the more basic adaptations of the community members, such as locomotion and feeding strategy. In his view,

> Ecological communities are merely an epiphenomenon of the overlap in distributional patterns of various organisms controlled primarily by the environmental framework. Interspecific interactions within a community are mostly weak and facultative, except for those developed through coevolution in predator-prey systems involving a negative feedback loop. (Hoffman, 1979, p 370)

Although the extent of participation may itself be a legitimate subject for investigation, man as a facultative carnivore clearly became a member of the Middle Pleistocene large-predator community in temperate Eurasia and part of the predator-prey system. That community, as we have seen, included lions, hyaenas, wolves and leopards, as well as lynxes, *Lynx* spp., and a whole host of smaller predators and at least two species of omnivorous bears, *Ursus* spp. The total food requirement of these carnivores was such that man could not have had an unrestricted choice from some pristine standing ungulate crop, but competed for a selection from an already exploited range of species. Clear appreciation of this point is no less important than the realization that many assemblages of animal bones assumed to be the result of hominid activity are likely to have resulted in whole or in part from the activities of man's meat-eating competitors.

## 4. COLONIZATION OF COLD ENVIRONMENTS

Human colonization of the arctic biomes produced by the Pleistocene glaciations is a relatively late phenomenon. It is only from deposits which date to the last glaciation in Eurasia that significant assemblages of stone tools are consistently found in association with cold-adapted mammal species and appropriate environmental indicators. Earlier evidence of man is generally associated with interglacial deposits at times when cold–temperate winters would have provided the most severe test of abilities. It is also notable that these first consistent appearances in association with cold-climate indicators occur in the more westerly parts of Eurasia, where oceanic effects had a moderating influence.

Human occupation of more easterly Eurasian areas which experience a continental climate, and thus the harshest winters, occurs later still. Klein (1975) argued that this late colonization, perhaps dating to 35–40 000 years ago and apparently coinciding with the appearance of Upper Palaeolithic technologies, marked the development of sufficient capabilities for coping with such conditions.

Certainly, adequate clothing and shelter, together with an efficient food procuring strategy, would have been essential for survival in arctic environments. But such requirements do not suddenly become essential only in extreme conditions, and it is likely that all last glaciation inhabitants of Eurasia were well-prepared for cold climates whether equipped with Middle or Upper Palaeolithic technologies. Clearly, the persistent popular view of Neanderthals prowling the landscape clad in no more than loin cloths, when animals such as mammoth, *Mammuthus primigenius*, woolly rhinoceros and reindeer were covered with extremely efficient cold-adapted protection, must be some way from the truth. Furthermore, if man was present in temperate Eurasia for several hundred thousands of years prior to the appearance of Upper Palaeolithic industries, then it seems highly unlikely that steps towards coping with at least seasonal cold had not brought considerable expertise. It is at least as likely that technological developments occurred in response to needs as that man was prevented from occupying arctic biomes by their absence.

But occupation of arctic environments would undoubtedly have had a major effect upon the subsistence base and upon the mode of extraction. Seasonal movements of ungulates, under pressure of seasonal variation in food supply, would have produced similar stimulus to movement in human groups. Recent studies show that, in carnivores of a given body weight, home range size is influenced by diet (Gittleman and Harvey, 1982), to the extent that species with a higher proportion of meat in their diet are required to exploit a larger area to satisfy intake needs. Further, species whose prey exhibit migratory habits must be equally mobile and therefore occupy a larger home range. The combination of relatively little vegetable food suitable for man in glacial Eurasia, together with seasonally migrant prey, would inevitably have resulted in considerable annual territories for Middle and Upper Palaeolithic groups. The only alternative, an option for well-organized hominids unavailable to other carnivores, would be *in situ* winter consumption of cached resources stockpiled following summer hunts.

Of course the only direct route to the New World lay via the Beringian land bridge between Eurasia and North America at times of glacial advance when sea levels were lowered sufficiently (Fig. 8.1). In a real sense, therefore, a prior requirement for dispersal of any terrestrial mammal to the New World was the occupation of northeastern Eurasia at times when the climate there was at or close to its harshest. In such circumstances, it is not hard to envisage large home ranges among the human groups leading almost inevitably to dispersal across Beringia.

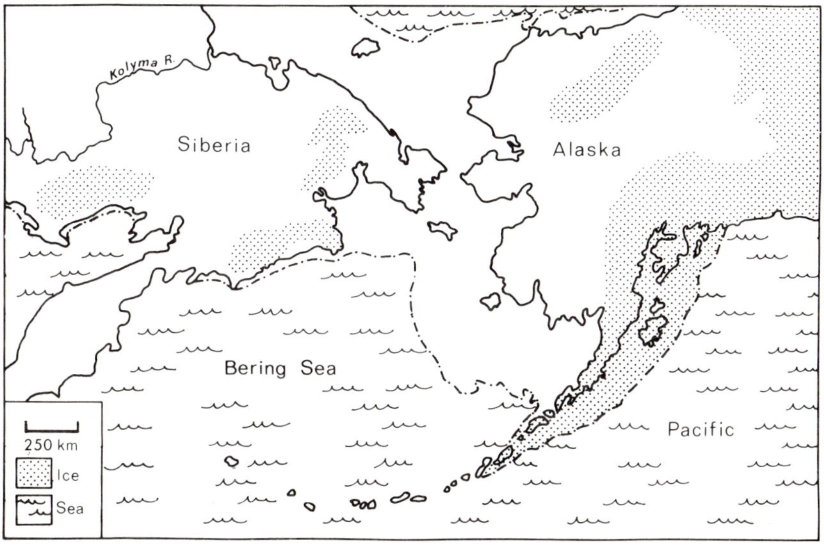


Fig. 8.1. Beringia at the height of the last glaciation, showing the extent of the land bridge and the position of the major ice sheets. Modern shorelines are shown in solid line (after Hopkins, 1967).

## 5. OCCUPATION OF THE NEW WORLD AND AUSTRALIA

Migrations of large mammals to the New World appear more complex than those into temperate Eurasia, but an excellent recent summary of the evidence is given by Kurtén and Anderson (1980).

The Beringian land bridge (Fig. 8.1) appears to have acted as a filter for large mammal movements (Kurtén, 1980). Good evidence for the precise time of human migration is lacking, but while most workers would accept that there is clear evidence for human occupation in the time range 12–15 000 B.P., opinion on claims for earlier dates is sharply divided. Doubtful associations between artifacts and dated material, and human activity claimed on the basis of bone "tools" which may well be the work of non-hominid predators, again cloud the issue (Binford, 1981; Haynes, in 1983). Klein (1975) pointed out that earliest clear presence of man in northeastern Asia tends to set a basal date of *ca.* 30 000 B.P. for human arrival in the New World. He also argued that a second constraint on interpretation tends to be set by the timing of any split in the ice sheet of the last North American glaciation sufficient to permit dispersal out of Beringia and into North America proper. There are thus two aspects to consider in any assessment of human colonization of the New World, given the existence of a land bridge: the timing of initial movements across the bridge into eastern Beringia and the subsequent movement southwards.

Fladmark (1979) has suggested that Alaska was first occupied by people with maritime abilities and marine exploitation interests, perhaps at a relatively early point in the last glaciation. He argued that such people may have continued to move southwards along the refugia afforded by the unglaciated coastal environments of northwestern America, eventually arriving south of the ice. In Fladmark's view, any gap in the North American ice sheet is likely to have existed in terrain which would have remained either climatically or hydrologically unsuitable for human movement, so that a coastal dispersal route becomes considerably more likely. A sea crossing to North America removes the emphasis on big-game hunting by the first invaders while also reducing the need to argue that they actually made use of the land bridge as other than a series of stopping-off points. It would also have made migration possible prior to the full exposure of the land. Neither author discusses the other members of the mammalian community at any length, although Klein points out that one species, the elk, *Alces alces*, has a delayed appearance south of the ice while two others, the saiga antelope, *Saiga tartarica*, and the yak, *Bos grunniens*, apparently never moved south out of Beringia. The evidence afforded by these latter species supports the view that an inland route to the south was difficult if not impossible, although it does not clarify the reasons for the barrier. What light can a consideration of other mammal dispersals perhaps shed on the issue of human colonization?

Woolly mammoth, the steppe bison, *Bison priscus*, the elk and the musk ox, *Ovibos moschatus*, are all known in Alaska from Illinoian and subsequent deposits. A striking absentee from this migrant fauna, however, is the woolly rhinoceros, since a species which appears to have originated in eastern Eurasia might be expected to have made the move to the New World during the Pleistocene, as did the reindeer. Indeed none of the Eurasian Pleistocene rhinoceros species seem to have managed to reach America, and the eastern boundary of woolly rhinoceros dispersal appears to have been approximately the Kolyma River (Fig. 8.1) (Kurtén, 1980).

Among the non-hominid predators the wolf migrated early, perhaps close to the Early–Middle Pleistocene boundary. The lion is first recorded in Beringian deposits from the last interglacial, probably as a result of migration during the previous cold stage, and from the evidence of its subsequent size and distribution flourished there and in North America, even reaching as far south as Peru. This distribution makes the lion probably the most widely dispersed of the large Pleistocene land mammals after man (Kurtén, 1980). The leopard, on the other hand, appears to have failed to disperse across Beringia, an odd failure on the part of a cat which is today well-adapted to a variety of habitats and now has one of the widest distributions of the large predators (Guggisberg, 1975). The spotted hyaena, like the leopard, also remained confined to the Old World, again presenting an apparent anomaly since this species is well-adapted to making a living as both a scavenger and a hunter (Kruuk, 1972).

That Beringia afforded a land bridge for Eurasian large mammals during the last glaciation has been elegantly demonstrated by Kurtén (1973a) in his discussion of size distributions among modern brown bears, *Ursus arctos*, either side of the Beringian Straits. The clines in size clearly support the idea of a once continuous population of brown bears during the last glaciation in the area of Beringia now covered by the sea. However, that species, like the elk, seems to have been unable to move into southern North America until after the end of the last glaciation, and this restriction offers further support for the idea of an actual barrier to movement south. It thus seems likely, as argued by Fladmark (1979), that late dates for human appearance south of the last glaciation ice sheet do not depend upon a late initial expansion into Beringia, and therefore do not assist with the problem of dating such initial movements. However, it seems that in any event these movements occurred substantially after the appearance of man in the Eurasian temperate zones. Klein's figure of 30 000 years as a basal date may therefore be largely correct.

Current evidence suggests that human colonization of Australia is also a relatively late phenomenon (Table 8.2). The earliest directly dated human remains are those from Lake Mungo in New South Wales at *ca.* 26 000 B.P., although various claims have been made for occupation by 50 000 B.P. on the basis of dated artifacts (Freedman and Lofgren, 1979; White and O'Connell, 1979). In the case of Australia a sea crossing is clearly implied, since the region has been cut off from southeast Asia throughout the Pleistocene even at the maximum fall in sea level. This isolation has meant that only man and the dingo, *Canis familiaris dingo*, managed to represent the larger placental mammals on the continent prior to European arrival (Keast, 1972).

What general patterns can be discerned among these large-mammal movements?

## 6. TOWARDS INTERPRETING DISPERSAL PATTERNS IN HOMINIDS AND LARGE PREDATORS: SOME POSSIBLE CAUSES AND CONSTRAINTS

For the purpose of this discussion I intend to restrict consideration to dispersal patterns of hominids and large predators. The carnivorous abilities of the hominids link them ecologically with the predators in a way which may be useful for palaeoecological reconstruction. Modern ethological studies have thrown light on the relationships between food availability and behavioural response in predators which permit valuable general conclusions to be drawn, of a kind which are particularly suitable for application to palaeontological enquiry. Such conclusions should be applicable to an investigation of hominid dispersal without requiring the assumption that meat was the sole source of hominid food. However, the advantages of meat as a food source in temperate Eurasia, and more especially in the arctic biomes during glacial advances, are difficult to ignore in any discussion of foods available to an intelligent animal equipped with a technology. As Foley (1982) has pointed out in the case of likely hominid predation in the tropics, the question to be asked in any dispute about human nutrition really concerns the structure of resource availability.

It may be appropriate to begin this discussion by considering the spotted hyaena, since the failure of this species to disperse to the New World seems to present the most glaring anomaly. The species appears to have been adequately adapted to cold conditions such as may have existed in northeastern Eurasia and in Beringia during times of lowered sea level. It is abundantly represented in Devensian Britain (Table 8.1) in quantities which suggests that it found conditions there highly suitable (Kurtén, 1963, 1969; Turner, 1981a, 1981b). Since most large predators exhibit wide environmental tolerance given adequate nourishment, absence of the hyaena from Beringia is unlikely to be a direct reflection of differences between that area and the British Isles. Neither is it likely that its absence is related to past availability of suitable food, since the great number of animal remains recovered from Upper Pleistocene deposits in Alaska suggest that the large-mammal fauna there was abundant and varied (Guthrie, 1968; Harington, 1978). This abundance would imply that a number of predator species could certainly have been supported, as indeed we know they were. Furthermore, as Kurtén and Anderson (1980, p 41) point out, finds of frozen carcases often exhibit immediate post-mortem decay, a fact which in conjunction with the quantities of bone recovered implies that considerable scavenging opportunities existed. But it is perhaps significant that the American dire wolf, *Canis dirus*, a large extinct canid with posterior teeth which point to considerable bone scavenging abilities perhaps not unlike those of the spotted hyaena, is present in North America from later Middle Pleistocene times onwards but also unknown in Beringia (Kurtén and Anderson, 1980, p 171). This dual absence suggests that the failure of the spotted hyaena to cross to the New World may reflect a basic feature of large predator-scavenger dispersal related to the behaviour of various species rather than simply to the availability of food.

The present-day larger predators exhibit varying degrees of socialization, from the essentially solitary cats through to the communal-living dogs and hyaenas. Within each species, the degree of socialization in a particular population appears to depend upon the abundance and distribution of food in the locality, but a sufficiently clear distinction between the upper limits of group size in each species seems to exist to enable broad yet useful generalizations to be made. In particular, both lions (Schaller, 1972) and wolves (Mech, 1970) appear able to tolerate and operate within large groups to a much lesser extent than the spotted hyaena (Kruuk, 1972). This observation would imply that the spotted hyaenas would be capable of existing in an area at a higher density than that accepted by the other two species, compounded by the fact that as a larger animal the lion would have greater food needs anyway. These differences in maximum group size seem to be an expression of each predator's ability to cooperate socially with other members of its species. Given generally similar behaviour patterns in the past, such a basic difference would have caused Pleistocene lion and wolf populations, over time, to seek territories in an ever greater area, by comparison with spotted hyaenas, as the social groups split up and sought separation from one another. Such tendencies to territorial expansion could explain the relatively early appearance of the wolf in the New World and the slightly later appearance of the lion. The wolf could be expected to have dispersed the earliest since it apparently originated closer to Eurasia at a time prior to the appearances there of the other two species. In this scheme, the spotted hyaena would be seen less as unable to migrate across Beringia as unpressured to do so. The absence of the leopard from the New World may be related to that animal's need for cover when hunting, a feature likely to have been missing from Beringia during times of lowered sea level.

Although an essentially solitary animal (Guggisberg, 1975; Schaller, 1972) and therefore likely at first sight to seek ever more remote and separate territories, it may also have been unable to compete with species such as the jaguar, *Panthera onca*, now confined to Middle and South America, which once had a Holarctic distribution (Kurtén, 1973b; Kurtén and Anderson, 1980). Furthermore, leopards tend to be excluded by lions and tigers today in areas where they overlap (Schaller, 1967, 1972; Sankhala, 1978), and the Beringian presence of lion may have acted as a major barrier to the migration of the species particularly in open terrain.

The one species which does not have a known analogous behaviour pattern to enable it to be fitted into this scheme is the extinct dire wolf. The model of behaviour proposed above would imply that it, like the spotted hyaena, was, however, capable of existing in rather large groups and was under comparatively little pressure to disperse in search of new territories. This interpretation would make sound ecological sense, since the larger the group within which a scavenger operates the more easily it can obtain food by driving off an original predator. In North America the dire wolf was presumably in competition for food with ordinary wolves, lions, jaguars and the sabre-toothed cat, *Smilodon fatalis*, and under considerable pressure to evolve not only appropriate physical adaptations but also suitable patterns of social behaviour. Such patterns, and the form which they might take, are nicely demonstrated by the relationship between lions and spotted hyaenas in the East African area studied by Kruuk (1972, 1975) and Schaller (1972). In that area, the degree of social cohesion among hyaenas appears to be a direct reflection of the availability of food and the extent of competition from lions, and is extremely flexible.

This interpretation of dire wolf behaviour receives considerable support from the work of Hemmer (1978) on the relationship between cephalization and sociality in predators. He found the dire wolf to have had the largest brain among the canids, extant or extinct, a feature which in living predators marks the most social members of a family, *Crocuta* in the case of the hyaenas, the lion in the case of the felids and the wolf in the case of the canids. (The tiger, equally large brained as the lion, appears temperamentally less suited to cooperative behaviour.) Furthermore, he found the Pleistocene American lion to have had a higher degree of encephalization than modern lions, whereas *Smilodon fatalis* fell among the non-social felids. The frequency of lesions in skeletal parts of *Smilodon* point to considerable intraspecific aggression compatible with such low implied sociality. Large groups of cooperatively hunting lions, and aggressive sabre-toothed cats unable to exploit carcases fully because of their specialized dentitions (Ewer, 1973) are likely to have provided considerable quantities of food for a scavenger which could muster sufficient numbers to ensure access to the potential supply. As a final point here, low sociality and considerable aggressiveness are likely to explain many features of the wide distribution of the machairodont cats if the indications based on *Smilodon* give a true picture of their general behaviour.

How does man fit into this scheme? The late expansion of man into eastern Eurasia and into the New World, summarized in Table 8.2, suggests a parallel with the pattern of spotted hyaena dispersal. Recent studies of Middle and Upper Palaeolithic sites and material in Germany (Sturdy, 1975; Gamble, 1978, 1979), southwestern France (Bahn, 1977; Speiss, 1979) and Britain (Turner, 1981a, 1981b) point to seasonal migrations by small groups of people rather than to large, permanent communities or to *in situ*, winter consumption of the fruits of summer hunting. These small, dispersed groups are likely to have come together during the winters in larger aggregations with clear social, biological and technological advantages of the kind outlined by Binford (1979) for modern Eskimo groups and proposed by Wobst (1974) as an overall feature of primitive human societies. If we can legitimately extrapolate these results to other human groups occupying temperate and arctic biomes in Eurasia, then the slow pace of human dispersal over time becomes an inevitable consequence. Large but mobile social groups, as opposed to permanent communities, would reduce the need for dispersal in search of discrete home ranges. It would also make sense to see such patterns of human behaviour having a considerable history, given the overall level of primate intelligence and potential for group activity. Such a history would imply a more direct relationship to environmental pressures as an ultimate mechanism although mediated through cultural capacities.

Seasonal change in group numbers is paralleled in the social organization of spotted hyaenas in Tanzania (Kruuk, 1972, 1975), where a large clan of up to 80 animals will split up into hunting parties of between two and 30 individual adults depending on the prey sought and local circumstances. The clan may be mobilized to defend territorial boundaries at times, and provides a secure barrier against threats to the young. To draw such parallels here is not to construct facile analogies between spotted hyaena and hominid social organizations, since precisely similar and purely mechanistic motivations need not be proposed for both cases. We are interested simply in the broad pattern of biogeographic response over a long timespan. Both species merely appear to develop similar patterns of behaviour in order to cope with pressures which may themselves be similar in the problems which they pose. Thus it is certainly not necessary to insist that hominid groups throughout the Pleistocene were simply scavengers just because they, like hyaenas, may have dispersed less rapidly than some other large predators. However, emphasis on scavenging seems a reasonable conclusion in the case of the dire wolf and may well reflect the exploitation strategies of hominid groups at various times in the past. That such complexities in aspects of economic behaviour may be discernible in the archaeological and palaeontological record has received its most recent and explicit statement from Binford (1981), who proposes a number of interesting and equally complex methods for their investigation. Any such efforts to discern economic behaviour must, however, give considerable attention to the problems of quantifying the components of a vertebrate assemblage (Fieller and Turner, 1982; Turner, 1983).

The specific abilities of the hominid and dire wolf groups would no doubt have produced particular patterns of resource exploitation and of social intercourse which would differ at organizational and intensity levels from those of the spotted hyaenas. That hominid groups, armed with technology, may have had the ability to maintain features of their social organization in the face of environmental shifts, while hyaena clans and dire wolf packs did not, is not at issue. However, it is difficult to escape the conclusion that the repertoire of behaviour in each species was ultimately a response to environmental circumstances.

One has to admit that fitting the human colonization of Australia into any overall pattern is not without difficulties. Certainly it is, like the dispersal to the New World, a relatively late event, but it is also a rather special case, perhaps, in view of the ocean crossing necessary to get there. But as Thorne (1980, p 96) has recently stressed, the long history of human occupation in southeast Asia, an area subjected to repeated changes in geography as the sea level altered during the Pleistocene, may be expected to have produced communities who had considerable familiarity with coastal environments. Colonization of Australia, in Thorne's view, is less likely to have been by accident than to have been "the product of a competent maritime technology". We may feel that the dispersal to Australia could not have taken place before the development of adequate maritime abilities, but we really do not know how early such developments occurred. We readily assume that other species are competent at what they do, and there seems no pressing reason to think that human expertise in coping with their environment was acquired in some sudden rush towards the light of knowledge at the end of the Pleistocene. If we may suppose a reasonable ancestry for perhaps modest maritime abilities among southeast Asian human groups, then the colonization of Australia may indeed fit the general pattern of human dispersal during the Pleistocene.

But the pattern of dispersal, and the palaeogeographic parallels between hominids and the large mammalian predators, may be of interest in yet another way. Vrba (1980) has pointed to the apparent correlation between high speciation rate and specialist environmental adaptation, whereby animals with particular food requirements or closely determined behaviour patterns will constantly encounter new environments and come under the influence of fresh selection pressures. The large predators appear to be more adapted to prey size than to prey species, and any dispersal is therefore unlikely to present them with dramatically new feeding environments of the kind likely to be encountered by a herbivore. This difference may explain the lower rates of speciation which tend to be observed in fossil carnivores in comparison with herbivores of similar body size, although it should perhaps be emphasized that herbivores receive the additional burden of predation pressure upon them which in itself could have a significant effect on rates of change. [Jerison (1973) showed an apparently greater rate of increase in brain size in North American herbivores compared with those of South America, and suggested that a correlation existed between rates of change and the numbers and species of large predators present in each area.]

We may expect that hominids fell prey to the other large predators from time to time, but it seems likely that they would have learned to cope with predation pressures at an early stage in their history. I therefore doubt whether such influences had a major bearing upon the course of human evolution in any but the earliest stages. Indeed, Brain (1981) has argued that the transition from prey status for hominids may actually be visible palaeontologically in the South African cave deposits of Swartkrans and Sterkfontein. The sequences of events represented by Members 1 and 2 at Swartkrans and by Members 5 and 6 at Sterkfontein each show a major fall-off in the number and proportion of hominids present in assemblages which appear to have accumulated largely through carnivore activity. But hominids, as facultative if not obligate predators fully able to scavenge from large-carnivore kills, are, like the large carnivores, unlikely to have encountered significantly new feeding environments during much of the Pleistocene. We might therefore expect a similar absence of related selection pressures to that experienced by the predators. How does this accord with the fossil record?

For the Pleistocene, the number of proposed species of hominids is relatively small (Cronin *et al.*, 1981). Vrba (1980, p 75) argued that, by comparison with other large mammals, the group has speciated slowly over the entire Plio-Pleistocene period. This low rate of speciation is in many respects intermediate between those of the mammalian herbivores and the carnivores, and would be in keeping with the omnivorous capabilities and behavioural plasticity of the hominids.

However, new environments are difficult to avoid entirely, and the Upper Pleistocene of Eurasia may afford an example of adaptation to fresh circumstances. Human colonization of the glacial arctic biomes, discussed in Section 4 above, is a relatively late event, coinciding to a great extent with the appearance of a Middle Palaeolithic technology. Gamble (1979; Chapter 10, this volume) has suggested that many features of Middle, and in particular Upper Palaeolithic material culture, may reflect an increasing concern with information exchange. Such exchange could well indicate attempts to maintain social identities and links, as much as to ensure efficient hunting strategies, preserving recognition between groups which would otherwise have tended to diversify and fragment as adaptations to more rigorous conditions developed. Problems of inter-group alienation may have become particularly acute during later stages of the Pleistocene, as language changes and development produced ever more marked barriers to social interaction. Paterson (1980) has proposed that sexually reproducing species each consist of populations which share a *specific-mate recognition system* (SMRS), which has evolved to ensure fertilization. Behavioural components of this recognition system may be expected to include communication at a number of levels, and a breakdown in verbal communication may lead to reproductive isolation. In a species entering a new and harsh environment such reproductive isolation could lead to genetic divergence resulting in further breaks in the recognition system. Non-verbal communication of the kind which Gamble suggests may be implied by the material culture would help to reduce linguistic barriers to contact and subsequent mating. There is of course no implication that the human groups involved were attempting to do more than maintain existing social links. But the "effect" of these maintained social links (*sensu* Vrba, 1980) is likely to have been to reduce the degree of genetic divergence between segments of the population below that to be expected in a species entering such a new and partitioned environment.

## 7. MOVEMENT INTO BRITAIN: A TEST CASE

Because Britain has been cut off and rejoined to the rest of Europe at various times during the Pleistocene as a result of fluctuations in sea level (West, 1977) it has a number of parallels with the New World as a scene of episodic colonization. At times when a sea crossing was possible, the dispersal of terrestrial mammals within temperate Eurasia also included this area (see Chapter 9, this volume).

In the recent detailed discussion of early man in Britain by Roe (1981) a number of mammal species receive mention, but little importance is attached to their presence. No consideration of their appearance in Britain is employed in the discussion of human movement patterns, yet many of these species are first recorded in Britain at or about the same time as the earliest evidence for human presence and had to face similar problems of fluctuating accessibility. Roe is at pains to point out the remoteness of Pleistocene Britain. If that were the case, then the colonization of the area by man and other large mammalian predators should be broadly predictable on the basis of the previous discussions.

As with the previous examples, dating is once again a key issue. Terminology used here follows that proposed by Mitchell *et al.* (1973), in full recognition that this probably provides a minimal view of the complexities of Pleistocene events in Britain (Table 8.1). Human occupation is well-attested by artifacts dated to the Hoxnian interglacial (Roe, 1981), but may also have occurred during phases of the preceding Anglian glaciation. Indeed, movements during times of lowered sea level resulting from ice build-up are inherently most likely. Man is almost unrepresented in deposits from the subsequent Ipswichian interglacial, and appears to have entered Britain in any significant numbers only during the latter half of the Devensian (last) glaciation, again on the basis of artifactual evidence. Lion and wolf first appear in deposits of the Cromerian interglacial and are present in all subsequent Pleistocene stages (Stuart, 1974). Spotted hyaena is also seen for the first time in Cromerian material [from West Runton (Stuart, 1975) ], but it is unrecorded from Hoxnian deposits and only becomes abundant in Ipswichian and Devensian times.

It is perhaps significant that few of the pre-Ipswichian vertebrate sites in Britain are cave deposits. This may reflect the paucity of caves available for deposit formation at that stage, or subsequent erosion of early accumulations, but may also be at least a partial function of the absence or scarcity of *Crocuta*, a noted modern-day collector of bones (Sutcliffe, 1970; Turner, 1981a–d). Stone-tool assemblages in cave deposits of that date are also scarce. Leopard, as elsewhere in Europe, is rarely found in Britain, and only clearly known from Ipswichian and/or Devensian deposits (Stuart, 1974; Turner, 1981a).

In broad terms, the Pleistocene colonization of Britain by man and large predators has a number of features in common with the pattern of dispersal into the New World. Wolf and lion represent the earliest phase, while man, and in this case the spotted hyaena, represent an essentially later stage. Leopard was perhaps hampered by the open vegetation of periods when low sea-level made migration possible. The parallels are of course not precise, in view of the Hoxnian and slight Ipswichian presence of man and the appearance of spotted hyaena, but still suggest an interesting degree of similarity. It is almost as though the early appearances of both man and the hyaena represent the colonization attempts of an early burst of dispersal, attempts for which no repetition occurred until later in the Pleistocene. The Hoxnian evidence for human presence need not contradict this impression, since although the quantity of Lower Palaeolithic implements found is considerable the number of known sites is small. The timespan of deposition is likely to be long, and any assumption of high human population density on the basis of such finds is therefore probably unwarranted.

Thus the dispersal of early man into Britain conforms to biogeographical expectations, and an understanding of the pattern requires a wider perspective than that which comes from a conventional study of the artifactual evidence alone.

## 8. CONCLUSIONS

Archaeologists may choose to argue that it is not their place to consider non-hominid behaviour in what is essentially an assessment of past human activity, but this is open to serious question. It may be worth suggesting that the archaeology of Pleistocene man might be subsumed within palaeontology, with the recognition that the archaeologist is studying one animal among a range of species, an animal which happened to make tools and leave them behind as an additional category of behavioural evidence. While for practical purposes it may be necessary to specialize in the analysis of artifacts, it would be short-sighted to interpret their patterning independent of palaeontological and palaeoecological evidence.

It is certainly unlikely that hominid dispersals had no unique features, since the ability to make implements, build shelters and pass on information must have played an important part in any behavioural pattern. But by looking at hominid dispersal during the Pleistocene as part of a broader pattern of change in large-mammal distributions, we begin to see that the appearances of human groups in various parts of the world were not isolated events. This wider perspective on Pleistocene dispersal is essential for any attempts to produce a more complete understanding of human evolution. It provides at the very least a background of normal processes in the development of the mammal fauna against which to assess the activities of the hominids. Without this basic contrast it is impossible to interpret features of hominid dispersal in other than particularistic terms which may be misleading and highly inappropriate. But, at the same time, it should be understood that the activities of other species do not provide a simple backdrop against which to view the human evolutionary progression. Man was a member of the mammal community. His activities impinged on those of the other community members and were in turn impinged upon.

It is probable that the broad pattern of hominid dispersal during the Middle and Upper Pleistocene has a number of features in common with that of some of the large predators. Man appears to have been a participant in a major biogeographic event at the beginning of the Middle Pleistocene, and to have moved into an area which other large predators also found attractive. His relatively late dispersal into the New World and, in any numbers, Britain, has many parallels with those of the dire wolf, spotted hyaena and leopard, and the pattern of dispersal in the case of these latter two species can be understood by reference to the ethology of modern representatives. The evidence of modern ethnography, archaeology and palaeogeography is in considerable accord for the Upper Pleistocene movements of man, and in turn makes sense when considered in the light of some features of modern predator/scavenger ethology. This Upper Pleistocene pattern of human dispersal may suggest the response of basic social organization to environmental circumstances, similar to that producing the repertoire of spotted hyaena behaviour. It seems inherently unlikely that behaviour has any less of an evolutionary history than physical appearance, or fails to respond to circumstances as we readily assume the body of an animal to have done. It should be perfectly possible to attempt a low resolution analysis of a species' social organization in relation to environment, that of man included, provided that suitable evidence has been preserved. To do so is not to claim that every facet of the behavioural repertoire is to be explained on that basis. It may also prove possible to assess features of the tempo and mode of hominid evolution by comparison with speciation rates in other large-mammal groups. The apparent intermediacy of Pleistocene hominid evolutionary rates between those of the herbivores and the large predators may be a reflection of intermediate feeding strategies and concomitant selection pressures.

While attention in this chapter has focused on the involvement of palaeontology in the study of early man, it must be stressed that the palaeontological information available is not without its difficulties (see Chapter 5, this volume). These result not so much from the common complaint that too little detail is available as from the relatively few major syntheses which have been attempted. Much work therefore remains to be done on the biogeography of large Pleistocene mammals, whether or not this proceeds from the perspective of community palaeoecology. Models of dispersal such as that proposed here may serve to focus attention on problems, point to shortcomings in the data and suggest specific questions which may be asked. One implication of the foregoing discussion, for instance, is that the lion may actually have appeared in Eurasia before the spotted hyaena or early man, but existing information does not appear to confirm this. Kurtén (Kurtén and Poulianos, 1977, p 124) showed that lion does in fact appear slightly earlier than the spotted hyaena in deposits from Petralona in Greece, but as he has also stressed (in. lit.) this may simply be an artifact of sampling. By adding their voice to questions of this sort, hitherto thought to be of minimal relevance to prehistory, archaeologists may help to promote a fuller understanding of Pleistocene palaeoecology which can only assist the study of early man.

## ACKNOWLEDGEMENTS

I am grateful to C. K. Brain, R. Foley, C. Gamble, D. R. Harris, B. Kurtén and E. S. Vrba for discussion and comments on topics covered in this chapter.

# REFERENCES


Bahn, P. (1977). Seasonal migration in south-west France during the late glacial period. *J. Archaeol. Sci.* **4**: 245–57.

Bar-Yosef, O. (1975). Archaeological occurrences in the Middle Pleistocene of Israel. In: K. W. Butzer and G. Ll. Isaac (eds). *After the Australopithecines*. The Hague: Mouton, pp. 571–604.

Binford, L. R. (1979). Organisation and formation processes: looking at curated technologies. *J. Anthrop. Res.* **35**(3): 255–72.

Binford, L. R. (1981). *Bones: Ancient Men and Modern Myths*. New York: Academic Press.

Bordes, F. and Thibault, C. (1977). Thoughts on the initial adaptation of hominids to European glacial climates. *Quat. Res.* **8**(1): 115–27.

Brain, C. K. (1981). *The Hunters or the Hunted?* Chicago: University of Chicago Press.

Campbell, B. G. (1972). Man for all seasons. In: B. G. Campbell (ed), *Sexual Selection and the Descent of Man*. Chicago: Aldine, pp. 40–58.

Coltorti, M. *et al.* (1982). Reversed magnetic polarity at an early Lower Palaeolithic site in Central Italy. *Nature* **300**: 173–76.

Cooke, H. B. S. (1972). The fossil mammal fauna of Africa. In: A. Keast, F. C. Erk and B. Glass (eds). *Evolution, Mammals and Southern Continents*. New York: State University of New York Press, pp. 89–139.

Cronin, J. E., Boaz, N. T., Stringer, C. B. and Rak, Y. (1981). Tempo and mode in hominid evolution. *Nature* **292**: 113–22.

Dermitzakis, M. D. and Sondaar, P. Y. (1978). The importance of fossil mammals in reconstructing paleogeography with special reference to the Pleistocene Aegean archipelago. *Annales Géologiques des Pays Helléniques* **29**: 808–40.

Eisenberg, J. F. (1981). *The Mammalian Radiations*. Chicago: University of Chicago Press.

Ewer, R. F. (1973). *The Carnivores*. London: Weidenfeld and Nicolson.

Fieller, N. R. J. and Turner, A. (1982). Number estimation in vertebrate samples. *J. Archaeol. Sci.* **9**: 49–62.

Fladmark, K. R. (1979). Routes: alternative migration corridors for early man in North America. *Am. Antiquity* **44**(1): 55–69.

Foley, R. (1982). A reconsideration of the role of predation on large mammals in tropical hunter-gatherer adaptation. *Man* **17**: 393–402.

Freedman, L. and Lofgren, M. (1979). The Cossack skull and a dihybrid origin of the Australian Aborigines. *Nature* **282**: 298–300.

Gamble, C. (1978). Resource exploitation and the spatial patterning of hunter-gatherers: a case study. *Brit. Archaeol. Rep.* **47**: 153–85.

Gamble, C. (1979). Hunting strategies in the central European Palaeolithic. *Proc. Prehist. Soc.* **45**: 35–52.

Gittleman, J. L. and Harvey, P. H. (1982). Carnivore homerange size, metabolic needs and ecology. *Behav. Ecol. Sociobiol.* **10**(1): 57–64.

Guggisberg, C. A. W. (1975). *Wild Cats of the World*. London: David and Charles.

Guthrie, R. D. (1968). Palaeoecology of the large-mammal community in interior Alaska during the late Pleistocene. *Am. Midland Naturalist* **79**: 346–63.

Harington, C. R. (1978). Quaternary vertebrate faunas of Canada and Alaska and their suggested chronological sequence. *Syllogeus* **15**: 105.

Haynes, G. (1983). Frequencies of spiral and green-bone fractures on ungulate limb bones in modern surface assemblages. *Am. Antiquity*. **48:** 102–14.

Hemmer, H. (1978). Considerations on sociality in fossil carnivores. *Carnivore* **1**: 105–7.
Heusser, L. E. and Shackleton, N. J. (1979). Direct marine-continental correlation: 150,000-year oxygen isotope-pollen record from the North Pacific. *Science* **204**: 837–9.
Hoffman, A. (1979). Community paleoecology as an epiphenomenal science. *Paleobiology* 5(4): 357–79.
Hopkins, D. M. (1967). The Cenozoic history of Beringia—a synthesis. In: D. M. Hopkins (ed). *The Bering Land Bridge*, Stanford: Stanford University Press, pp. 451–84.
Horowitz, A. (1979). *The Quaternary of Israel*. New York: Academic Press.
Howells, W. W. (1980). *Homo erectus*—who, when and where: a survey. *Yearbook Phys. Anthropol.* **23**: 1–23.
Isaac, G. Ll. (1975). Sorting out the muddle in the middle. In: K. W. Butzer and G. Ll. Isaac (eds). *After the Australopithecines*. The Hague: Mouton, pp. 875–87.
Isaac, G. Ll. (1978). Early man reviewed. *Nature* **273**: 588–9.
Jerison, H. J. (1973). *Evolution of the Brain and Intelligence*. New York: Academic Press.
Keast, A. (1972). Australian mammals: zoogeography and evolution. In A. Keast, F. C. Erk and B. Glass (eds). *Evolution, Mammals and Southern Continents*. New York: State University of New York Press, pp. 195–246.
Klein, R. G. (1975). The relevance of Old World archaeology to the first entry of man into the New World. *Quat. Res.* 5: 391–4.
Kruuk, H. (1972). *The Spotted Hyaena*. Chicago: University of Chicago Press.
Kruuk, H. (1975). *Hyaena*. London: Oxford University Press.
Kurtén, B. (1963). The cave hyaena, an essay in statistics. In: 1st edn, D. Brothwell and E. S. Higgs (eds). *Science in Archaeology*. London: Thames and Hudson, pp. 224–34.
Kurtén, B. (1968). *Pleistocene Mammals of Europe*. London: Weidenfeld and Nicolson.
Kurtén, B. (1969). Evolution at the population level: a statistical approach. In, 2nd edn., D. Brothwell and E. S. Higgs (eds). *Science in Archaeology*, London: Thames and Hudson, pp. 661–68.
Kurtén, B. (1971), *The Age of Mammals*. London: Weidenfeld and Nicolson.
Kurtén, B. (1973a). Transberingian relationships of *Ursus arctos* Linné (brown and grizzly bears). *Commentationes Biologicae* **65**: 10.
Kurtén, B. (1973b). Pleistocene jaguars in North America. *Commentationes Biologicae* **62**: 3–23.
Kurtén, B. (1980). Pleistocene mammal faunas of the Holarctic region. In L-K. Königsson (ed.) *Current Argument on Early Man*, Proceedings of a Nobel Symposium, May 21–27 1978, Karlskoga, Sweden. Oxford: Pergamon Press, pp. 145–51.
Kurtén, B. and Anderson, E. (1980). *Pleistocene Mammals of North America*. New York: Columbia University Press.
Kurtén, B. and Poulianos, A. N. (1977). New stratigraphic and faunal material from Petralona Cave with special reference to the carnivora. *Anthropos* **4** (1–2): 47–130.
Maglio, V. J. (1975). Pleistocene faunal evolution in Africa and Eurasia. In K. W. Butzer and G. Ll. Isaac (eds). *After the Australopithecines*. The Hague: Mouton, pp. 419–76.
Maglio, V. J. (1978). Patterns of faunal evolution. In V. J. Maglio and H. B. S. Cooke (eds). *Evolution of African Mammals*. Harvard: Harvard University Press, pp. 603–19.

Mech. L. D. (1970). *The Wolf*. New York: Natural History Press.

Mitchell, G. F., Penny, L. F., Shotton, F. W. and West, R. G. (1973). A correlation of Quaternary deposits in the British Isles. *Special Report of the Geological Society of London* **4**: 99.

Molleson, T. I. and Oakely, K. P. (1966). Relative antiquity of the 'Ubeidiya hominid. *Nature* **209**: 1268.

Paterson, H. E. H. (1980). A comment on 'Mate Recognition Systems'. *Evolution* **34**: 330–1.

Repenning, C. A. and Fejfar, O. (1982). Evidence for earlier date of 'Ubeidiya, Israel, hominid site. *Nature* **299**: 344–7.

Roe, D. A. (1981). *The Lower and Middle Palaeolithic Periods in Britain*. London: Routledge and Kegan Paul.

Sankhala, K. (1978). *Tiger*. London: Collins.

Schaller, G. B. (1967). *The Deer and the Tiger*. Chicago: University of Chicago Press.

Schaller, G. B. (1972). *The Serengeti Lion*. Chicago: University of Chicago Press.

Shackleton, N. J. (1975). The stratigraphic record of deep-sea cores and its implications for the assessment of glacials, interglacials, stadials and interstadials in the Mid-Pleistocene. In: K. W. Butzer and G. Ll. Isaac (eds). *After the Australopithecines*. The Hague: Mouton, pp. 1–24.

Speiss, A. E. (1979). *Reindeer and Caribou Hunters*. New York: Academic Press.

Stringer, C. B. (1980). The phylogenetic position of the Petralona cranium. *Anthropos* **7**: 81–93.

Stringer, C. B. (1981). The dating of European Middle Pleistocene hominids and the existence of *Homo erectus* in Europe. *Anthropologie* **19**(1): 3–14.

Stringer, C. B., Howell, F. C. and Melentis, J. K. (1979). The significance of the fossil hominid skull from Petralona, Greece. *J. Archaeol. Sci.* **6**: 235–53.

Stuart, A. J. (1974). Pleistocene history of the British vertebrate fauna. *Biol. Rev. Cambridge Philos. Soc.* **49**: 225–66.

Stuart, A. J. (1975). The vertebrate fauna of the type Cromerian. *Boreas* **4**: 63–76.

Sturdy, D. A. (1975). Some reindeer economies in prehistoric Europe. In E. S. Higgs (ed). *Palaeoeconomy*. Cambridge: Cambridge University Press, pp. 55–95.

Sutcliffe, A. J. (1970). Spotted hyaena: crusher, gnawer, digester and collector of bones. *Nature* **227**: 1110–3.

Thorne, A. (1980). The arrival of man in Australia. In A. Sherratt (ed.) *The Cambridge Encyclopedia of Archaeology*. Cambridge: Cambridge University Press, 96–100.

Tobias, P. V. (1966). Fossil hominid remains from 'Ubeidiya, Israel. *Nature* **211**: 130–3.

Turner, A. (1981a) Unpublished Ph.D. Thesis. University of Sheffield.

Turner, A. (1981b). Predation and palaeolithic man in northern England. In G. W. W. Barker (ed.), *Prehistoric Communities in Northern England*. Sheffield: University of Sheffield, Department of Prehistory and Archaeology, pp. 11–26.

Turner, A. (1981c). Ipswichian mammal faunas, cave deposits and hyaena activity. *Quat. Newsletter* **33**: 17–22.

Turner, A. (1981d). Ipswichian mammal faunas—a reply to Stringer and Currant. *Quat. Newsletter* **35**: 18–24.

Turner, A. (1983). The quantification of relative abundances in fossil and sub-fossil bone assemblages. *Ann. Transvaal Museum*, **33**: 311–12.

Vrba, E. S. (1980). Evolution, species and fossils: how does life evolve? *South Afr. J. Sci.* **76**, 61–84.

Vrba, E. S. (in press). Palaeoecology of early Hominidae, with special reference to

Sterkfontein, Swartkrans and Kromdraai. In *L'Environment des Hominidés*, Y. Coppens (ed.).

West, R. G. (1977). *Pleistocene Geology and Biology*, 2nd edn, London: Longmans.

White, J. P. and O'Connell, J. F. (1979). Australian prehistory: new aspects of antiquity. *Science* **203**: 21–8.

Wobst, H. M. (1974). Boundary conditions for palaeolithic social systems: a simulation approach. *Am. Antiquity* **39**(2): 147–78.

# 9 *Hunter-Gatherers and Large Mammals in Glacial Britain*

*Katharine Scott*

## 1. INTRODUCTION

Although man evidently ventured into western Europe at the end of the Lower Pleistocene, possibly as long ago as 0·9 million years (Butzer and Isaac, 1975; de Lumley, 1975 and see Chapter 8, this volume), it was to be another 600 000 years before he reached Britain, where the earliest plausible evidence for hominid presence, at Westbury-sub-Mendip, is believed to date to *ca.* 300 000 years ago (Bishop, 1975; Wymer, 1977).

No doubt, such a comparatively late arrival in Britain is partly due to the geographical situation of these islands in that they were always the north-west limit of possible hominid occupation. However, environmental conditions, and climate in particular, probably represented the greatest limiting factor to hominid expansion into this region.

Europe underwent numerous climatic and biological changes during the Pleistocene, so that although much of hominid occupation of the western continent in the past 900 000 years or so would have been in temperate interglacial or interstadial conditions, during glacial episodes the northern regions marginal to the ice sheets would have been affected by extremely cold, even arctic, conditions for considerable periods. Apparently unable to endure the environmental conditions of the periglacial zones, hominids seem to have remained in the central and southern regions for the earlier part of their occupation of Europe, where they would have been less directly affected by the growth of continental ice sheets and accompanying deterioration of the climate and environment. Even after their arrival in the northwestern regions some 300 000 years ago, the relative scarcity of archaeological material until the second half of the last (Devensian) cold stage (see Table 9.1), suggests that occupation was sporadic.

The fact that human activity in Britain becomes more archaeologically visible in the latter half of the Devensian is interesting for a number of reasons.

HOMINID EVOLUTION
ISBN 0-12-261920-X

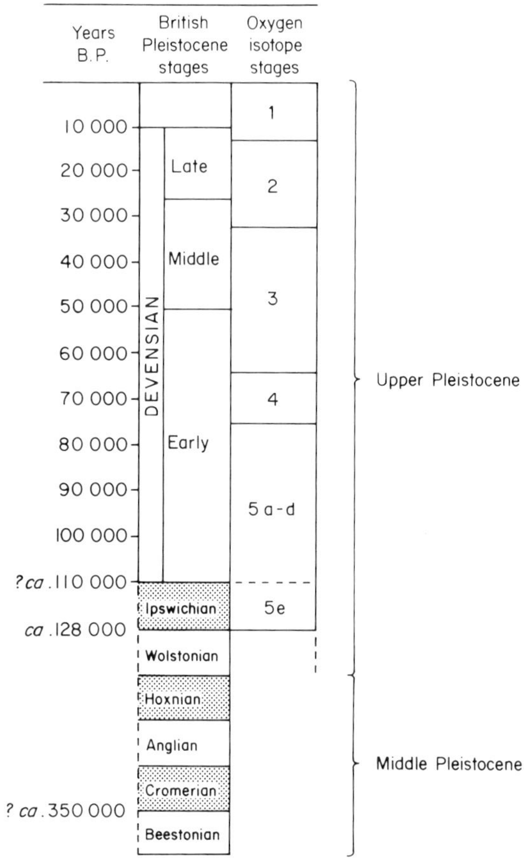


Table 9.1. British Middle and Upper Pleistocene stages. The interglacial stages are shaded. Note that the chronological scale applies to the Devensian only (after Shotton 1977; Stuart 1982). The Ipswichian and Devensian are correlated with oxygen isotope stages 1–5 on the basis of the estimated ages of the stage boundaries in core V28–238 (data from Shackleton, 1975).

First, as mean annual temperatures are estimated to have been several degrees centrigrade below zero (Watson, 1977; see also Table 9.2), hominids must have adapted to, or developed protection against, the cold. Secondly, as more biological and climatic data is available for the Middle and Late Devensian than is available for any preceding cold stage, it is possible to examine the environmental constraints operating on human adaptation in relation to the rest of the biological community. Thirdly, this period has the added advantage of falling within the range of radiocarbon dating, which in turn makes it

Table 9.2. Estimated Devensian temperatures (°C) based on permafrost features (from Watson, 1977).

| Years B.P. | Mean air temperatures annual | Mean air temperatures July | Fall from present annual | Fall from present July |
|---|---|---|---|---|
| 10 500 | −5 | 10 | 15 | 5 |
| 11 500 | −4 | 12/13 | 14 | 3 |
| 13 000 | −7 | 10 | 17 | 5 |
| 18 000 | −15 | 5 | 25 | 10 |
| 55 000 | −7 | 11 | 17 | 4 |

possible that the movements of animals (including man) in response to changing environmental conditions might be defined more precisely.

This paper thus considers the evidence for hominid occupation of Britain against the ecological background of the last (Devensian) cold stage. It discusses the possibility that man was sufficiently adapted, by the Late Pleistocene at least, to withstand winter conditions in this region. Finally, it examines the resource base of man in Britain during the Devensian, and his consequent adaptation.

## 2. HOMINID OCCUPATION OF BRITAIN

The fact that, right until the late Upper Pleistocene, there is comparatively little evidence of hominid presence in Britain may be attributed principally to the geographical situation of these islands. Not only would they have been the farthest regions for colonization from the south and east, but they would have been markedly affected by the numerous fluctuations of climate and sea-level that characterize the Pleistocene.

During glacial periods of low sea-level, Britain could have been approached from the continental mainland. Simultaneously, however, it would have been subjected to the southerly advances of the ice sheets, so that habitable areas would be confined to the central and southern regions, depending on the extent of glacial oscillation. With the retreat of the ice and the onset of warmer, interglacial conditions, when Britain would have provided a thermally congenial environment in which to live, any further population expansion from the south and east would have been prevented by the interglacial rise in sea-level and consequent separation of the British Isles from the continental mainland (Fig. 9.1). Yet despite the fact that Britain was often peripheral to ice sheets, and at times inaccessible, artifacts from dateable horizons indicate that sporadic colonization continued (Wymer, 1981; Roe, 1981).

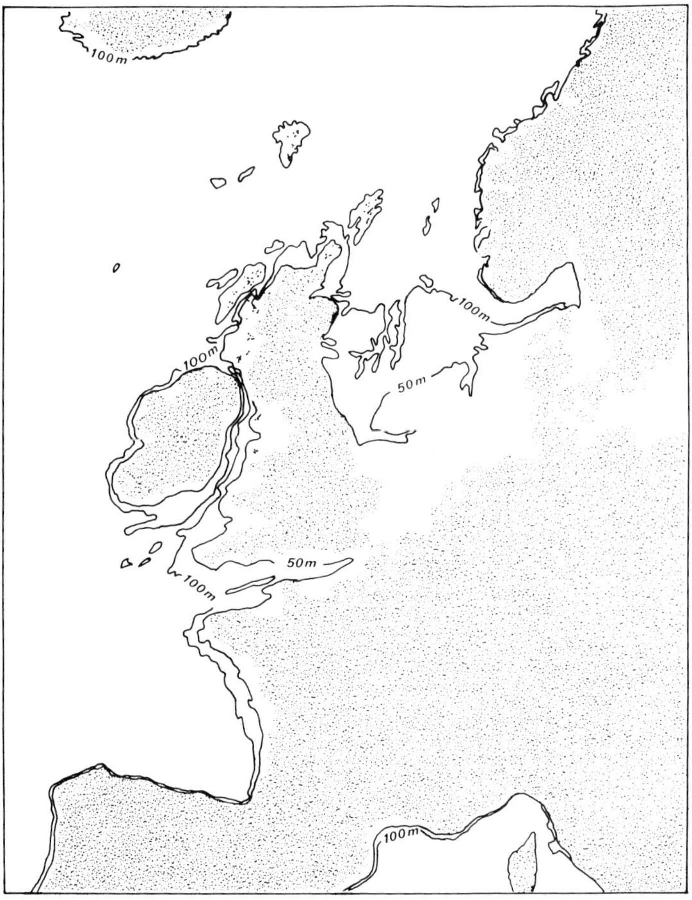


Fig. 9.1. Britain in relation to the north-west of continental Europe (from Roe, 1981). During glacial episodes a lowering of the sea-level would have connected the British Isles to the continental mainland to a greater or lesser extent as indicated by the modern marine contours. The 100 m contour gives an idea of the extent of the north-west European peninsula during the Devensian (Weichselian) Cold Stage.

The artifacts from Westbury-sub-Mendip (possibly of Cromerian age) and others from Hoxnian and Ipswichian deposits testify to hominid adaptation to interglacial conditions in Britain, but the evidence from the intervening cold stages is far less abundant. There are no archaeological finds positively attributable to the Anglian Cold Stage, and few from the Wolstonian Cold Stage. Of course, it is possible that what evidence there might have been of hominid presence during the Anglian was destroyed by the more extreme southerly advance of the Wolstonian ice sheets, but it may also be possible that hominids scarcely ever ventured so far to the north-west during these glacial episodes. In fact, even in the earlier part of the Devensian, the evidence for man's presence is remarkably scant, and consists chiefly of artifacts with Mousterian affinities from poorly-stratified or poorly-dated contexts. Human settlement becomes more apparent in the early part of the Middle Devensian (after about 50 000 B.P.) in the form of the industries of the so-called "Earlier Upper Palaeolithic", and even more so in the Late Devensian in the form of the "Later Upper Palaeolithic", as defined by Campbell (1977) and Jacobi (1980). By comparison with the rest of western Europe, the quantity of archaeological material representing hominid activity in the Devensian is small, but its presence at least is evidence of human adaptation to a generally very cold environment.

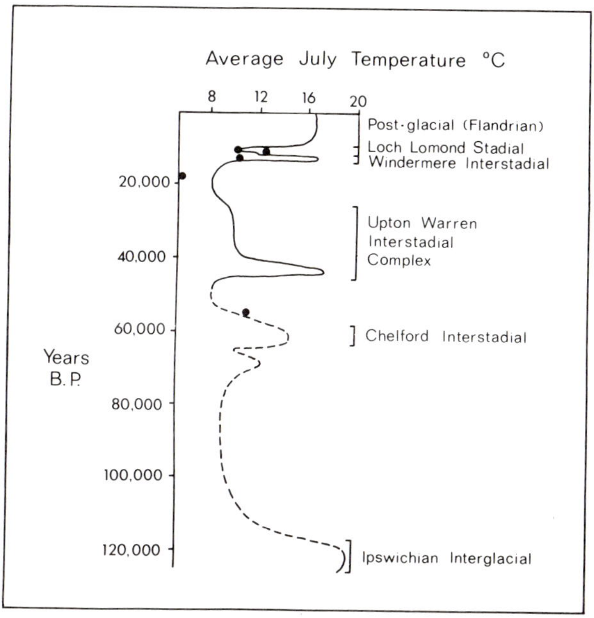


Fig. 9.2. Variations in the average July temperatures in lowland areas of the southern and central British Isles since the Last (Ipswichian) Interglacial. The temperature curve is based on the interpretation of fossil beetle faunas (from Coope, 1977). The black dots (•) are estimated July temperatures based on permafrost features (data from Watson, 1977).

## 3. THE ENVIRONMENT OF THE MIDDLE AND LATE DEVENSIAN

The palaeobiological evidence leaves little doubt that the climate in the early part of the Middle Devensian (*ca.* 50 000 B.P.) was of arctic severity. These extremely cold conditions were interrupted by the sudden climatic amelioration of the Upton Warren Interstadial, the thermal maximum being attained between about 44 000 and 42 000 B.P. It is suggested, on the basis of fossil beetle assemblages, that summer temperatures at this time were higher than those of the present day (Fig. 9.2; Coope, 1977).

After this thermal maximum, climatic conditions began to deteriorate, and for the rest of the period included in the Upton Warren Interstadial Complex (until about 25 000 B.P.), the climate was generally very cold indeed. Evidence from both ice wedges and the fauna and flora indicate a permafrost environment, fluctuating between tundra and subpolar desert conditions. The landscape was treeless and fairly arid, the prevailing vegetation being dominated by plants of subarctic or montane distribution today, such as *Betula* (birch), *Salix* (willow), and various weeds, ruderals and grasses, indicative of disturbed and open ground (Godwin, 1975; West, 1977; Watson, 1977). The vertebrate fauna is represented by species with a present day circumpolar distribution, adapted to tundra and boreal forest environments—for example, arctic fox, reindeer, arctic lemming. The prevalence of open grassland is indicated by a number of large grazers such as the horse, bison, and the extinct mammoth and woolly rhinoceros, both found principally in cold stage deposits (Stuart, 1977, 1982). Fossil beetle assemblages indicate that average July temperatures might have been at or below 10°C, with average winter temperatures below −20°C, somewhat similar temperatures to those experienced near the forest limit of the Siberian arctic today (Coope, 1977).

Certainly man was in Britain during this period. The artifactual material shares many characteristics of the French and Belgian lithic assemblages of this age, although the British material lacks the stratigraphic continuity evident on the continental mainland (Jacobi, 1980). This suggests sporadic rather than long-term occupation, but the archaeological evidence is insufficient as yet to allow more than speculation on this point. Even the evidence of subsistence in the form of animal remains is uncertain. The "Earlier Upper Palaeolithic" artifacts recovered from open context are not associated with faunal remains, and those recovered from caves are associated with bones that have evidently derived from a number of sources: some of these may represent the prey of man, but others are undoubtedly the result of carnivore activity, particularly that of the spotted hyaena *Crocuta crocuta* (Scott, in prep.; see Chapter 10, this volume).

The visits of hominids to Britain apparently ceased with the further climatic deterioration preceding the Devensian glacial maximum (*ca.* 18 000–15 000 B.P.). What is interesting is that a number of animals apparently survived in Britain as late as 18 000 B.P. (Shotton and Williams, 1973; Stuart, 1982) whereas man did not. In fact, there are no radiocarbon dates convincingly associated with evidence for human presence in the late Middle Devensian, and Jacobi (1980) has commented on the absence in Britain of any artifact types characteristic of those in northwest Europe at this time that may confidently be assigned to the period between about 27 000 and 22 000 B.P. As he points out, the culturally undiagnostic artifacts from Caldey Island that are supposed to indicate that man was indeed in the British Isles as late as *ca.* 22 000 B.P., cannot be demonstrated to be contemporary with the woolly rhinoceros bone from which the date is derived.

During the main glaciation of the Devensian (*ca.* 18 000–15 000 B.P.) there is no evidence that man, or indeed any large mammals, occupied the British Isles. There are no plant records from the ice maximum, and the evidence from permafrost features suggests an arid, continental climate. Full polar desert conditions probably existed in most of England south of the ice sheets, with mean annual temperatures in the order of $-16°$ to $-17°$C (Coope, 1977; Watson, 1977; Pennington, 1977).

As man appears to have disappeared from Britain considerably in advance of the rest of the mammalian fauna at the onset of the climatic extremes of the Devensian ice maximum, so he appears to have taken longer to return. Radiocarbon dates based on bone collagen of reindeer (BM-729), suggest that this animal had returned to Britain by about 15 000 B.P., although dates for other large mammals such as horse and giant deer indicate later re-entry. Several species evidently did not return—such as the mammoth, woolly rhinoceros, and spotted hyaena—and may even have become extinct in western Europe by this time (Stuart, 1982).

The earliest convincing evidence of man's reappearance in these islands is some time between about 13 500 and 12 000 B.P. (Jacobi, 1981), a period for which the palaeobiological evidence implies a climate similar to the present day. After 12 000 B.P., the climate deteriorated gradually and irregularly until the return of the arctic conditions of the Loch Lomond Stadial between about 11 000 and 10 000 B.P. This period provides an interesting point at which to discuss hominid adaptation to the tundra environment, not only because the palaeoenvironmental evidence is good, but because there is some interesting and detailed evidence available on man's presence in Britain at this time.

## 4. HOMINID ADAPTATION TO ARCTIC CONDITIONS IN BRITAIN DURING THE LOCH LOMOND STADIAL

The palaeobiological evidence for the period encompassed by the Loch

Lomond Stadial implies an environment similar to that of present day arctic tundra regions. The comparison is valid in that tundra regions have many of the cold-adapted plant and animal species common to the Loch Lomond Stadial and, by inference from permafrost features and fossil beetle assemblages, similar annual temperatures (Watson, 1977; Coope, 1977). A problem is encountered, however, when comparisons are drawn between present day human adaptation to periglacial regions (such as parts of Alaska and northern Siberia) and human adaptation to the conditions in periglacial regions of Late Glacial Europe. The contention expressed here, that modern ethnographic observation is of limited value in the interpretation of prehistoric adaptation, is based on a consideration of the *differences* as well as similarities in the two environments. These differences would have had a marked effect on the food resources potentially available to prehistoric man, and would surely have resulted in adaptive strategies different from those observed today.

The palaeobiological evidence makes one thing quite clear: that there is no modern equivalent of the Pleistocene tundra. One reason why periglacial regions differ now from those of the Pleistocene concerns the complex interrelationship of fauna and flora. Most significantly, species diversity of both plants and animals is considerably reduced in tundra regions today. During the Pleistocene, a variety of large ungulates is consistently represented in faunas from periglacial deposits right across northern Europe, the Ukraine, Siberia, Beringia and south-central Asia. These include the members of the so-called "Mammoth Complex": mammoth, woolly rhinoceros, horse, reindeer and bison (Matthews, 1979). Today the only large ungulates are reindeer and musk ox, but even if the others were not extinct in circumpolar regions today, the fact is that modern tundra does not have a high enough level of productivity to support any additional species (Bliss *et al*, 1975). However, as Matthews (1979) points out, the character of modern tundra—its moss-dominated, poorly drained substrates and abundance of heaths—may be partly due to the absence of certain key ungulate species. Grazing, trampling, and manuring are known to increase productivity and nutrient cycling in modern tundra, and these feedback loops must have been enhanced when the ungulate community contained the greater variety of grazers present in the Pleistocene periglacial environment. The pollen spectra for the Late Devensian show a great diversity of vegetation with region. There were communities of dwarf shrubs such as birch and, in particular, arctic willow, and there is ample evidence for fen and marsh communities associated with streams and shallow pools, but in general, the highest values (between 80 and 95%) are for non-arboreal pollen (West, 1977). The Late Devensian landscape was characterized by herb vegetation dominated by grasses, and adequate to support a much higher diversity of ungulates than is presently the case in tundra regions.

Temperatures throughout Britain during the Loch Lomond Stadial appear to have been similar to those of the arctic tundra today. On the basis of permafrost features and fossil beetle assemblages, average summer temperatures are estimated at 10°C, only a few degrees lower than those of the present day in Britain. However, winter temperatures might have been as low as −20°C, some 10–15 degrees lower than at present (Coope, 1977; Watson, 1977).

The particular geographical situation of the British Isles in relation to the Continental mainland and to the Atlantic Ocean would have had an effect on the Devensian climate and environment. As Lockwood (1979) points out, the fact that Britain has a more southerly latitude than present-day tundra areas means that glacial tundra climates would have differed in a number of important aspects from existing tundra climates. The great expansion westwards of the Siberian permafrost, accompanied by an increase in continentality, would have resulted in cold, dry winters. Ice-cap associated high pressure would have been stronger during winter, making winter precipitation generally light. This winter high pressure, combined with a considerable drop in surface temperature of the North Atlantic during cold stages, would have resulted in most precipitation falling as rain during moderately cool summers (Bonatti, 1966; West, 1977; Lockwood, 1979). The presence of ice wedges in Britain also indicate that winter snowfall was lighter during the Devensian than in circumpolar regions today; ice wedges will not crack, and therefore will not grow if the snow cover is thick (Williams, 1975). The widespread occurrence of *Artemesia* in the Devensian pollen spectra implies that snow cover was localized to upland areas of the north and west, where a more snow-tolerant vegetation predominated (Godwin, 1975; Pennington, 1977).

The likelihood that winter snowfall was considerably lighter than in present-day tundra regions raises some particularly relevant points with respect to Pleistocene hominid—or indeed any other mammal—adaptation. Animals in circumpolar regions today adapt to cold, and particularly to the extremes of winter, in a variety of ways (Batzli *et al.*, 1981). Some combat winter through a lowering of basal metabolic rate, thus reducing their energy requirements, food intake and water loss. Some hibernate, and others, such as the arctic lemming, continue to live in passages dug deep below the snow. Most store fat reserves during the summer months on which they draw during the winter, and all increase pelage. Either the animal must adapt through physiological and behavioural mechanisms, or it must migrate.

Particularly well-known are the migrations of the reindeer-caribou. These animals sometimes cover hundreds of miles per annum in search of seasonal pastures and breeding grounds, and it is commonly supposed that reindeer made similar seasonal migrations in Europe in the Pleistocene, moving from the periglacial northern regions at the end of the summer to central and southern Europe for the duration of the winter. It is also frequently suggested that prehistoric man made such migrations in pursuit of this prey. Indeed, the archaeological evidence from hundreds of sites in northwest Europe indicates that reindeer and hunters were in particular regions at certain times of the year (Sturdy, 1975). However, it may be that a model in which humans follow reindeer over vast distances on a regular annual basis is too simplistic. Reindeer migration is far more complex than is generally realized. Although reindeer are on the move year round in search of particular vegetation, shelter and breeding grounds, migrations are not necessarily over the same routes from year to year. In fact, recent studies show that particular routes and

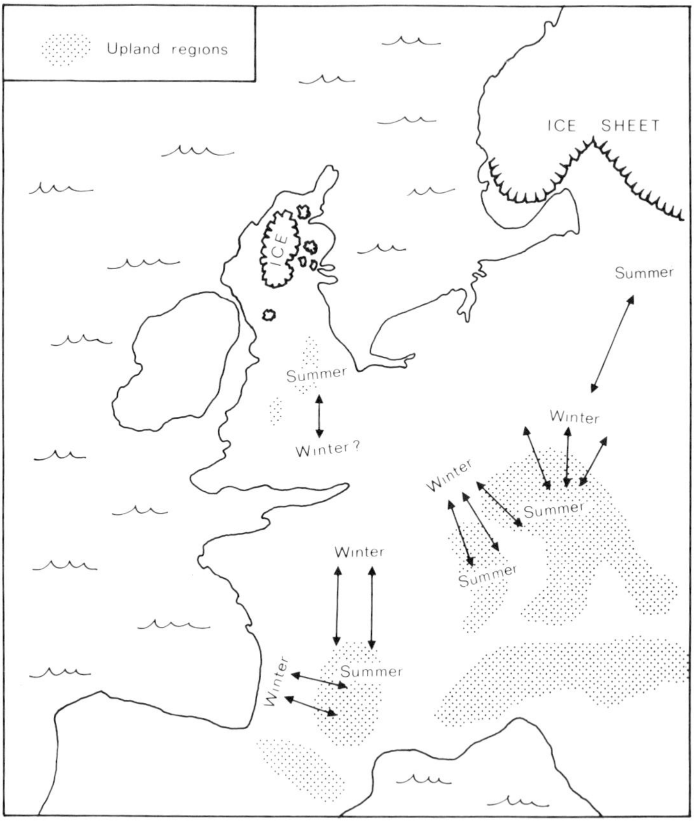


Fig. 9.3. Suggested reindeer seasonal movement in north-west Europe based on Late Glacial archaeological sites (data from Sturdy, 1975; Scott, in prep.). Arrows indicate general directions rather than actual routes.

summer and winter grounds may only be used three or four years out of every ten (White, 1981). Furthermore, although some herds do migrate vast distances, it is not the distance travelled that is important to the reindeer, but the location of certain requirements. Among some herds, if the right conditions are within close proximity, migrations are comparatively limited (Banfield, 1974). In terms of the possible dependance of prehistoric hunters on reindeer as a source of food, it needs to be stressed that reindeer movement is unpredictable, and governed from year to year by local conditions such as the prevalence of biting flies and mosquitoes, wind and weather conditions, but most especially on the depth and condition of the snow. This is particularly relevant to a consideration of possible migration of reindeer to and from Britain during the Devensian. If, as the palaeoclimatic evidence suggests, there was considerably less snow than in the arctic tundra today, then it might not have been necessary for large herbivores to leave Britain during the winter for regions far to the south.

To what extent large mammals such as mammoth, bison, horse or giant deer, migrated seasonally during the Devensian is not known, but it is clear that some reindeer at least were in Britain during the winter months. For example, in deposits of early Devensian age, at Wretton in Norfolk, in a fauna including bison, reindeer and mammoth, there are male reindeer shed antlers (Stuart, 1982). As males shed their antlers in winter, these obviously indicate that some reindeer had spent the winter at Wretton. Shed male antlers are also present in Middle Devensian deposits at Tattershall Castle, Lincolnshire, and Isleworth, Middlesex, again indicating the winter presence of this species (Stuart, 1982). However, it is also interesting that carrion and dung beetles recovered from the same sites indicate that large herbivores were in those regions in the summer months as well (Coope and Angus, 1975; Coope, 1977). Arguably, these examples illustrate large mammal adaptation to winter conditions that might not have been as extreme as those of the Loch Lomond Stadial. However, there is evidence from a small site in the Peak District of central England that not only animals, but hominids, were in the region in the late winter of this extremely cold climatic episode. Ossom's Cave, a small shelter in the Manifold Valley, appears to represent a short-term occupation for the purpose of butchering several reindeer. These reindeer remains are associated with a number of artifacts, a small amount of charcoal, and micromammals including arctic and Norway lemmings (Bramwell *et al.*, in prep.). The fauna, and a radiocarbon date on reindeer antler of *ca.* 10 590 B.P., place this occupation or visit well within the Loch Lomond Stadial (pollen zone De III). What is particularly interesting is that there are virtually complete mandibles and maxillae of several young reindeer from which it is possible to estimate the time of year at which the region was visited.

It is common practice among archaeologists to use dental eruption and wear to determine the age and season of death of animals. This is especially facilitated if the species concerned has a restricted season of birth. In most reindeer herds 95% of all animals are born within a two-week period, mid-May being the mean birth date. Speiss (1979) found a very close correlation between tooth eruption and age in hundreds of reindeer-caribou from several populations in Canada. Among a mid-May calving population, the eruption of the second molar (M2) begins around mid-March to mid-May. A group of individuals with M2 one quarter or less erupted is placed at 10–11 months of age—that is to say, mid-March to mid-April. All the young reindeer from Ossom's Cave are at exactly this stage of dental eruption (Fig. 9.4). In addition, there are several very young antlers—the first year spikes—still attached to skulls. In both males and females these are shed in late April and May, which further suggests a March/April (or, at any rate, early Spring) killing of the Ossom's Cave reindeer (Scott, in prep.).

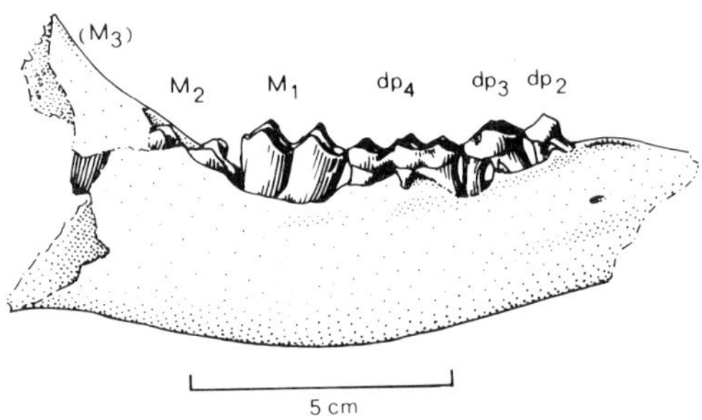


Fig. 9.4. Right mandible of reindeer from Ossom's Cave (Specimen Number O.V.C./4′8″) showing well-worn deciduous premolars, first molar in wear, and second molar erupting. Part of the unerupted third molar is visible.

Whether the hunters followed the reindeer to this upland region in the Spring, or waited for them to pass through the valley on annual migration, is debatable. Reindeer are constantly on the move, also travelling at night, which suggests that it is unlikely that hunter-gatherers (including young and old) kept up with them over great distances. Given the relatively slow speed at which humans move on foot (which presumably they were), it also seems unlikely that a group that had reached central England by the early Spring could have spent the winter very far away. However, irrespective of where this group might have spent the winter, it is interesting in terms of hominid adaptation to the periglacial environment that they had reached central England by the late Winter/early Spring. This would have necessitated travelling across, or remaining in, a region devoid of natural shelter in the form of caves in the extreme cold of winter.

In all probability, the storage of subcutaneous fat and an increase in hair cover aided survival in such conditions, and no doubt, the control of fire was vital to hominids in the periglacial regions of north-west Europe. Another possibility of course is that with the increase in technological skill evident by the Late Pleistocene, hominids were able to protect themselves by constructing shelter. However, apart from such cultural adaptations, it is evident from extensive studies made on groups of primitive people that live naked or poorly clothed in recurrently very cold conditions, that man behaves like other species by making metabolic changes to combat the effects of cold. The aborigines of central Australia and the san of the Kalahari Desert, for example, sleep with little or no clothing at temperatures of 0°C or below, which they are able to do, not by producing more heat as people unused to enduring the cold are observed to do, but by a lowering of the metabolic rate and body temperature, and by conserving heat through a reduction in shivering (Wyndham, 1970; LeBlanc, 1975).

Thus far this paper has dwelt on hominid adaptation to cold, which was a major but, of course, not the only factor affecting his survival. The availability of food would have controlled his movements equally as rigorously. A variety of large and small mammals was abundantly available in north-west Europe during the cold stages, so that of hunting and/or scavenging we can be sure. Gathering, it seems, could not have played a significant part in hominid adaptation for the simple reason that there was apparently scarcely anything available in the periglacial tundra that the human body could process for energy.

Various botanists have described the vegetation of periglacial regions and the adaptations that plants make in order to survive in certain climatic conditions (see Godwin, 1975; Bliss *et al.*, 1981; Pruitt, 1978; Grime, 1979). If a plant is to be edible to a hunter-gatherer it must store nutrients in the form of berries, bulbs, etc., the storage of which material above and below the ground is limited by certain external factors. These are classified by Grime (1979) into two categories: (1) *stress*, which includes such factors as sub-optimal temperatures, or the shortage of light, mineral nutrients or water, and (2) *disturbance*, such as brought about by trampling, mowing, ploughing or high winds.

The strategies evolved by plants to cope with these factors fall into three extremes of evolutionary specialization, best summarized with reference to Fig. 9.5. In this figure, Grime (1979) presents an equilateral triangle in which variation in the relative importance of competition, stress and disturbance as determinants of the vegetation is indicated by three sets of contours. At their respective corners of the triangle, competitors (C), stress-tolerators (S), and ruderals (R) become the exclusive constituent of the vegetation:

(1) *The competitors*, subject to low stress and low disturbance, have high competitive ability which enables them to acquire *and store* resources (my italics). They are characterized by productive, crowded vegetation above and

below ground.
(2) *The stress-tolerators* (stress in this context being principally environmental —extreme cold, short summers) adapt by slow growth and reduced stature, longevity, infrequent flowering, and the maintenance of the evergreen habit, a mechanism which obviates the necessity to spend food resources on a wholly new photosynthetic apparatus every year.
(3) *The ruderals* are subject to sufficient disturbance to cause their partial or total destruction regularly. Such disturbance includes many activities such as ploughing, flooding, high winds, extreme climatic fluctuation, and the activities of herbivores.

These are the extreme consequences of three types of response, between which are a number of strategies that have evolved in habitats experiencing intermediate intensities of competition, stress and disturbance. On this model, Dr. J. Birks (University of Cambridge), has plotted the range of strategies encompassed by floras from glacial and interglacial periods. Those plants with storage organs—corms, bulbs, roots—fall within the Competitor zone. It is this category of plants that comprises the plant foods of modern hunter-gatherers. The plants of the Devensian, however, fall within the range of strategies (along the base of the triangle, Fig. 9.5) where storage is not a viable adaptation. Birks suggests therefore (pers. comm.), that apart from a few unpalatable berries, little plant food would have been available for human consumption in the cold, disturbed periglacial environment of north-west Europe.

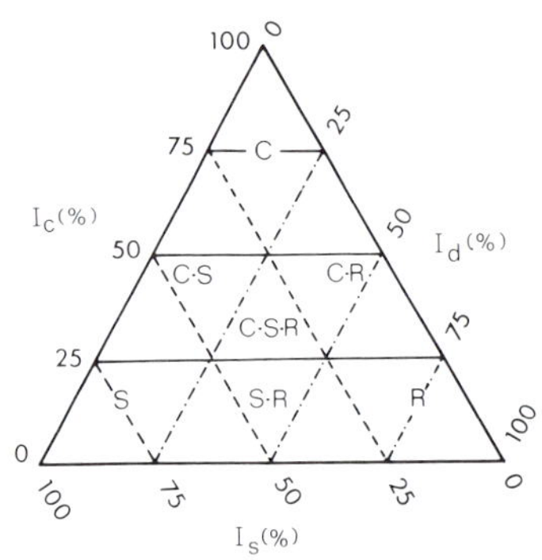


Fig. 9.5. Model describing the various equilibria between competition, stress, and disturbance in vegetation and the location of primary and secondary strategies. $I_c$ = relative importance of competition (———); $I_s$ = relative importance of stress (– – – –); $I_d$ = relative importance of disturbance (. — . — . — .). The symbols for the strategies are described in the text. Redrawn with permission from Grime (1979).

## 5. CONCLUSIONS

Despite the harsh environment, it is clear from the archaeological evidence that hominid presence in Britain increased during the Devensian, particularly after the glacial maximum. However, the detailed reconstruction of the pattern of human presence in this region is unlikely ever to be achieved. Principally this is because the depth of stratified archaeological deposits, so common in the caves of western Europe, is lacking in Britain. One reason for this is probably that hominid occupation of this north-west outpost was in any case intermittent in response to climatic fluctuation. Another almost certainly is that most research into palaeolithic activity has focused on deposits from caves. In general, these are in the central and south-western regions of the country, and relatively few are more than shallow or narrow shelters. Few are spacious enough to provide shelter for a large group, and, as ethnographic observation suggests, where hunting is an important strategy, the band is necessarily fairly large. Thus the fact that British caves are generally restricted, both in size and location, perhaps adds an important dimension to our understanding of human adaptation in the Late Pleistocene: the western regions were being reached in extremely cold, even winter, conditions by traversing the south and east—a vast area devoid of natural shelter in the form of caves. Settlements or camps made en route have little chance of coming to light considering the extensive trampling, ploughing, cultivation and building of the past few thousand years. Nevertheless, even without these, it is evident that a milestone had been reached in hominid evolutionary or technological development, or both. Almost certainly man had made physiological adaptations, but above all, it must have been an increase in technological skills—the control of fire and the construction of clothing and shelter—that enabled him, by modifying the effects of his environment, to endure the rigours of life in periglacial Britain.

## REFERENCES

Banfield, A. W. F. (1974). *The Mammals of Canada*. Toronto: Toronto University Press.

Batzli, G. O., White, R. G. and Bunnel, F. L. (1981). Herbivory: a strategy of tundra consumers. In: L. C. Bliss, O. W. Heal and J. J. Moore, (eds). *Tundra Ecosystems*. Cambridge: Cambridge University Press, pp. 359–376.

Bishop, M. J. (1975). Earliest record of man's presence in Britain. *Nature* **253**: 95–7.

Bliss, L. C., Heal, O. W. and Moore, J. J. (eds) (1981). *Tundra Ecosystems*. Cambridge: Cambridge University Press.

Bonatti, E. (1966). North Mediterranean climate during the last Würm glaciation. *Nature* **209**: 984–5.

Bramwell, D., Jacobi, R. M., Scott, K. and Stuart, A. J. (in prep.). Ossom's Cave: a Late Glacial reindeer butchery site in Staffordshire, Great Britain.

Butzer, K. and Isaac, G. L. (eds) (1975). *After the Australopithecines*. Hague: Mouton.
Campbell, J. 1977. *The Upper Palaeolithic of Britain. A Study of Man and Nature in the Late Ice Age*. (2 vols). Oxford: Clarendon Press.
Coope, G. R. (1977). Fossil coleopteran assemblages as sensitive indicators of climatic changes during the Devensian (Last) cold stage. *Phil. Trans. Roy. Soc. Lond. B*. **280**: 313–40.
Coope, G. R. and Angus, R. B. (1975). An ecological study of a temperate interlude in the middle of the last glaciation, based on fossil Coleoptera from Isleworth, Middlesex. *J. Anim. Ecol.* **44**: 365–91.
Godwin, H. (1975). *The History of the British Flora* (2nd ed.). Cambridge: Cambridge University Press.
Grime, J. P. 1979. *Plant Strategies and Vegetation Processes*. Chichester: Wiley.
Jacobi, R. M. (1980). The Upper Palaeolithic in Britain, with special reference to Wales. In: J. A. Taylor (ed.) *Culture and Environment in Prehistoric Wales*. Oxford: British Archaeological Reports, No. 76 pp. 15–100.
Jacobi, R. M. (1981). The Late Weichselian peopling of Britain and North-West Europe. *Archaeologia Interregionalis* Vol I (Krakow).
LeBlanc, J. (1975). *Man in the Cold*. American Lecture Series Publication No. 986. Springfield: Thomas.
Lockwood, J. G. (1979). Water balance of Britain, 50,000 yr. B.P. to the present day. *Quat. Res.* **12**: 297–310.
Lumley, H. de (1975). Cultural evolution in France and its palaeoecological setting. In: K. Butzer and G. Isaac (eds). *After the Australopithecines*. Hague: Mouton, pp. 745–808.
Matthews, J. V. (1982). Arctic steppe – an extenct biome. In: D. M. Hopkins, J. V. Matthews Jr., C. E. Schweger and S. B. Young (eds). Palaeoecology of Beringia. New York: Academic Press.
Pennington, W. (1977). The Late Devensian flora and vegetation of Britain. *Phil. Trans. Roy. Soc. Lond. B*. **280**: 247–271.
Pruitt, W. O. Jnr. (1978). *Boreal Ecology*. Institute of Biology Studies in Biology 91. London: Arnold.
Roe, D. A. (1981). The Lower and Middle Palaeolithic Periods in Britain. London: Routledge and Kegan Paul.
Scott, K. (in prep.). Hominids and other predators in Britain during the last (Devensian) Cold stage.
Shackleton, N. J. (1975). The stratigraphic record of deep sea cores. In: K. Butzer and G. Isaac (eds). *After the Australopithecines*. Hague: Mouton, pp. 1–24.
Shackleton, N. J. (1977). The oxygen isotope stratigraphic record of the Late Pleistocene. *Phil Trans. Roy. Soc. Lond. B*. **280**: 169–182.
Shotton, F. W. (1977). Chronology, climate and marine record. The Devensian stage: its development, limits and substages. *Phil. Trans. R. Soc. Lond. B*. **280**: 107–118.
Shotton, F. W. and Williams, R. E. G. (1973). Birmingham University radiocarbon dates VII. *Radiocarbon* **15**: 451–68.
Spiess, A. E. (1979). *Reindeer and Caribou Hunters: an Archaeological Study*. New York: Academic Press.
Starkel, L. (1977). Palaeogeography of Europe. *Phil. Trans. Roy. Soc. Lond. B*. **280**: 351–372.
Stuart, A. J. (1982). *Pleistocene Vertebrates in the British Isles*. London: Longman.
Sturdy, D. A. (1975). Some reindeer economies in prehistoric Europe. In: E. S. Higgs (ed). *Palaeoeconomy*. Cambridge: Cambridge University Press, pp. 55–95.

Van der Hammen, T. Wijmstra, T. A. and Zagwijn, W. H. (1971). The floral record of the Late Cenozoic of Europe. In: K. K. Turekian (ed). *Late Cenozoic Glacial Ages*. New Haven: Yale University Press, pp. 391–424.

Watson, E. (1977). The postglacial environment of Great Britain during the Devensian. *Phil. Trans. Roy. Soc. Lond. B*. **280**: 183–198.

West, R. G. (1977). Early and Middle Devensian flora and vegetation. *Phil. Trans. Roy. Soc. Lond. Ser. B*. **280**: 229–46.

White, R. G. (1981). Ungulates on Arctic ranges. In: L. C. Bliss, O. W. Heal and J. J. Moore (eds). *Tundra Ecosystems*. Cambridge: Cambridge University Press.

Williams, R. G. B. (1975). The British climate during the last glaciation: an interpretation based on periglacial phenomena. In: A. E. Wright and F. Mosely (eds). *Ice Ages: Ancient and Modern*. Geological Journal Special Issue 6, pp. 95–120.

Wymer, J. J. (1977). The archaeology of man in the British Quaternary. In: F. W. Shotton (ed.) *British Quaternary Studies: Recent Advances*. Oxford: Clarendon.

Wymer, J. J. (1981). *The Palaeolithic Age*. London: Croom Helm.

Wyndham, C. H. (1970). Adaptation to heat and cold. In: D. H. K. Lee and D. Mynard (eds). *Physiology, Environment and Man*. New York: Academic Press, pp. 177–204.

# 10 *Regional Variation in Hunter-Gatherer Strategy in the Upper Pleistocene of Europe*

*Clive Gamble*

## 1. INTRODUCTION

An archaeological discovery is often proof enough that a cultural system functioned at a level that ensured long-term survival. We do not need a complicated methodology to infer from the recovery of flints, bones and dwelling structures that palaeolithic man had adequately solved the survival problems connected with continuing existence. If man was present in an area then he must have been eating and reproducing! To go beyond this simple level of recovery as evidence of existence we must confront the problem of accounting for variation within and between cultural systems. Earlier work has largely sidestepped this issue and concentrated instead upon detailing and comparing the contents of past cultural systems in a qualitative manner. However, since the *survivors* are marked by their own paleolithic record it becomes difficult to see how we are to understand any variation in these records, since they are all representative, by their own lights at least, of successful adaptations. In other words, why should variability exist in the palaeolithic record, and can it be accounted for in terms of alternative strategies adapted to comparable environments and the resources they contain?

Most approaches accounting for variability in the patterns of archaeological materials are based on judgements about the adaptive significance of changes in those patterns. For example according to Klein (1973, pp 122–123)

> In the long term view of culture history, the upper palaeolithic appears to constitute a quantum advance over the Mouterian. This advance was probably manifest not only in material culture and technology . . . but also in social organisation.

HOMINID EVOLUTION
ISBN 0-12-261920-X

These judgements stem directly from the way in which the palaeolithic record has been partitioned up into analytical units for the purposes of studying culture history.

As Dunnell (1978, p 193) has pointed out, the major achievement of the culture history approach has been to demonstrate that the range of cultural variation in the past is greater than could be imagined from a consideration of contemporary and historically documented cultures. The contents of each cultural partition have on occasion been described by a series of labels which hint at an adaptive rather than cultural flavour. Aside from standard descriptions such as "Acheulian hunters", we find big game hunters (Pfeiffer, 1978), wild harvesters (Freeman, 1973), broad spectrum hunting (Straus, 1977) generalized and specialized strategies (Gamble, 1978) as well as the many descriptions that result from the dominance of a particular species—reindeer, horse, cave bear and so forth—in a collection of animal bones. Any variation in the description on the label is often seen as a mute reflection of what was available in the environment. If the environments appear to be similar in terms of climate and species composition, but the exploited resources differ, then technological competence on the part of palaeolithic man is frequently invoked to account for variations in the faunal menus. In these instances emphasis is laid upon the associated cultural material as a medium for explaining aspects of past adaptive strategies.

This brief review suggests that the study of adaptive strategies in palaeolithic Europe has been concerned with detailing what was eaten and observing how other environmental factors, cold and the presence of ice sheets for example, were adapted to by the acquisition of fire and the construction of huts. However, the lack of a common unit for making comparisons between cultures—analagous to that of "fitness" or "reproductive success" within the evolutionary sciences—has limited the value of the findings of the culture history approach. The conclusion to be drawn from this is that we cannot begin to understand variation in past cultural systems until we have developed, as a first step, a framework of measurement. Before we can make judgements about "quantum advances", as Klein (1973) has done, we need to be able to measure the variable output in the materials of the archaeological record along axes which we know enable valid comparisons of the evidence for the past that we can recover. The second step follows a point made by Kirch (1980, p 115) that in any archaeological study of adaptation the problem is not only one of measuring variation but also in relating such observations to the probable selection pressures of the environment, both social and physical.

In considering this second step of relating adaptation to the environmental circumstances that condition it, the term adaptation may mean both the process of conforming to the environment and also the extent to which the organism does conform to the environment. Our measurements of adaptation must allow us to specify both meanings.

The measurement framework favoured here is based upon the differential distribution of energy. This can be observed at a global scale (Leith and Whittaker, 1975) and used to explain variation in contemporary hunter-gatherer adaptive systems in relation to settlement systems, mobility (Binford, 1980), technology (Oswalt, 1976; Torrence, 1983) and the dependence upon animal products (Foley, 1982). Energy models can predict environmental structure, and in turn, past adaptive structures via the character of the archaeological record. Such models fulfil the requirements of the first step outlined above in that they provide a common measurement that may be applied to all archaeological contexts; and of the second step by relating adaptation to the selective constraints of energy distribution.

This chapter will attempt to relate the distribution of energy in past environments to the variable patterns of archaeological data within the upper Pleistocene of palaeolithic Europe. This is not an easy task. Archaeologists have always had difficulties in providing measures of ecological variables and environmental structure (Foley, 1977). It is well worth pointing out that our sister discipline, palaeoecology, has had similar difficulties (Gifford, 1981). It is still not possible to transform a pollen diagram or a collection of fossil insects into a measurement of ecological productivity, biomass or any other useful statement about past ecological relationships. At best we are provided with a "thermostat" model of the past 0·7 Myr which some mighty hand periodically turned from cold (treeless), through temperate (gallery woods and shade loving snails) to warm (oak mixed forests and hippos by the Thames).

Butzer (1980) has reacted to this state of affairs by suggesting that the multivarious strands of human decision making may be related to environmental concepts that can then be measured by the battery of environmental techniques now at our disposal. The conceptual themes he identifies include the properties of space, complexity, interaction, scale and stability of past human ecosystems. While this is an important theoretical step it fails to show how we are to link patterns in the archaeological record, which are inert, with the dynamic interplay we believe they represent in terms of past action. In my own work (Gamble, 1978) I have tried to measure variation in hunting strategies by contrasting the diversity of species in faunal samples from separate chronological and environmental segments of the Upper Pleistocene of Europe. While such measures may appear to conform to ecological assessments of species diversity and dominance, and hence to the characterization of different structural relationships in past ecosystems, this conclusion sadly ignores the conditions which created a faunal community of animal bones in a series of archaeological sites. As a result the faunal record of the palaeolithic was read as a book about ecology and not, as it should have been, as a book of behaviour by a whole host of authors, who each contributed a chapter or two.

This dissatisfaction will help to explain the approach that follows. The pressing need in palaeolithic studies is not for more palaeoeconomic studies using more palaeoecological detail in more precise ways to illustrate more sophisticated models culled from ethnoarchaeological observations. Instead we need to document, in the materials of the palaeolithic record, variation which we can understand as adaptation to environmental conditions. This approach does not involve working towards definitions of economic or adaptive types but rather looks at the variable signatures (Gould, 1980, p 113) contained in the record of the past. These signatures reflect the impact of the environment on adaptation. This implies a spectrum of variation among past adaptations which can be measured via the archaeological record: a pattern of flexible responses (Hatter, 1982) both within the mosaic of environments encountered at a continental scale and between them as climatic change leads to alterations in environmental structure. This perspective and approach is complementary to a regional framework of analysis and interpretation where we must expect flexibility of response among adaptations and so utilize appropriate modes of description. This is of course in stark contrast to site-based approaches to the study of past adaptations. In these studies (e.g. Klein, 1973; Straus, 1977) the concept of the region is nothing more than the sum total of all the sites from a particular time block contained within a manageable geographical area. As a result the study of adaptation is partitioned up into parcels which are determined by the archaeological strategy of investigation rather than by the behavioural phenomenon being investigated.

In the rest of this chapter I will outline a regional model for the investigation of adaptation in Upper Pleistocene Europe. This will then be examined for the relations between energy, adaptation and the resultant signatures in the palaeolithic record. The model will then be elaborated to consider the significance of differences in the palaeolithic records from a single region under comparable interglacial conditions. This leads us to consider exactly what aspects of these past adaptive systems are subject to selection pressure from the environment, both social and physical.

## 2. A REGIONAL MODEL OF UPPER PLEISTOCENE EUROPE

In order to put some of this into practice I have constructed a simple regional model of Europe with which to study cultural adaptations during the Upper Pleistocene. This has taken into account the following criteria.

(1) Hunter-gatherer adaptations need to be studied at a very extensive spatial scale (Foley, 1978). We require analytical units which are larger than any

single local and regional population since we cannot expect to understand inter-population variation unless sufficient populations are incorporated to show the wider network of constraints that shape decisions. In this regard the continent of Europe is an appropriate scale for the investigation of variation in palaeolithic systems. This assertion is considered in greater detail below. A further discussion of analytical concepts for the study of hunter-gatherer regional adaptations can be found in Wobst (1974, 1976, 1978), Hill (1978), Bender (1978), Gamble (1982, 1983a) and in the earlier works of Birdsell (1953, 1958) and Steward (1938).

(2) Within such a framework we also require analytical units which can serve to investigate local variation. Foley (1981) has argued that mobile hunter-gatherer behaviour should be viewed as home range specific, which further emphasizes that our analytical concepts need to be based on spatial criteria. Equipped with such a concept we can begin to organize archaeological material collected in specified areas in order to investigate its attributes of aggregation, dispersal, frequency and visibility within the landscape. These factors will vary according to the behaviour that created an archaeological record. They will also be affected by post-depositional processes of sorting and erosion.

(3) While it would be possible to define arbitrarily the limits of a region, it has been noted on several occasions (Peterson, 1976; Jochim, 1976; R. Jones, 1971) that the boundaries of regional populations of hunters and gatherers frequently coincide with the area marked by a drainage basin. This observation will be used to help select appropriate regions within the continent.

(4) Other criteria for dividing up the European continent into analytical units for the purpose of providing a framework for measuring variation in palaeolithic materials require even firmer points of reference. The "eternal objects" that A. N. Whitehead (1932, p 188) referred to as a necessity for investigating the past have more recently been dusted off by Binford (1982), transformed into *intellectual anchors* and cast into the murky waters of archaeological methodology. These reference points provide a basis for strong inference when investigating past behaviour. They form a benchmark against which to observe variation. In this instance our points of reference are provided by the constants of latitude, longitude and relief. These may not seem very strong anchors on which to moor our regional model but a moment's consideration will reveal that they are all we have. They act as independent benchmarks against which we can use reconstructions of past vegetation zones (Frenzel, 1973), the distribution of permafrost features (Kaiser, 1960) and faunal distributions (Kurtén, 1968) as indicators of their impact on environmental structure. Latitude, longitude and relief serve as a basis for predicting the distribution of energy since the amount of incoming solar radiation and precipitation budgets are in large part governed by the position of any locality on the earth's surface. These factors help condition the length of the growing

season (Lockwood, 1974) and the productivity of the vegetation layer (Jones, 1979). While soil factors (Bridges, 1978) also play an important role in vegetation productivity they vary in the same manner as those palaeo-environmental indicators that have already been mentioned, and hence cannot be regarded as a constant.

The model (Fig. 10.1) based on these four criteria recognizes three provinces which have a zonal distribution. The Mediterranean province with its characteristic summer droughts is separated from the southern province by high mountain chains. The southern province contains the uplands and mountains of Europe while above latitude 48°N lie the extensive plains of the northern province.

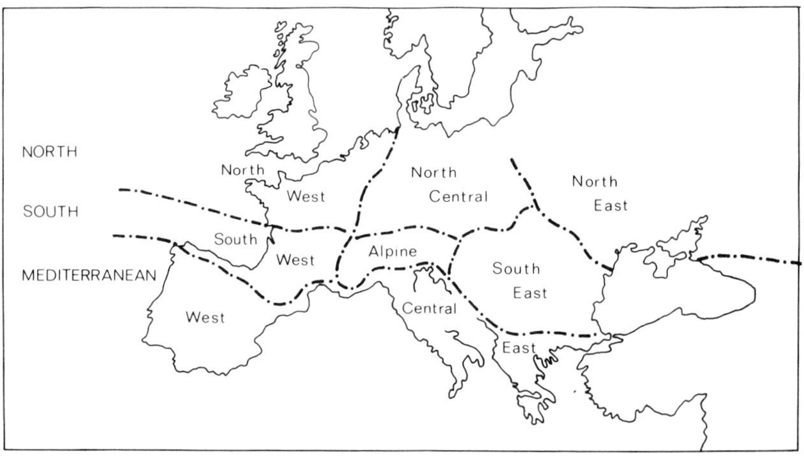


Fig. 10.1. A regional model for the study of variation in Palaeolithic Europe.

These three provinces are divided into nine regions which tend to follow the major drainage basins. In the northern province the regional boundaries take into account the increasing continentality in climate in an eastern traverse. A similar distinction is drawn for the southern province between the south-west and south-east regions which are separated by the Alps. The south-east region is a complex mosaic of environments that reflect not only its geographical position within the continent but also the varied pattern of local relief. In the south-west region the Cantabrian coast of Spain is included with the Pyrenees and southern France. This takes into account the high altitude of this coastal area which produces marked environmental differences from more southerly areas in the Iberian peninsula. The three regions in the Mediterranean province have been divided into a western, central and eastern region for ease of reference.

The purpose of any regional model is to provide a framework for tracing the outcome of variable environmental conditions on a set of behaviours. In this sense the regions in my model are nothing more than a convenient means of summarizing a complex set of environmental conditions which imposed different selection pressures upon human adaptations within the continent.

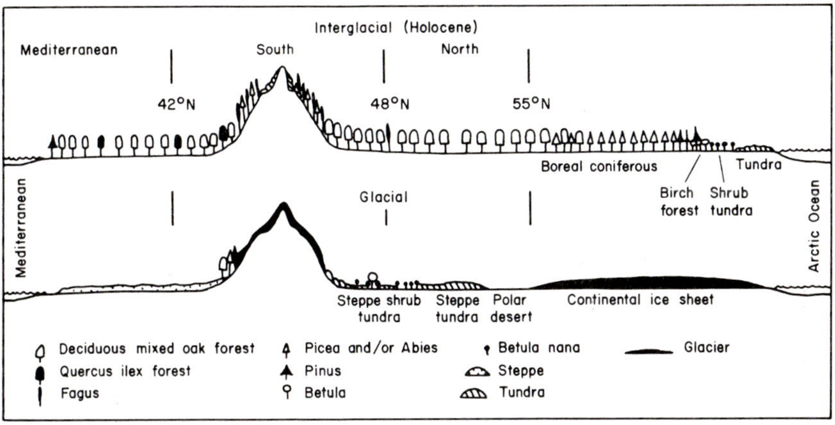


Fig. 10.2. A schematic transect through Europe under extreme interglacial and glacial conditions showing variation in vegetation zonation. The transect contrasts in general terms the potential of the three provinces. Under the full glacial conditions it is expected that animal biomass would be concentrated in the southern and Mediterranean provinces as a reflection of environmental productivity over the continent (based on Van der Hammen *et al.*, 1971).

An objection to the model might be that there is insufficient contrast in the range of longitude and latitude in Europe to result in differences in the distribution of energy which might affect adaptation. The latitudinal range under glacial conditions is between 35–55°N. While this produces a contrast in modern Europe between Mediterranean and temperate environments we cannot assume that this is enough to affect patterning in the palaeolithic record. Moreover the principal effect of decreased solar radiation during the cold stages of the Pleistocene could have been to simplify further the limited contrasts in this continental pattern. In Fig. 10.2 the three provinces are contrasted in terms of resources and energy during full glacial conditions. The data suggest sufficient variation as well as determining which regions would have been more or less advantageous for occupation. However, such a qualitative assessment is unhelpful in using the model as a basis for measuring selection pressure upon past adaptations. We might for example always be tempted to characterize the northern province as harsh and inhospitable, which it no doubt was, but if we are to understand variation in adaptive strategies then this needs to be measured rather than assumed. This is the test of the model, and should establish whether: (1) the continent of Europe is an appropriate scale (criteria 1) for measuring variation in human adaptations under the climatic conditions of the Upper Pleistocene; (2) the materials left behind by a set of behaviours do vary in relation to the regional model and its implications for energy and archaeological structure; (3) this state of affairs, if demonstrated, is itself a product of relative selection pressures operating on long-term survival strategies.

## 3. ENERGY VARIATION, ADAPTIVE STRATEGY AND THE PALAEOLITHIC RECORD

The regional model provides a framework for analysis based upon the assumption that the distribution of energy will affect the pattern of residues left behind by palaeolithic adaptations. We are, however, still left with the problem that all we might be doing is setting up a self-fulfilling model where we cannot fail to find variation between the provinces and regions in the materials of the palaeolithic record, and so claim that the analytical framework has value. In the first instance we need an independent check to examine this relationship between energy–adaptation–residues, and so produce methodological links between dynamics and statics.

One way to tackle the problem is to look at the behaviour of other species that form a part of the archaeological and palaeontological records. These species experienced the same glacial–interglacial environments as man and their behaviour patterns were subject to environmental selection. The large carnivores have been known for a long time to create their own records (Brain, 1981; see Chapter 5, this volume) through denning behaviour. The activities involved in such a general set of behaviours varies between and within species. Denning behaviour in bears is most commonly a response to the lack of winter forage (Herrero, 1972), while their use of caves and rock shelters is a reflection of permafrost conditions that makes den digging difficult. Wolves and hyaenas use den sites to protect their young. This behaviour involves the transportation of food to such safe sites. Dens are also used as refuges for finishing meals undisturbed by other carnivores and is well illustrated by the leopard which caches food in trees (Brain, 1981).

Denning behaviour is therefore part of a carnivore survival strategy. It is a means by which feeding competition is averted and a way for the smaller carnivores to avoid predation by larger species (Gifford, 1981, p 413). While it is a part of many carnivores behavioural repertoire the intensity of denning behaviour appears to vary as a function of competition. It was this variation that led to the controversy over the role of the hyaena as a bone collecting agent

(Hughes, 1954; Brain, 1981; see Chapter 5, this volume). It is reasonable to suggest that the creation of carnivore residues will reflect the intensity of competition among carnivore species and that the static material residues that result from this coping strategy are thus a measure of past dynamic behaviour.

An example of this variability is provided by Binford (1981, p 221) where the treatment of prey carcasses by African and Alaskan carnivores is contrasted. In the former area where competition among carnivores is high there is generally a greater incidence of the carnivores transporting parts of the prey away from the kill site and back to a lair. In contemporary Alaska the wolf is the only large carnivore, and the pattern of food transport is very different.

The large carnivores of Pleistocene Europe combined both arctic and sub-tropical species—lion, hyaena, wolf, leopard, bear as well as lynx, fox, wolverine and other smaller forms. The eurytopic nature of several of these species must have led to inter-specific competition, which would, however, have varied as a function of carnivore to prey densities. Variation in the incidence of denning behaviour by carnivores could therefore be explained in terms of competition for prey resources between the large carnivores.

If this is the case it then follows that a high incidence of den residues is a result of intense selective pressures placed by the environment on the predators. Carnivore residues are therefore informing us about the contexts and conditions that shape variation in an aspect of behaviour.

The Table 10.1 shows the mean number of carnivore species for various periods during the Upper Pleistocene. Bear, lion, hyaena, wolf and foxes are the most commonly represented species. The presence of these carnivores in cave and rock shelter deposits is assumed to reflect mortalities that occurred during denning activity. Only in rare instances might man be responsible for such accumulations. While the sample is largely restricted to the limestone areas of Europe it quite clearly points up a major difference between the northern and southern provinces for which data are available. In the north-west and central regions the carnivores are very common as shown by the larger mean values for the chronological periods. By contrast the sample from the south-west and Mediterranean regions show a much lower incidence of carnivore recovery from caves and rock shelters. Elsewhere (Gamble 1983b) the relative abundance of these carnivore species in these same faunal samples is presented and this complements the pattern reported here. Moreover, the quantities of faunal specimens attributed to the main carnivores also points to a much greater representation in the central region when compared with faunal collections from the south-west region. This would confirm the hypothesis that environmental selection differed in the three provinces. Competition was higher in the three northern regions and this is demonstrated by the data on the mean number of large carnivore species per stratigraphical unit as well as by the abundance of their remains.

Table 10.1. Carnivore representation in cave and rock shelter deposits during the last glacial. The figures refer to the mean number of carnivore species per stratigraphic level. The figures in parentheses indicate the number of faunal assemblages from which the mean is calculated.

| | Middle Palaeolithic 118ka–35ka early last glacial and pleniglacial | Early Upper Palaeolithic 35ka–20ka pleniglacial | Late Upper Palaeolithic 20ka–10ka pleniglacial and late glacial |
|---|---|---|---|
| NORTHERN PROVINCE | | | |
| England | | 4·1 (15) | 2·3 (33) |
| Southern Germany | 4·02 (49) | 4·05 (20) | 1·87 (47) |
| Bükk/Hungary | | 4·6 (20) | |
| Crimea | 3·3 (13) | | |
| SOUTHERN PROVINCE | | | |
| South-west France | 1·06 (59) | 1·6 (46) | 1·56 (70) |
| Cantabria | 0·84 (19) | 1·0 (31) | 0·94 (54) |
| Switzerland | 2·57 (14) | | |
| MEDITERRANEAN | | | |
| Languedoc/Roussillon | | | |
| Provence | 2·89 (38) | | |
| Istria/Yugoslavia | | 2·9 (13) | |

1. Carnivores included in the table are; bear, lion, hyaena, wolf, red fox, arctic fox, leopard, lynx, wolverine, badger and wild cat.
2. Sources: Campbell (1977); Gamble (1978); Mottl (1941); Klein (1965); Delpech (1975); Freeman (1973), Straus (1977); Müller-Beck (1968); Pillard (1972); Musil (1980–1).

Another test to this hypothesis may be made by examining the distribution of cave bear (*Ursus spelaeus*) within the continent. A recent comprehensive review by Musil (1980–1981) has listed the occurrences of this fossil species. In his view the majority of finds of this species date to the last glaciation. While the overall distribution reflects that of karst we can none the less see some marked variation in the frequency with which the cave bear has been identified in the nine regions (Table 10.2). Furthermore, not only does the frequency of cave bear representation vary, but there are differences in the quantities of cave bear bones recovered from sites in the various regions, a pattern that complements that already noted for the large carnivores. Straus (1982, p 87) makes a particular point that the Mousterian levels at the Cantabrian site of Lezetxiki comprise between 60 and 92% cave bear bones for the faunal collections. As he shows this is exceptional for the south-west region. It is, however, commonplace in the central region at sites such as the Weinberghöhlen (Zotz, 1955), Stadel (Gamble, 1979), Istállöskö (Vértes, 1955), Pod hradem (Musil, 1965) or the Drachenhöhle in Austria where Ehrenberg (1951) estimated that cave bear made up some 99% of the 225 000 bones and teeth dug out of this high altitude cave. At the open site of Érd (Gábori-Csánk, 1968) outside Budapest, and which in my opinion is also a hibernation den, cave bear formed 89% of the 14 871 identified large mammal remains. Thus the distribution of carnivores and cave bear in both archaeological and palaeontological sites confirm the hypothesis, tested through the regional model, that a link between the distribution of energy, inter-specific competition and variation in the record of the past can indeed be traced.

Table 10.2. The proportion of sites by region which have yielded remains of the cave bear (*Ursus spelaeus*). The sample comprises 792 sites from Musil (1980–1). His survey deals only with *U. spelaeus* and other forms such as the brown bear (*U. arctos*) are not included.

| | | | | |
|---|---|---|---|---|
| | | *NORTH* | | |
| West | | Central | | East |
| 4[a] | % | 29 | % | 8[b] |
| | | *SOUTH* | | |
| West | | Alpine | | East |
| 14 | | 14 | | 17 |
| | | *MEDITERRANEAN* | | |
| West | | Central | | East |
| 2 | | 4 | | 7 |

[a] Musil regards nearly all the specimens from England as arctoid forms, which are common in this part of the north west region.
[b] Sample mainly from the Caucasus.

These regional studies of the large carnivores are examples of how we can put back the dynamics of behaviour into the inert facts of the palaeoecological record. The measures used are very simple but the variation can be explained, partially at least, in behavioural terms. Moreover these studies show that the continent of Europe under the glacial and pleniglacial phases of the full interglacial/glacial cycle (Kukla, 1977) presented a mosaic of conditions which at a regional and provincial scale were of sufficient size and diversity to condition behaviour. The prospects are therefore good for conducting similar exercises with the archaeological record and establishing a comparable link between energy, here represented by the position of the region within the continent, behaviour, and a variable material product.

What outcomes, though, can we predict and test for human adaptations and the formation of residues? The denning model is hardly applicable to the human use of caves and rock shelters. In order to account for the variable use of rock shelters by human groups we would require a more subtle hypothesis linking the availability of such facilities with the pattern of resources and the settlement system strategy designed to exploit the environment (Binford 1978, p 492).

An example of the difficulties involved in measuring the intensity of rock shelter use is provided by Straus (1982, p 92) in his study of Vasco-Cantabrian caves. He uses the occurrence of the carnivores as a measure of the intensity of use by man of these same locations. He interprets his data as indicating a shift from "time sharing" of these facilities by both carnivores and man to "more full-time occupation" by human groups who were apparently primarily interested in trapping carnivores for furs in the late palaeolithic of the area. The lack of carnivore remains in caves which also contain traces of human occupation is therefore taken as a measure of the intensity of human use. This measure is not, however, based on any contemporary understanding that the variable use of the same caves by man and the large carnivores produces this sort of outcome. In this instance we do not understand the conditions which result in patterning in our data. While Straus' argument is plausible his data do not provide a measure of the factor, intensity of cave occupation by man, which in fact he wishes to determine.

An alternative way to investigate the regional framework for information about variation in human adaptations resulting from selection pressure imposed by the environment is to compare the settlement histories of the different regions. The results of this approach are presented elsewhere (Gamble, 1983a, 1983b); they indicate a distinction between continuous settlement during the last glaciation (75 Kyr–10 Kyr) from local areas within the south-west region (Laville *et al.*, 1980), and the discontinuous settlement record in the central (Hahn, 1976) and western (Stuart, 1982; Fig. 10.1) regions of the northern province. This may be described by means of an ebb and flow model of settlement and abandonment until coping strategies evolved that enable uninterrupted occupation and population maintainance. Abandonment most probably took the form of local extinction although others (Jochim, 1983) have argued for the movement of population out of such areas and into refuge zones.

The advantage of this approach is that we do not need initially to specify the type of settlement system and its implications for other aspects of the adaptive strategies. The preliminary studies suggest that during the Upper Pleistocene the three provinces have different settlement histories with, in particular, and as would be expected, the northern province showing the most marked effects of changing climate upon occupation. In other words we have a comparable link between energy, selection upon aspects of behaviour and a material outcome in the archaeological record as we have previously demonstrated for the carnivores.

At a very general level the pattern of settlement history reflects the risks involved in animal exploitation strategies in specialized glacial environments. While such settlement histories are an unsophisticated measure of the relationship between energy and adaptive strategies they none the less serve to show that regional variation is to be expected in the palaeolithic records from within the continent. It would, however, be unwise to assume that the variation in this record, although linked to latitude, longitude and relief was just a reflection of the gross distribution of energy in the continent. On the contrary, all regional environments at all times during the European Upper Pleistocene had enough energy to sustain some palaeolithic groups. Energy availability was not the limiting factor, nor the reason why settlement ebbed and flowed. Rather than absolute amounts of energy it was the way it was structured within these regions in the form of parcel size, protein return, timing and so forth. Therefore at a more detailed and local level of analysis we need to bring into play the models of optimal foraging theory (Winterhalder and Smith, 1981) since they provide us with a framework for the detailed assessment of resource choice, based upon energy costs and risks.

One purpose of these models is to break down the monolithic characterizations of energy distribution that so far in this paper have taken the guise of a regional framework. Let us initially assume that local settlement ebbs and flows as a response to the costs involved in an exploitation strategy, leading to variation in the archaeological record. These factors resulting in settlement abandonment can of course be overcome, for instance through increased technological efficiency to reduce certain critical exploitation costs. Torrence (1983) has elegantly demonstrated that time stress places a selective pressure upon technology. This occurs, for example, when key resources are only available in a group's home range for very limited periods of the year. Failure to get sufficient of these resources during these brief moments would lead to the breakdown of the strategy. As a result the technology is finely honed to minimize risks and provide the best possible chances of success. These resources are mainly large mammals which are costly to exploit in terms of search, pursuit and handling time.

The benefits, however are vast, particularly in the context of few alternative resources. The increasing complexity of technologies that depend upon such resources is well demonstrated in Oswalt's (1976) survey of food-procuring equipment. Time saving facilities such as traps and storage strategies would also form very necessary elements in the development of adaptations that could cope with sets of resources that posed such exploitation costs in the form of time-budgeting decisions.

These mammal-based strategies have often been contrasted with plant gathering strategies (Lee, 1968) where such costs seem much less. This view has recently been questioned by Hawkes and O'Connell (1981) where they point out that the high handling/preparation costs for many plant foods often means that they are left out of an exploitation strategy even though they may be very abundant in the environment. The preparation of certain species of nuts and seeds does not always require a complex technology but often involves the expenditure of a great deal of time and hence cost.

While resource availability, and hence the environment, may determine the level of costs the social system will ultimately determine what constitutes acceptable costs. The environment does not set the level of demand for the appropriation of resources although it does establish a framework which requires a flexible response from the set of social relations. In this sense the social system is dominant but the environment determinant in any adaptive situation (Ingold, 1981). Social solutions to environmental problems are as important as any technological advance. For example information sharing networks are needed to cope with the intricate adjustment of individuals and groups to variable resource conditions. These networks can only be carried on a framework of alliances (Gamble 1982, 1983a) which serve to articulate social relations over extensive spatial areas.

## 4. INTERGLACIAL CONDITIONS

Many of these points can be illustrated with an example from the north-west region. This region is well represented by large mammal faunas dating from the last warm period, particularly in southern England (Stuart, 1976, 1982). These Ipswichian faunas, as they are known, contained hippopotamus (Table 10.3) and a recent $^{230}Th/^{234}U$ date on flowstone in the Victoria Cave, Yorkshire, suggests that this characteristic fauna is dated to 120 Kyr (Gascoyne *et al.*, 1981). This suggests a tie in with the marine isotope record stage 5, and in particular with the first sub-stage (5e) where the $^{18}O$ curve indicates the smallest volume of continental ice at any time during the Middle and Upper Pleistocene (see Chapter 2, this volume). Shackleton (1969) has argued that this short-lived sub-stage of some 10 Kyr duration, and beginning at 128 Kyr (Shackleton and Opdyke, 1973), should be regarded as the interglacial, and support for this comes from the pollen profile of Grande Pile (Woillard, 1978) in the Voges.

While the precise correlation of deep sea core sub- stages 5e, c and a with palaeoecological evidence from terrestrial sources is difficult, it is none the less striking that a number of authors (Wymer, 1981, p 67; Evans, 1975, p 33; Stuart, 1982, p 176) have commented on the paucity of archaeological materials associated either with sub-stage 5e or any part of this general warm

Table 10.3. A comparison of the large mammal faunas attributed to the last interglacial (Ipswichian) and post glacial (Flandrian) in the British Isles (after Stuart, 1982).

| | Ipswichian | Flandrian |
|---|---|---|
| Carnivora | | |
| *Canis lupus* (wolf) | * | * |
| *Ursus* sp. (bear) | * | * |
| *Meles meles* (badger) | * | * |
| *Crocuta crocuta* (hyaena) | * | |
| *Panthera leo* (lion) | * | |
| *Vulpes vulpes* (red fox) | * | * |
| *Felis sylvestris* (wild cat) | * | * |
| Proboscidea | | |
| *Palaeoloxodon antiquus* (straight tusked elephant) | * | |
| *Mammuthus primigenius* (mammoth) | * | |
| Perissodactyla | | |
| *Equus* sp. (horse) | * | * |
| *Dicerorhinus hemitoechus* (narrow nosed steppe rhino) | * | |
| *Dicerorhinus kirchbergensis* (Merck's rhino) | * | |
| Artiodactyla | | |
| *Hippopotamus amphibius* (hippo) | * | |
| *Megaloceros* sp. (giant deer) | * | |
| *Dama dama* (fallow deer) | * | |
| *Cervus elaphus* (red deer) | * | * |
| *Bos primigenius* (aurochs) | * | * |
| *Bison priscus* (steppe bison) | * | |
| *Sus scrofa* (wild pig) | * | * |
| *Alces alces* (elk) | | * |
| *Capreolus capreolus* (roe deer) | * | * |

stage. This is in striking contrast to earlier but less temperate interglacials. Given the warm conditions at 125 Kyr, the rich Ipswichian vegetation, and the large mammal fauna this state of affairs is perhaps contrary to the expectation that the cold was the primary limiting factor for palaeolithic populations.

A very different pattern indeed is shown by the settlement record from the same area of southern England during the post-glacial. The $^{18}O$ curves indicate many points of environmental comparability between this stage and the last interglacial. At this time, however, the landscape of southern England is covered with mesolithic sites (Wymer, 1977) even though the faunal communities appear somewhat impoverished (Table 10.3) compared with those assigned to the various stages of the Ipswichian (Stuart, 1976, 1982).

If we allow, for present purposes that the two environments were roughly equivalent in terms of the energy they contained then why should these differences exist in the archaeological record of the two periods? Destruction of middle palaeolithic, Ipswichian age, sites on a massive scale does not seem possible given the substantial recovery of earlier archaeological material (Wymer, 1981). Neither does differential recovery seem a plausible explanation. The paucity of evidence of man in England during the Ipswichian is therefore neither a result of lack of energy or food in the environment nor of taphonomic factors. This particular Garden of Eden must have imposed heavy cost penalties on exploitation strategies. The animals did not come in large aggregations and the forest conditions favoured smaller body size species than would have been the case on the steppe tundras of the early glacial and pleniglacial phases. Even the forests which were rich storehouses of potential food (Clarke, 1976) only yielded up their harvest at a cost that involved the long preparation of individually small plants, nuts and fruits. Therefore, man was having to invest energy in processing food for his own consumption that in more open environments is effort expended by the herbivores themselves.

The contrast in mesolithic and middle palaeolithic technologies brings out this point. The mesolithic consists of small standardized lithic segments that can be mounted in a variety of ways to perform a wide range of reaping, cutting, piercing tasks. Digging stick weights and axes are also present in the lithic inventories (Wymer, 1977, Fig. 2, xiii) and represent part of the technological solution to the problem of getting enough of a harvest in from the woods. This involved intricate scheduling (Jochim, 1976) of resource use and the flexible adjustment of personnel to resources with the partitioning of the environment into territories staked with their social markers such as cemeteries (Albrethsen and Brinch-Petersen, 1976). The accent is on a longer working day and an intensive level of exploitation to which the technology is geared.

The middle palaeolithic lacked this emphasis on technological complexity. Rather than gathering a wide range of resources under a labour and information intensive system, the strategy was to attach groups to locally abundant resources (see Chapter 9, this volume) and to take by shock and intercept tactics what in the mesolithic was gained by stealth and traps. As a result widescale information sharing and the necessary framework which carried it (Gamble, 1983a, 1983b) was not required.

This strategy of "going for broke" is suited to ecologically specialized environments and the animal resources they contain (Wilmsen, 1973). Any changes in such environments which led to increased labour costs as measured by search, pursuit, processing time, and which required more investment in acquiring information on which to base a resource use schedule, would impose their own ceiling above which costs were too high for continued exploitation without developments in adaptive strategies. It should therefore come as no surprise that the cultural inventories of the mesolithic are markedly different from those of the middle palaeolithic which might have been found in the last interglacial, and that the settlement histories for the same region during the two periods should be so distinctive.

These contrasted settlement histories and technologies are an example of different archaeological signatures on comparable environmental documents. The variation can be understood in terms of increased labour effort directed towards procuring foods with high exploitation costs. The general principle is that groups move down the rank of resources, thereby incurring these higher exploitation costs, rather than moving out of the area. The basic difference between these two adaptive strategies separated by some 100 Kyr hinges on the use of costly resources. The environment was not a limiting factor on adaptive strategies in the sense that it did not contain enough food to sustain populations. It was, however, the structure of those food parcels, the way in which they were distributed around the environment, that imposed unacceptable costs for existing adaptive strategies.

These high cost tariffs are however unlikely to have been broken for the first time with the resumption of temperate conditions at the beginning of the post glacial. Indeed there must have been many occasions during the Middle and Upper Pleistocene when glacial conditions led to a similar set of problems in the exploitation of the environment. The ebb and flow model is applicable to the climatic minima as well as to the optima.

The build-up prior to the glacial maximum at 18 Kyr (Peterson *et al.*, 1979) is a period of change in the material cultural inventories of Europe. At 25 Kyr we see the development of spatially extensive technocomplexes (Otte, 1981) and the material expression of equally extensive information sharing networks (Gamble, 1982). The use of art, ornament, display and the properties of utilitarian objects to carry information via stylistic messaging are pointers to the adaptive changes that have taken place. At 18 Kyr there is abandonment in much of the northern province (Gamble 1983a, 1983b) as the ice sheets reached their full extent forming a glacial corridor described as polar desert (Van der Hammen *et al.*, 1971). Two regions where abandonment did not take place were the south-west and west Mediterranean. The distinctive Solutrean lithic assemblages (Smith, 1966), now well dated to around 18 Kyr (Davidson, 1974), are found in the Périgord, Cantabria and south-east Spain within these two regions.

At this time Jochim (1983) has pointed out in a challenging paper that the compression of ocean currents at latitude 42°N would have resulted in a very rich marine environment off the coast of northern Spain and southern France. He suggests that this would have made available another important resource in the form of salmon runs up the rivers of these same areas. He argues that refugee populations from the northern province could have been accommodated in these more southerly regions by utilizing these extra resources. This dense packing of population to resources would have needed alternative ways to assist group integration and the allocation of spatially restricted, but important resources, to groups. One consequence, he maintains, would be the spectacular rise of cave art, acting as a social marker in an arena of increasing social complexity.

However, this same set of marine circumstances presumably existed during earlier glacial cycles without leading to such results. There is no evidence for the earlier use of salmon in these areas of the south-west region. It therefore seems unlikely that population pressure forms part of the explanation for the use of this "new" resource. Rather than refugee populations arriving in the southern areas it is more likely that the abandonment of the northern province represents the local extinction of population in those regions at 18 Kyr.

I would agree, however, that the outcome in terms of settlement continuity and the appearance of social markers is indicative of a major change in adaptive strategy. This refers to solving the high cost tariffs involved in exploiting these environments. While the salmon may not have required a complex technology in order to catch them in the rivers of the Dordogne and Cantabria they were no doubt expensive in terms of processing and storage time (Schalk, 1977, pp 232–3). Most importantly the use of this resource would have required an intricate schedule to key in its exploitation with other important species such as reindeer and red deer. In this instance overcoming the cost tariffs is not a result of population pressure acting as a selective force on adaptation. It is rather the case that the system of appropriation, which specified that increased costs were now acceptable, was already in existence and long before any population pressure became a factor in the system. While this brings us no closer to understanding why adaptive strategies change, it does focus on those areas of the entire social/cultural package which act as restraints on the status of being adapted to a set of conditions.

In this context it is also interesting to note the growing body of evidence for a shift to lower rank, more expensive foods at the 18 Kyr climatic minima. This has been demonstrated in Australia by O'Connell and Hawkes (1981, p 115) with the appearance of grinding equipment in the archaeological record, and a shift to the use of seed resources. These had always been abundant in the Australian environment but costly to process for consumption. They were only used as less expensive resources fell away under increasingly rigorous climatic conditions. They still remained expensive to use. Neither did their use at 18

Kyr lead to some form of Antipodean agricultural revolution as soon as the climate and other less expensive resources improved. Their use formed part of an adaptive strategy with new levels of cost limitations on exploitation. Agriculture implies a very different social system for the appropriation of resources. Overcoming exploitation costs, working harder for a greater return, is only an aspect and not *the* major obstacle in the transition from one productive system to another.

Grinding equipment together with the intensive use of cat-fish have also been dated to 18 Kyr in the Wadi Kubbaniya in the Nile valley (Wendorf *et al.*, 1980). I would suggest that this "intensification" is comparable to that in the Périgord and Cantabria although the resources involved are very different. In both cases the adaptive strategies of the respective regions went further down the rank of resources. The populations worked harder, and required more complex information systems to deal with intricate scheduling decisions. It is not surprising that the material record produced by these adpatations appears so very different from much earlier archaeological traces (White, 1982). In the case of the Nile valley the subsequent history of this use at 18 Kyr of seed resources still has to be documented in detail. However, Wendorf *et al.* (1980, p 278) note that in the terminal palaeolithic of the area, *ca.* 12 Kyr, grains disappear and grinding equipment becomes very rare in this same area of the Nile.

We shall probably find more and more instances where the later palaeolithic record of the Upper Pleistocene contains material evidence which mimics the picture of agricultural origins and developments usually associated with later prehistory. This mimicry arises from the fact that the use of some resources as food, by whatever type of social formation, carries a commitment of higher exploitation costs. At some point in the palaeolithic this obstacle to exploitation was overcome. This differentiates the later adaptive strategies of the Upper Pleistocene from those which came before. However, changes in adaptive strategies during this period in no way fore-shadowed, as a consequence, the agricultural "revolution".

## 5. CONCLUSIONS

In this paper I have attempted to show that in order to investigate past human adaptations we need systems of measurement if we are to understand variation in the palaeolithic record in terms of alternative strategies. Energy and space provide the framework for examining variation in past behaviour and these have been converted into regional units of analysis and applied to European data. These units provide a basis for comparison between the variable archaeological signatures that they contain and which, on a synchronous basis, can be understood in terms of different distributions of energy and resources within the continent. In this regard energy is seen as a determining factor in occupation and behaviour. An independent check is provided by looking at the fellow travellers with palaeolithic man (see Chapter 8, this volume), the large carnivores, and observing that variation in their residues can likewise be interpreted by tracing the link between energy, adaptive behaviour and the creation of residues for preservation in deposits.

The measures discussed here are a first attempt to understand the European palaeolithic record in terms of behaviour rather than as a succession of descriptive labels. From a diachronic perspective the energy space framework has been chosen to explore the relationship between the probable selection pressures on adaptation of both the social and physical environments. By using the assumptions of optimal foraging models it is possible to see that one trend in long-term adaptive change has been the use of resources which carry a high tariff of exploitation costs. The change to these resources has resulted in a lessening of the ebb and flow model of settlement in the northern latitudes of Europe. During the Upper Pleistocene occupation of these regions becomes less interrupted under both glacial and interglacial conditions. It is argued that this stems from changes in the social environment which devised appropriate ways by which these costs could be defrayed. However, the consequences of surmounting these cost barriers is not an automatic agricultural revolution once the climate improves. The selection pressure on adaptive change therefore stems from the social environment. The environment acts only as a determinant factor within this wider dominant framework. We must remember that our perspective of the past is such that energy and the environment have *not* been limiting for the establishment of hunting and gathering adaptations in the process of global colonization. Through this process we have an indication of the long term evolutionary success of the species. From this perspective it will always be the delay in this process, rather than its attainment, which will be the central interest in the study of the evolution of human adaptations.

## REFERENCES

Albrethsen, S. E. and Brinch-Petersen, E. (1976). Excavation of a mesolithic cemetery at Vedbaek, Denmark. *Acta Archaeologica* **47**: 1–28.

Bender, B. (1978). Gatherer-hunter to farmer: a social perspective. *World Archaeology* **10**: 204–22.

Binford, L. R. (1978). *Nunamiut ethnoarchaeology*. New York: Academic Press.

Binford, L. R. (1980). Willow smoke and dogs' tails: hunter-gatherer settlement systems and archaeological formation processes. *American Antiquity* **45**(1): 4–20.

Binford, L. R. (1981). *Bones: Ancient Men and Modern Myths*. New York: Academic Press.

Binford, L. R. (1982). Meaning, inference and the material record. In Renfrew, C., and Shennan, S. J. (eds). *Ranking, Resource and Exchange*. Cambridge: University Press, pp. 160–3.

Birdsell, J. B. (1953). Some environmental and cultural factors influencing the structure of Australian aboriginal populations. *Am. Naturalist* **87**: 171–207.

Birdsell, J. B. (1958). On population structure in generalised hunting and collecting populations. *Evolution* **12**: 189–205.

Brain, C. K. (1981). *The Hunters or the Hunted?* Chicago: Chicago University Press.

Bridges, E. M. (1978). *World Soils*. Cambridge: Cambridge University Press.

Butzer, K. W. (1980). Context in archaeology: an alternative perspective. *J. Field Archaeol.* 7: 417–22.

Campbell, J. B. (1977). *The Upper Palaeolithic of Britain: A Study of Man and Nature in the Late Ice Age (2 vols)*. Oxford: Clarendon Press.

Clarke, D. L. (1976). *Mesolithic Europe: The Economic Basis*. London: Duckworth.

Davidson, I. (1974). Radiocarbon dates for the Spanish Solutrean. *Antiquity* **48**: 63–5.

Delpech, F. (1975). *Les faunes du paléolithique supérieur dans le sud-ouest de la France*. These de Doctorat d'Etat ès Sciences naturelles. Université de Bordeaux No. 479.

Dunnell, R. C. (1978). Style and function: a fundamental dichotomy. *Am. Antiquity* **43**: 192–202.

Ehrenberg, K. (1951). 30 Jahre palaobiologischer Forschung in Österreichischen Höhlen. *Quartär* 5.

Evans, J. G. (1975). *The Environment of Early Man in the British Isles*. London: Elek.

Foley, R. (1977). Space and energy: a method for analysing habitat value and utilisation in relation to archaeological sites. In: Clarke, D. L. (ed.) *Spatial Archaeology*. London: Academic Press, pp. 163–188.

Foley, R. (1978). Incorporating sampling into initial research design: some aspects of spatial archaeology. In: Cherry, J. F., Gamble, C. S. and Shennan, S. J. (eds). *Sampling in Contemporary British Archaeology*. Oxford: British Archaeological Reports (British Series 50), pp. 49–66.

Foley, R. (1981). *Off-site Archaeology and Human Adaptation in Eastern Africa*. Oxford: British Archaeological Reports (International Series 97).

Foley, R. (1982). A reconsideration of the role of predation on large mammals in tropical hunter-gatherer adaptation. *Man* 17: 393–402.

Freeman, L. G. (1973). The significance of mammalian faunas from palaeolithic occupations in Cantabrian Spain. *Am. Antiquity* **38**: 3–44.

Frenzel, B. (1973). *Climatic Fluctuations of the Ice Age*. Cleveland and London: Case Western Reserve University.

Gábori-Csánk, V. (1968). *La station du paléolithique moyen d'Érd, Hongrie*. Budapest: Kiado.

Gamble, C. S. (1978). Resource exploitation and the spatial patterning of hunter-gatherers: a case study. In: Green, D., Haselgrove, C. and Spriggs, M. (eds). *Social Organisation and Settlement*. Oxford: British Archaeological Reports (International Series 47), pp. 153–185.

Gamble, C. S. (1979). Hunting strategies in the central European palaeolithic. *Proc. Prehist. Soc.* **45**: 35–52.

Gamble, C. S. (1982). Interaction and alliance in palaeolithic society. *Man* **17**: 92–107.

Gamble, C. S. (1983a). Culture and society in the upper palaeolithic of Europe. In: Bailey, G. N. (ed.) *Hunter-Gatherer Economy in Prehistory*. Cambridge: Cambridge University Press, pp. 201–211.

Gamble C. S. (1983b). Caves and faunas from last glacial Europe. In Grigson, C. and

Clutton-Brock, J. (eds). *Animals and Archaeology: Hunters and their Prey*. Oxford: British Archeological Reports. (International Series 163), pp. 163–172.

Gascoyne, M., Currant, A. P. and Lord, T. C. (1981). Ipswichian fauna of Victoria Cave and the marine paleoeoclimatic record. *Nature* **294**: 652–4.

Gifford, D. P. (1981). Taphonomy and palaecology: a critical review of archeology's sister disciplines. In: Schiffer, M. B. (ed.) *Adv. Archeol. Method and Theory Vol 4*. New York: Academic Press, pp. 365–438.

Gould, R. A. (1980). *Living Archaeology*. Cambridge: Cambridge University Press.

Hahn, J. (1976). Das gravettian im westlichen Mitteleuropa. Nice: U.I.S.P.P. IX Congrès, Colloque XV *Perigordian et gravettian en Europe*.

Hatter, C. S. (1982). *Some Problems of Archaeological Explanation*. B. A. Dissertation. Sheffield University: Department of Prehistory.

Hawkes, K. and O'Connell, J. F. (1981). Affluent hunters? Some comments in light of the Alyawara case. *Am. Anthropol.* **83**: 622–26.

Herrero, S. (ed.) (1972). *Bears—Their Biology and Management*. Morges: International Union for Conservation of Nature and Natural Resources.

Hill, J. R. (1978). Language contact systems and human adaptations. *J. Anthropol. Res.* **34**: 1–26.

Hughes, A. R. (1954). Hyaenas versus australopithecines as agents of bone accumulation. *Am. J. Phys. Anthrop.* **12**: 467–86.

Ingold, T. (1981). The hunter and his spear: notes on the cultural mediation of social and ecological systems. In: Sheridan, A. and Bailey, G. N. (eds). *Economic Archaeology*. Oxford: British Archaeological Reports (International Series 96), pp. 119–130.

Jochim, M. A. (1976). *Hunter-Gatherer Subsistence and Settlement: a Predictive Model*. New York: Academic Press.

Jochim, M. A. (1983). Palaeolithic cave art in ecological perspective. In: Bailey, G. N. (ed.) *Hunter-Gatherer Economy in Prehistory*. Cambridge: Cambridge University Press, pp. 212–219.

Jones, G. (1979). *Vegetation Productivity*. London: Longman.

Jones, R. (1971). The demography of hunters and farmers in Tasmania. In Mulvaney, D. J. and Golson J. (eds). *Aboriginal Man and Environment in Australia*. Canberra: Australian National University Press, pp. 271–287.

Kaiser, K. (1960). Klimazeugen des periglazialen Dauerfrostbodens in Mittel- und Westeuropa. *Eiszeitalter und Gegenwart* **11**: 121–41.

Kirch, P. V. (1980). The archaeological study of adaptation: theoretical and methodological issues. In Schiffer, M. B. (ed.) *Advances in Archeological Method and Theory Vol 3*. New York: Academic Press, pp. 101–156.

Klein, R. G. (1965). The middle palaeolithic of the Crimea. *Arctic Anthropol.* **3**: 34–68.

Klein, R. G. (1973). *Ice Age Hunters of the Ukraine*. Chicago: Chicago University Press.

Kukla, G. J. (1977). Pleistocene land-sea correlations I. Europe. *Earth Sci. Rev.* **13**: 307–74.

Kurtén, B. (1968). *Pleistocene Mammals of Europe*. London: Weidenfeld and Nicolson.

Laville, H., Rigaud J. P. and Sackett, J. R. (1980). *Rock Shelters of the Périgord*. New York: Academic Press.

Lee, R. B. (1968). What hunters do for a living, or how to make out on scarce resources. In: Lee, R. B. and DeVore, I. (eds). *Man the Hunter*. Chicago: Aldine, pp. 30–48.

Leith, H. and Whittaker, R. H. (eds). (1975). *Primary Productivity of the Biosphere*. Berlin: Springer Verlag.

Lockwood, J. C. (1974) *World Climatology*. London: Edward Arnold.

Mottl, M. (1941). Die Interglacial und Interstadialzeiten im Lichte der Ungarischen Säugertierfaunen. *Mitteilungen aus dem Jahrbüche der kgl. ungar. Geolögischen Anstalt* XXXV: 1–33.

Müller-Beck, H-J. (1968). *Die Altere und Mittlere Steinzeit.* Basel: Ur- und Frühgeschichtliche Archäologie der Schweiz Band 1.

Musil, R. (1980–1). *Ursus spelaeus—Der Höhlenba.* Vols. I–III. Weimar: Monographien zur Ur- und Frühgeschichte.

O'Connell, J. F. and Hawkes, K. (1981). Alyawara plant use and optimal foraging theory. In Winterhalder, B. and Smith, E. A. (eds). *Hunter-Gatherer Foraging Strategies.* Chicago: Chicago University Press, pp. 99–125.

Oswalt, W. H. (1976). *An Anthropological Analysis of Food-Getting Technology.* New York: John Wiley.

Otte M. (1981). *Le Gravettien en Europe centrale.* Dissertationes Archaeologicae Gandenses Vol XX de Tempel: Brugge.

Peterson, G. M., Webb, T., Kutzbach, J. E., Van der Hammen, T., Wijmstra, T. A. and Street, F. A. (1979) The continental record of environmental conditions at 18,000 yr B.P.: an initial evaluation. *Quat. Res.* **12**: 47–82.

Peterson, N. (1976). The natural and cultural areas of Aboriginal Australia. In: Peterson, N. (ed.) *Tribes and Boundaries in Australia.* Canberra: Australian Institute of Aboriginal Studies, pp. 50–71.

Pfeiffer, J. E. (1978). *The Emergence of Man.* New York: Harper and Row.

Pillard, B. (1972). La faune des grands mammifères du Würmien II. In de Lumley, H. (ed.) *La grotte de l'Hortus.* Marseille: Études Quaternaires I.

Schalk, R. F. (1977). The structure of an anadromous fish resource. In: Binford, L. R. (ed.) *For Theory Building in Archaeology.* New York: Academic Press, pp. 207–249.

Shackleton, N. J. (1969). The last interglacial in the marine and terrestrial records. *Proc. Roy. Soc. Lond. B* **174**: 246–56.

Shackleton, N. J. and Opdyke, N. D. (1973). Oxygen isotope and palaeomagnetic stratigraphy of Equatorial Pacific core V28–238. *Quat. Res.* **3**: 39–55.

Smith, P. E. L. (1966). *Le solutréen en France.* Bordeaux: Institut de préhistoire, memoire 5.

Steward, J. H. (1938). *Basin Plateau Aboriginal Sociopolitical Groups.* Washington: Smithsonian Institution, Bureau of American Ethnology Bulletin 120.

Straus, L. G. (1977). Of deerslayers and mountain men: palaeolithic faunal exploitation in Cantabrian Spain. In: Binford, L. R. (ed). *For Theory Building in Archaeology.* New York: Academic Press, pp. 41–76.

Straus, L. G. (1982). Carnivores and cave sites in Cantabrian Spain. *J. Anthropol. Res.* **38**: 75–96.

Stuart, A. J. (1976). The history of the mammal fauna during the Ipswichian/last interglacial in England. *Phil. Trans. Roy. Soc. Lond. B* **276**: 221–50.

Stuart, A. J. (1982). *Pleistocene Vertebrates in the British Isles.* London: Longman.

Torrence, R. (1983). Time budgeting and hunter-gatherer technology. In: Bailey, G. N. (ed). *Hunter-Gatherer Economy in Prehistory.* Cambridge: University Press, pp. 11–22.

Van der Hammen, T. C., Wijmstra, T. A. and Zagwijn, W. H. (1971). The floral record of the late Cenozoic of Europe. In: Turekian, K. K. (ed.) *Late Cenozoic Glacial Ages.* New Haven: Yale University Press, pp. 391–424.

Vértes, L. (1955). Neuere Ausgrabungen und paläolithische Funde in der Höhle von Istállöskö. *Acta Archaeologica Hungarica* 5: 111–181.

Wendorf, F., Schild, R. and Close, A. E. (1980). *Loaves and Fishes: The Prehistory of*

*Wadi Kubbaniya*. Dallas: Department of Anthropology, Southern Methodist University.
White, R. (1982). Rethinking the middle/upper palaeolithic transition. *Curr. Anthropol.* **23**: 169–192.
Whitehead, A. N. (1932). *Science and the Modern World*. Cambridge: Cambridge University Press.
Wilmsen, E. N. (1973). Interaction, spacing behaviour, and the organisation of hunting bands. *J. Anthropol. Res.* **29**: 1–31.
Winterhalder, B. and Smith, E. A. (eds) (1981). *Hunter-Gatherer Foraging Strategies*. Chicago: Chicago University Press.
Wobst, H. M. (1974). Boundary conditions for paleolithic social systems: a simulation approach. *Am. Antiquity* **39**: 147–178.
Wobst, H. M. (1976). Locational relationships in palaeolithic society. *J. Hum. Evol.* **5**: 49–58.
Wobst, H. M. (1978). The archaeo-ethnology of hunter-gatherers or the tyranny of the ethnographic record in archaeology. *Am. Antiquity* **43**: 303–9.
Woillard, G. M. (1978). Grande Pile peat bog: a continuous pollen record for the last 140,000 years. *Quat. Res.* **9**: 1–21.
Wymer, J. J. (ed.) (1977). *Gazetteer of Mesolithic Sites in England and Wales*. London: Council for British Archaeology, Research Report 20.
Wymer, J. J. (1981). The palaeolithic. In: Simmons, I. G. and Tooley, M. J. (eds). *The Environment in British Prehistory*. London: Duckworth, pp. 49–81.
Zotz, L. (1955). *Das Paläolithikum in den Weinberghöhlen bei Mauern*. Bonn: Quartär Bibliothek 2.

# 11 Community Ecology and Pleistocene Extinctions in the Levant

*Andrew N. Garrard*

## 1. INTRODUCTION

During the Upper Pleistocene, many areas of the world were affected by a wave of animal extinctions and this appears to have had a particular impact on the large mammal community. In the mid-1960s a conference was held to discuss the reasons for these extinctions (Martin and Wright, 1967) and whilst some of the participants stressed environmental causes, others regarded excessive human predation as the principal factor.

Within the Near East there were fewer extinctions than in many areas, but similar agencies may well have been responsible. A factor which remained largely undiscussed at the conference was the role of human interspecific competition in the extinctions. In this chapter an attempt is made to reconstruct the role of man in the biotic community of the Levant during the Upper Pleistocene, and the model is then used as a framework for assessing the likelihood of human interspecific competition being responsible for local extinctions. An assessment is also made of the role of human predation and environmental change on the extinction pattern in this area.

## 2. PAST THOUGHT ON PLEISTOCENE EXTINCTIONS

The models presented at the conference on "Pleistocene Extinctions" fell into two categories; those which regarded human overkill as the major factor and those which attributed them to climatic change. The chief protagonist of the former was Martin (1967), who noted that the Upper Pleistocene wave of extinctions affected over 200 genera of large terrestrial herbivores and their ecologically dependent carnivores and scavengers, and that in many cases there appeared to be no phyletic or ecological replacements. Martin noted that the

HOMINID EVOLUTION
ISBN 0-12-261920-X

time of the extinctions varied in different parts of the world, from 40 to 50 000 B.P. in Africa and southern Asia, to 11–13 000 B.P. in northern Eurasia, to 13 000 B.P. in Australia, to 11 000 B.P. in North America, to 10 000 B.P. in South America. Since the extinctions appeared to strike only the larger animals (in excess of 50 kg) and since they did not occur simultaneously around the world, Martin suggested that they were caused by a non-climatic factor. He noted that the extinctions in Africa and southern Asia occurred shortly after the advent of big game hunting and that the extinctions elsewhere shortly after initial human colonization. From this he inferred that man was responsible and that he caused the catastrophe by excessive predation or the overkill of large game.

Jelinek (1967) put forward a similar argument. He noticed that extinctions were particularly common in Old World periglacial areas, where a lack of plant resources stimulated a technology geared towards the exploitation of animals. Towards the end of the glacial, the megafauna may have disappeared as a result of hunting in increasingly restricted habitats. By contrast, the New World fauna, developed independent of human contact and the intrusion of man as a formidable predator into this natural community, unprepared for such stress, may have led to the disappearance of many large mammals.

There are a number of problems with the overkill hypothesis. First, there is little physical evidence of the species which became extinct from hunting camps of the period and in fact many of the species which formed a large component of the human diet survived until the late Holocene. Secondly, the colonization of Australia and the New World is now known to have begun considerably in advance of the extinctions, and big game hunting in Africa and southern Asia probably has a history running back into the Middle Pleistocene. Thirdly, many historically recorded hunter-gatherers have possessed game laws or traditions which have acted consciously or unconsciously to conserve favoured game species. For example, many such societies have had laws or taboos governing the number of animals killed, which have safeguarded pregnant females, and have prevented hunting during the courtship seasons and which have left areas fallow for varying periods (Heizer, 1955).

Other researchers such as Guilday (1967), Slaughter (1967), Hester (1967), Kowalski (1967) and Wilkinson (1972) have noted a strong coincidence between the time of the final withdrawal of the Wisconsin and Würm ice sheets and the extinctions in North America and northern Eurasia, and have suggested that periodic climatic changes may also have been responsible for extinctions elsewhere. Guilday suggested that post-Pleistocene desiccation may have been a factor and that large herbivores were most affected by virtue of their greater demands on the system for space, food and cover. Slaughter argued that wider temperature fluctuations in the late glacial, together with inflexible mating habits established under conditions no longer existing, may have upset the balance. Grayson (1977) has recently provided further evidence in support of an environmental cause, by demonstrating that the wave of extinctions in North America affected not only large vertebrates, but also small species including at least ten genera of birds. He found it difficult to credit that human overkill was responsible for the loss of these smaller more elusive species.

A factor which was not discussed at the conference was the role of human interspecific competition in Pleistocene extinctions—that is, competition between hominids and the now extinct species for a shared resource. This may

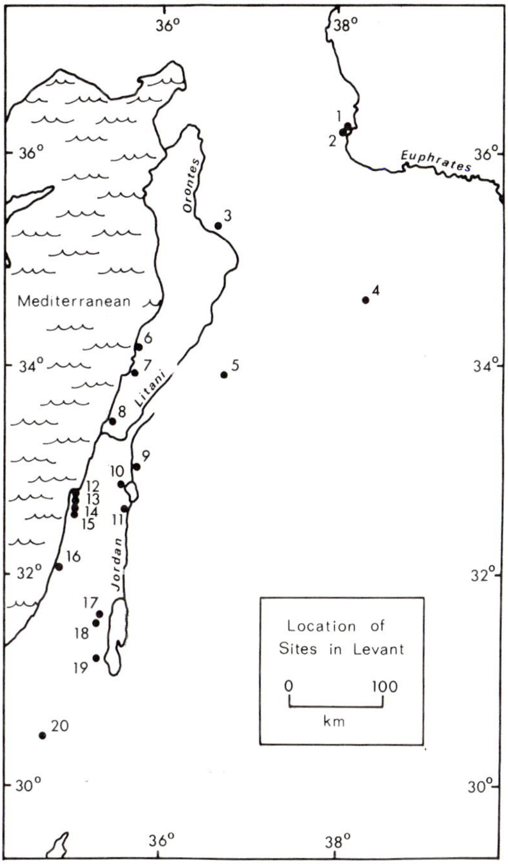


Fig. 11.1. Location of archaeological sites in the Levant mentioned in the text. (1) Tell Hadidi; (2) Ta'as; (3) Latamne; (4) Douara Cave; (5) Yabrud; (6) El-Masloukh; (7) Ksar Akil; (8) Mugharet el-Bezez; (9) Jisr Banat Yaqub; (10) Mugharet el-Emireh, Mugharet es-Zuttiyeh; (11) Ubeidiya; (12) Geulah Caves; (13) Nahel Oren; (14) Mugharet et-Tabun, Mugharet es-Skhul, Mugharet el-Wad; (15) Mugharet el-Kebara; (16) Tell Qassile; (17) Umm Qatafa, Erq el-Ahmar; (18) Sahba; (19) Arad; (20) Rosh Horesha.

result from the desire of participants to find a universal cause and direct competition is only likely to have affected those species closest to man in the biotic community. However it seems likely that a whole complex of factors were involved in the extinctions, and this chapter examines the role of interspecific competition as well as overpredation and climatic change.

## 3. THE PATTERN OF EXTINCTIONS IN THE LEVANT

Turning to the Levant (Lebanon, Syria, Jordan and Palestine), Bate (1937) reported finding evidence for a "faunal break" or wave of extinctions at a similar time to that recorded by Martin (1967) in Africa and southern Asia (50–40 000 B.P.). Some of her species were replaced by modern relatives whilst others had no phyletic replacements. Bate, from her study of the faunal assemblages at the Wadi el-Mughara Caves in Palestine, recorded the following species as disappearing.

(1) Five species of large herbivore: *Phacochoerus garrodae*, *Hippopotamus amphibius*, *Rhinocerus* cf. *hemitoechus*, *Sus gadarensis* and *Elephas* sp.

(2) Three species of large carnivore: *Hyaena prisca*, *Canis lupaster* and *Vulpes vinetorum*.

(3) Three species of rodent: *Philistomys roachi*, *Ellobius pedorychus* and *Microtus machintoni*.

(4) Three species of insectivore: *Erinaceus sharonis*, *Crocidura samaritana* and *Crocidura katinka*.

(5) One species of bat: *Myotis* cf. *baranensis*.

(6) Two species of large reptile: *Trionyx* sp. and *Crocodilus* sp.

However, since her publication, *Hippopotamus amphibius*, *Philistomys* (=*Myominus*) *roachi*, *Trionyx* sp. and *Crocodilus* sp. have turned up in post-glacial deposits (Haas, 1953, 1959; Bytinsky-Salz, 1965) and there are historical reports that elephant occurred in the wild in first millennium B.C. Mesopotamia (Bodenheimer, 1960). Also, *Sus gadarensis*, *Hyaena prisca*, *Canis lupaster*, *Vulpes vinetorum*, *Erinaceus europaeus* and *Crocidura samaritana* have been found to fall within the size range of the surviving *Sus scrofa*, *Hyaena hyaena*, *Canis aureus* or *lupus*, *Vulpes vulpes*, *Erinaceus europaeus* and *Crocidura leucodon* (Hooijer, 1959; Kurtén, 1965; Heller, 1970 and the author). Thus, apart from two species of rodent, one species of insectivore and one species of bat, which have not been reexamined since the original study, the only species which definitely disappeared are *Phacochoerus garrodae* (=*aethiopicus*) and *Rhinoceros* cf. *hemitoechus* (=*Dicerorhinus mercki*) (Hooijer, 1961b).

Although the Middle Pleistocene evidence is fairly limited, up-to-date information (Garrard, 1980) suggests that extinctions occurred steadily throughout the Pleistocene of the Levant and there is little evidence of a sudden increase in the Upper Pleistocene or in fact prior to the appearance of firearms, vehicles and the wholesale habitat destruction of the late Holocene

Table 11.1. Large mammals which are known from fossil or recent evidence to have inhabited the Levant during the Upper Pleistocene or Holocene (hare size and larger). The list excludes exclusively marine and human introduction and domesticates. All dates B.P.; L = Lebanon; P = Palestine; S = Syria; J = Jordan; H = Harrison (1964–72); Sur = Surviving.

| Species | Earliest record | Site and country | Reference | Latest Record | Site and country | Reference |
|---|---|---|---|---|---|---|
| *ARTIODACTYLA* | | | | | | |
| Bovidae | | | | | | |
| *Bos primigenius* | Pre-70 000 | M. et-Tabun, P | Garrard (1980) | 1 400–500 | Ta'as, S. | Clason & B. (1978) |
| *Oryx leucoryx* | Recent | | | 50 | J. | Stewart (1963) |
| *Alcelaphus buselaphus* | Pre-70 000 | M. et-Tabun, P. | Garrard (1980) | 100 | P. | Tristram (1884) |
| *Gazella gazella* | Pre-70 000 | M. et-Tabun, P. | Garrard (1980) | Sur. | | H. |
| *Gazella subgutturosa* | Recent | | | Sur. | | H. |
| *Gazella dorcas* | | R. Horesha, P. | Butler *et al.* (1977) | Sur. | | H. |
| *Capra ibex* | Pre-70 000 | Umm Qatafa, P. | Vaufrey (1951) | Sur. | | H. |
| *Capra aegagrus* | Pre-70 000 | Masloukh, L. | Gautier (1970) | Sur? | | H. |
| *Ovis orientalis* | 70–40 000 | Douara, S. | Payne p.c. | Sur? | | H. |
| Cervidae | | | | | | |
| *Cervus elaphus* | Pre-70 000 | Umm Qatafa, P. | Vaufrey (1951) | 9–8 000 | Nahel Oren, P. | Noy *et al.* (1973) |
| *Dama mesopotamica* | Pre-70 000 | Umm Qatafa, P. | Vaufrey (1951) | 50 | P. | H. |
| *Capreolus capreolus* | Pre-70 000 | M. et-Tabun, P. | Bate (1937) | Sur? | | H. |
| Suidae | | | | | | |
| *Sus scrofa* | Pre-70 000 | M. et-Tabun, P. | Bate (1937) | Sur. | | H. |
| *Phacochoerus aethiopicus* | 70–40 000 | M. es-Skhul, P. | Bate (1937) | 70–40 000 | ibid. | ibid. |
| Hippopotamidae | | | | | | |
| *Hippopotamus amphibius* | Pre–70 000 | Ubeidiya, P. | Haas (1966) | 2200 | Tell Qassile, P. | Haas (1959) |
| Camelidae | | | | | | |
| *Camelus dromedarius* | 70–40 000 | Sahba, P. | Vaufrey (1951) | 5–4700 | Arad, P. | Lernau (1978) |

| | | | | | | |
|---|---|---|---|---|---|---|
| *PERISSODACTYLA* | | | | | | |
| Equidae | | | | | | |
| *Equus caballus* | Pre-70 000 | Jisr B. Yaqub, P. | Hooijer (1959) | Pre-19 000 | M. el-Wad, P. | Bate (1937) |
| *Equus hemionus* | Pre-70 000 | M. et-Tabun, P. | Bate (1937) | 50 | S. | Groves (1974) |
| *Equus hydruntinus* | 70–40 000 | M. et-Tabun, P. | Bate (1937) | Pre-19 000 | M. el-Wad, P. | Garrard (1980) |
| Rhinoceratinae | | | | | | |
| *Dicerorhinus mercki* | Pre-70 000 | Jisr B. Yaqub, P. | Hooijer (1959) | 70–40 000 | Ksar Akil, L. | Hooijer (1961b) |
| *Dicerorhinus hemitoechus* | Pre-70 000 | Yabrud IV, S. | Perkins (1968) | 70–40 000 | M. el-Emireh, P. | Bate (1927a) |
| *PROBOSCIDAE* | | | | | | |
| *Elaphas maximus* | Pre-70 000? | M. et-Tabun, P. | Garrard (1980) | 2800 | S. | Bodenheimer (1960) |
| *HYRACOIDEA* | | | | | | |
| *Procavia capensis* | Pre-70 000 | Masloukh, L. | Gautier (1970) | Sur. | | H. |
| *LAGOMORPHA* | | | | | | |
| *Lepus capensis* | Pre-70 000 | Umm Qatafa, P. | Vaufrey (1951) | Sur. | | H. |
| *RODENTIA* | | | | | | |
| *Castor fiber* | 1400–500 | Tell Hadidi, S. | Clason & B. (1978) | 1400–500 | ibid. | ibid. |
| *Hystrix indica* | Pre-70 000 | M.et-Tabun, P. | Garrard (1980) | Sur. | | H. |
| *CARNIVORA* | | | | | | |
| Felidae | | | | | | |
| *Panthera leo* | Pre-70 000 | Yabrud IV, S. | Perkins (1968) | 800–700 | P. | H. |
| *Panthera pardus* | Pre-70 000 | Umm Qatafa, P. | Vaufrey (1951) | Sur. | | H. |
| *Acinonyx jubatus* | 70–40 000 | Geulah Caves, P. | Haas (1967) | 50 | J. | H. |
| *Caracal caracal* | Recent | | | Sur. | | H. |
| *Lynx lynx* | 70–40 000 | M. el-Bezez, L. | Garrard (1980) | 70–40 000 | ibid. | ibid. |
| *Felis silvestris* | Pre-70 000 | Umm Qatafa, P. | Vaufrey (1951) | Sur. | | H. |
| *Felis chaus* | Pre-70 000 | M. es-Zuttiyeh, P. | Bate (1927b) | Sur. | | H. |
| Canidae | | | | | | |
| *Canis lupus* | 70–40 000 | M. et-Tabun, P. | Garrard (1980) | Sur. | | H. |
| *Canis aureus* | Pre-70 000 | M. et-Tabun, P. | Garrard (1980) | Sur. | | H. |
| *Vulpes vulpes* | Pre-70 000 | M. et-Tabun, P. | Bate (1937) | Sur. | | H. |
| *Vulpes rupelli* | Recent | | | Sur. | | H. |
| *Fennecus zerda* | Recent | | | Sur. | | H. |

| | | | | | | |
|---|---|---|---|---|---|---|
| Hyaenidae | | | | | | |
| *Crocuta crocuta* | Pre-70 000 | Latamne, S. | Hooijer (1961a) | 10 500 | M. el-Kebara | Saxon (1974) |
| *Hyaena hyaena* | Pre-70 000 | Umm Qatafa, P. | Vaufrey (1951) | Sur. | | H. |
| Ursidae | | | | | | |
| *Ursus arctos* | Pre-70 000 | Umm Qatafa, P. | Vaufrey (1951) | Sur? | | H. |
| Mustelidae | | | | | | |
| *Meles meles* | 70–40 000 | Geulah Caves, P. | Heller (1970) | Sur. | | H. |
| *Mellivora capensis* | 40–19 000 | Erq el-Ahmar, P. | Vaufrey (1951) | sur. | | H. |
| *Lutra lutra* | Recent | | | Sur. | | H. |
| *Martes foina* | 70–40 000 | Geulah Caves, P. | Heller (1970) | Sur. | | H. |
| *Vormela peregusna* | Pre-70 000 | M.es-Zuttiyeh, P. | Bate (1927b) | Sur. | | H. |
| *Mustela nivalis* | Recent | | | Sur. | | H. |
| Viverridae | | | | | | |
| *Herpestes ichneumon* | Pre-70 000 | M. es-Zuttiyeh, P. | Bate (1927b) | Sur. | | H. |
| *Genetta genetta* | Recent | | | Sur. | | H. |
| *PINNIPEDIA* | | | | | | |
| *Monachus monachus* | Recent | | | Sur. | | Bodenheimer (1935) |

(see Table 11.1). The following species are known to have disappeared.
(1) Four large mammals in the first half of the Würm glacial (*ca.* 70–40 000 B.P.): *Phacochoerus aethiopicus*, *Dicerorhinus mercki*, *D. hemitoechus* and *Lynx lynx*.
(2) Three species in the second half of the Würm glacial (*ca.* 40–10 000 B.P.): *Equus caballus*, *Equus hydruntinus* and *Crocuta crocuta*.
(3) One species in the first half of the Holocene (10–5 000 B.P.) *Cervus elephas*.
(4) Six species in the succeeding period up until the last century (5000–100 B.P.): *Camelus dromedarius*, *Elephas maximus*, *Hippopotamus amphibius*, *Bos primigenius*, *Castor fiber* and *Panthera leo*.
(5) Five species in the last one hundred years (100–0 B.P.): *Alcelaphus buselaphus*, *Dama mesopotamica*, *Equus hemionus*, *Oryx leucoryx*, *Acinonyx jubatus*. Several others may currently be on the point of extinction.

The aim of the following sections is to assess the extent to which human interspecific competition, human overpredation and environmental change may have been responsible for the extinctions which occurred during the Pleistocene of the Levant.

## 4. EXTINCTION AS A RESULT OF INTERSPECIFIC COMPETITION WITH MAN

Man is an omnivore and in fact the only Near Eastern mammal to combine a large vegetable intake with the killing and eating of other large mammals. Unfortunately, little is known of his early vegetable intake, but his inability to digest large quantities of cellulose makes it unlikely that he was in direct competition with any of the large herbivores in this region, prior to the start of intensive cereal utilization and animal herding when it would have become necessary to protect certain areas against the attention of undesirable species of herbivore. The discussion which follows is therefore limited to man's relationship with the carnivores and omnivores living in the region.

In Table 11.2 an attempt has been made to compare the habitat preferences, hunting behaviour and food preferences of man with those of the other large carnivores and omnivores of *ca.* 2 kg or more which inhabited the Levant during the Upper Pleistocene and/or Holocene. The species listed are either known from fossil and historical evidence, or still inhabit the area (see Table 11.1). The ecological and behavioural data is based on recent observations and thus there is a necessity for some caution in extrapolating the data backwards through time. Habitat preferences may have changed as a result of human predatory pressure and loss of optimal habitats due to man's interference with the landscape, and activity time may also have altered due to predatory pressure as well as factors such as diurnal temperature range. Food preferences

Table 11.2. Habitat preferences, hunting behaviour and food preferences of Levantine carnivores and omnivores. *X* indicates preferred habitat or food if known. Sources: Bodenheimer (1935); Harrison (1964–72); Simon (1966); van der Brink (1967); Kruuk and Turner (1967); Wymer (1967); Schaller (1968); Dorst and Dandelot (1970); Mech (1970); Chaani (1972); Schaller (1972); Kruuk (1972); Kistchinski (1972); Hufnagl (1972); Ewer (1973); Fradrich (1974); Eaton (1974); Guggisberg (1975); Kruuk (1976); Neal (1977); Roberts (1977).

| Species | Average adult wt. (kg) | Preferred Habitat: Forest | Open Wood | Steppe | Hammada | Erg | Crags | Riverine | Marine | Size of hunting or gathering group (excludes young) | Activity time (m=mainly) | Food preferences (V = vertebrate): Plant | Invertebrates | Small V. –10 kg | Med. V. 10–100 kg | Large V. 100+kg |
|---|---|---|---|---|---|---|---|---|---|---|---|---|---|---|---|---|
| *Monachus monachus* | $c^a$ 300 | | | | | | | | *X* | | | | | x | x | |
| *Ursus arctos* | 105–265 | x | x | | | | | x | | 1 | M. Nocturnal | *X* | x | x | | |
| *Panthera leo* | 110–180 | | *X* | x | | | | | | 1–15± Cooperative | Nocturnal | | | | x | *X* |
| *Sus scrofa* | 35–175 | x | *X* | x | | | | *X* | | 1+ | Diurnal | *X* | x | x | | |
| *HOMO SAPIENS* | 58–70 | | x | x | | | | | | 1+ Cooperative | Diurnal | *X* | x | x | *X* | *X* |
| *Crocuta crocuta* | 45–70 | | x | *X* | | | | | | 1–30± Cooperative | M. Nocturnal | | | | x | x |
| *Panthera pardus* | 35–55 | x | *X* | x | | | x | *X* | | 1 | Nocturnal | | | x | *X* | |
| *Acinomyx jubatus* | 35–55 | | | x | x | | | | | 1 rarely 2 | Diurnal | | | x | *X* | |
| *Canis lupus* | 35–45 | x | x | x | x | | | x | | 1–12± Cooperative | M. Nocturnal | | | x | *X* | *X* |
| Hyaena hyaena | 35–43 | | | x | x | | | | | 1 rarely 2 | Nocturnal | x | x | *X* | x | |
| *Lynx lynx* | 18–38 | *X* | x | | | | x | x | | 1 | M. Nocturnal | | | *X* | x | |
| *Meles meles* | 10–22 | | *X* | | | | | | | 1+ | Nocturnal | *X* | x | x | | |
| *Caracal caracal* | 15–18 | | x | x | | | | x | | 1 | M. Nocturnal | | | *X* | | |
| *Mellivora capensis* | *ca.* 12 | x | x | x | x | | | x | | 1+ | M. Nocturnal | x | x | *X* | | |
| *Felis chaus* | 6–12 | x | x | | | | | *X* | | 1 | Nocturnal | x? | x | *X* | | |
| *Felis silvestris* | 5–10 | x | x | x | | | x | x | | 1 | Nocturnal | x | x | *X* | | |
| *Lutra lutra* | 5–10 | | | | | | | *X* | | 1 | M. Nocturnal | | x | *X* | | |
| *Canis aureus* | 5–9 | x | x | x | x | | | x | | 1–2 | M. Nocturnal | x | x | *X* | | |
| *Vulpes vulpes* | 5–9 | x | x | x | x | | | x | | 1 rarely 2 | M. Nocturnal | x | x | *X* | | |
| *Vulpes rüpelli* | 2–5 | | | | x | x | | | | 1 rarely 2 | Nocturnal | x? | x | *X* | | |
| *Herpestes ichneumon* | 2·5–4 | x | x | | | | | *X* | | 1 rarely 2 | M. Nocturnal | x | x | x | | |
| *Fennecus zerda* | 1·5–3 | | | | x | *X* | | | | 1 rarely 2 | Nocturnal | x? | x | *X* | | |
| *Martes foina* | 1·3–2·3 | x | x | | | | x | x | | 1 | Nocturnal | x | x | x | | |
| *Genetta genetta* | *ca.* 2·2 | x | x | | | | x | *X* | | 1–2 | Nocturnal | | x | *X* | | |
| *Vormela peregusna* | 1–2 | x | x | | | | | | | 1 | M. Nocturnal | x | x | *X* | | |
| *Mustela nivalis* | 0·45–1·3 | x | x | x | | | | x | | 1 | M. Nocturnal | | x | *X* | | |

and hunting group sizes may also have varied according to the local availability of food and the population density of the predators. However, by drawing data from as wide a range of habitats and geographical regions as possible, it is hoped that the variability has been compensated for and that the table provides a useful guide to the niches occupied by each of the species described.

It will be seen that the species have been arranged according to size. Bourlière (1963) has suggested that a relationship may exist between the size of a predator and the size of its prey and that the two are usually near enough equal. This appears to be true in many cases, but for species with a large plant element in their diet, the average prey is often considerably smaller. At the other extreme, the more carnivorous species, which hunt in groups and on a co-operative basis, often succeed in catching prey of a larger size than themselves (Schaller, 1972; Ewer, 1973).

In the table, man has been presented as preferring an open habitat; either steppe or open woodland. This would seem a reasonable hypothesis, as group cohesion is difficult in thick forest and it is not easy for a large group to subsist on the food normally available beneath a dense arboreal canopy. Man is also not physiologically well adapted to life in the extreme arid zone, as frequent access to water is required. This is not to say that man was never found in these biotopes, but that his preferences are likely to have been for steppic and open woodland habitats.

The table continues by listing the size of the average hunting and gathering group and the time of day when this activity is most common. It will be seen that only a few species such as *Panthera leo*, *Crocuta crocuta* and *Canis lupus* regularly co-operate in their hunting activities, although there are several advantages to co-operative hunting when sufficient prey is available. First, as mentioned above, it often allows the predator to catch larger prey than is normally possible when hunting as an individual. Secondly, it allows a more efficient utilization of harvested food and provides the opportunity for division of labour. Thirdly, operating in a group helps to raise the status of a population in the general predator dominance hierarchy. In other words, it reduces the chance of a kill being seized by other populations of predators and it also reduces the likelihood of the predator himself falling prey to other populations (Schaller, 1972). Man is included with the co-operative hunters although it is recognized that he may often hunt as an individual. Co-operative group hunting and the use of traps or weapons have enabled him to overcome the physical disadvantage he suffers from relatively slow speed on the ground.

Man's primate heritage suggests that his hunting and gathering activities have normally been diurnal. It will be seen from the table that most other Levantine carnivores and omnivores are mainly or exclusively noctural in their food collection activities. This would mean that man was largely free from competition in his daytime hunting activities and also free from the danger of being preyed upon by other large carnivores at this time (Schaller and Lowther, 1969).

In terms of food preferences, both the prehistoric evidence and recent analogy suggests that man was extensively reliant on medium and large vertebrates and probably vegetation for food. His reliance on small vertebrates or invertebrates is likely to have been minimal, at least in terms of nutritional yield. The majority of carnivorous animals are known to include at least a little plant food in their diet, but there are very few that utilize this resource to any extent (Ewer, 1973). The exceptions in the Levant are *Ursus arctos*, *Meles meles* and *Sus scrofa* and to a lesser extent *Martes foina* and *Herpestes ichneumon*. However, there is unlikely to have been any direct competition between these species and man, as first, they are all woodland or forest creatures and secondly, their meat intake is largely restricted to invertebrates and small vertebrates.

The species which are most likely to have been in direct competition with man are those which lived on medium or large vertebrates in open country or light woodland habitats. This would include the three co-operative hunters, *Panthera leo*, *Crocuta crocuta* and *Canis lupus*. Of these, *Panthera leo* has a slight preference for open woodland, *Crocuta crocuta* for open steppe and *Canis lupus* seems equally at home in either. Although all three are nocturnal hunters, their food tastes appear to be more specialized than man's and would consequently be the first to suffer if there was a depletion in the population of large game. *Canis lupus* has survived, but it is interesting to note that this was the first species of carnivore to be domesticated by man and as Eaton (1969) and others have suggested, this liaison may have resulted from living off the same game in the same habitat and from the development of co-operative hunting. On the other hand, *Crocuta crocuta* and *Panthera leo* have both become extinct, the former in the late glacial and the latter in the Mediaeval period (see Table 11.1). It is possible that their extinction resulted from competition with man for the same food resources.

*Acinonyx jubatus* is also an open steppe dweller, but has the advantage over man of speed and thus can manage to catch prey in the absence of suitable cover for stalking. It is also able to live in a semi-desert environment and at extremely low population densities. *Hyaena hyaena* is likewise found in open steppe and semi-desert conditions, but is less carnivorous and thus more versatile in food requirements than the other species so far mentioned. *Panthera pardus* and *Lynx lynx* are both solitary hunters and occupy habitats where they are unlikely to have come into direct competition with man (i.e. dense riverine thickets and forest ravines respectively), and thus the disappearance of the latter in the early Würm is likely to have resulted from other causes.

There is less likely to have been serious competition between the small carnivores and man (i.e. of *Caracal caracal* size and smaller) because of differences in food interests. Slight differences in hunting method, habitat and food preferences probably account for the lack of competition between the seemingly large number of species subsisting on small vertebrates and invertebrates.

Thus, out of the species which became extinct during the Upper Pleistocene of the Levant, *Crocuta crocuta* (the spotted hyaena) may have disappeared as a result of interspecific competition with man. Kurtén (1965) has suggested that there was a relationship between the disappearance of *Crocuta crocuta* and the emergence of *Hyaena hyaena*. However, the two species have different food preferences as well as hunting strategies (see Table 11.2; Kruuk, 1972, 1976) and it is therefore unlikely that they were in competition.

## 5. EXTINCTION FROM OTHER CAUSES

As was mentioned in the introduction, several researchers have blamed overpredation by man for the extinctions which occurred in the Pleistocene. However, of the prey species which became extinct in the Upper Pleistocene of the Levant, *Lynx lynx* and *Phacochoerus aethiopicus* are only known from one find, and the others rarely account for more than 10% and never more than 20% of the specimens at any one site (Garrard, 1980). Unless the animals lived at very low population levels, it seems unlikely that they became extinct simply as a consequence of overpredation by man. Similarly, the absence of any new herbivores or carnivores in the community, suggests that that they did not disappear as a result of displacement by a new species. The most likely cause in most cases seems therefore to have been environmental change or interspecific competition within the non-hominid community during a period of changing conditions.

Palaeoclimatic evidence for this region is available from a number of sources including pollen, deep-sea cores and lake and wadi geomorphology (Garrard, 1980). The data from wadis draining into the Jordan-Arava valley (Bar Yosef *et al.*, 1974; Goldberg, 1976) suggests that the early Würm glacial, until *ca.* 60 000 B.P., was a relatively moist period and that this was followed by a dry period lasting until the end of the Middle Palaeolithic at *ca.* 40 000 B.P. During the Upper Palaeolithic, between *ca.* 40 000 and 19 000 B.P. there is considerable evidence for a moist phase in the southern and presently more arid regions of the Levant. It was marked by aggradation in the wadis draining into the Jordan-Arava valley (Bar Yosef *et al.*, 1974; Goldberg, 1976), high lake levels on the Syrian-Arabian plateau (Huckriede and Wiesemann, 1968; Kaiser *et al.*, 1973; Garrard *et al.*, 1981), by brecciation, slumping and fluvial erosion in the Palestine and Lebanon caves (Bar Yosef and Vandermeersch, 1971) and by high percentages of arboreal pollen in northern Palestine

(Horowitz, 1971). This was followed by an intensely cold and dry period. The palynological evidence from northern Syria on the one hand (Niklewski and van Zeist, 1970) and from northern Palestine on the other (Horowitz, 1971), suggests that it may have been drier in the north than the south. During the late glacial and early Holocene the climate steadily ameliorated. The pollen diagrams from the north suggest that temperature and precipitation reached their present state at that time, whilst the data from the south suggests the climate was temporarily more humid than at present.

Returning to the fauna, all the living Dicerorhininae require frequent access to water (Talbot, 1960; Schenkel and Schenkel-Huluger, 1969; Owen-Smith, 1974), and so it is possible that increased aridity during the early Würm led to the loss of *Dicerorhinus hemitoechus* and *D. mercki*. On the other hand *Hippopotamus amphibius*, *Crocodilus* and the large turtle *Trionyx*, which are only found in and adjacent to freshwater, are found in both early Würm and early Holocene sites, and so either they survived the late glacial in this region or they managed to recolonize from elsewhere during the early post-glacial. The loss of the two equid species, *Equus caballus* and *Equus hydruntinus*, seems to coincide with the start of the late glacial arid phase (*ca.* 19 000 B.P.), but since the fossil evidence from Europe and central Asia suggests that they were both well adapted to cold and steppic conditions (Kurten, 1968, Groves, 1974), this seems an inadequate explanation. Another possibility, is that interspecific competition between the three Levantine equids during a period of deteriorating environmental conditions, caused reduced diversity through the extinction of two of the species. Unfortunately too little is known of the ecological preferences of the extinct *Equus hydruntinus* or of the ecological relationship between the small surviving populations of *Equus caballus* and *Equus hemionus* for this to be determined. The only species for which there appears to be a straightforward climatic explanation is the temperate woodland species *Cervus elaphus* whose northern retreat coincides with the post-glacial temperature rise. More refined dating and climatic chronologies will be required before further light can be shed on these extinctions.

## 6. CONCLUSIONS

In this chapter an attempt has been made to assess the role of human competition, human predation and environmental change in the extinction of large mammals which occurred in the Levant during the last glacial cycle. Four species are known to have disappeared in the first half of the Würm cycle, namely *Phacochoerus aethiopicus*, *Dicerorhinus mercki*, *D. hemitoechus* and *Lynx lynx* and three in the second half, namely *Equus caballus*, *E. hydruntinus* and *Crocuta crocuta*. A reconstruction has been presented of man's role in the biotic community in this area and period and it would appear that the species which are most likely to have suffered from human interspecific competition, i.e. those which also lived on medium and large vertebrates in open country or light woodland habitats, were *Canis lupus*, *Crocuta crocuta* and *Panthera leo*. The first species has in fact survived, however the other two became extinct, the former in the late glacial and the latter in the late Holocene. It is possible therefore that human interspecific competition played a role in Pleistocene and Holocene extinctions in the Near East. From a community viewpoint it is significant that wolf was the earliest species to be domesticated in many areas of the world. Both man and wolf are co-operative hunters within their own societies and an interspecific co-operative relationship may have developed between them, as a result of sharing the same habitat and the same resources.

It seems unlikely that over-predation by man was a significant factor in Near Eastern extinctions, as the species which disappeared rarely account for more than 10% and never for more than 20% of the specimens at individual sites. It is more likely that environmental change played a role in their disappearance and it is suggested, although not conclusively, that early and late Würm desiccation may have been a contributing factor.

Although few species became extinct in this area relative to other parts of the world, it is possible that similar factors were responsible. The use of a model based on community ecology has provided a useful tool for identifying why certain species disappeared in the Near East and could be usefully applied to other regions.

## REFERENCES

Bar-Josef, O. and Vandermeersch, B. (1971). The stratigraphical and cultural problems of the passage from Middle to Upper Palaeolithic in Palestinian caves. In: UNESCO. (ed.) *The Origin of Homo Sapiens*. Paris: UNESCO, pp. 221–225.

Bar-Yosef, O., Goldberg, P. and Leveson, T. (1974). Kebaran and Natufian sites in Wadi Fazael, Jordan Valley, Israel. *Paléorient* 2: 415–428.

Bate, D. M. A. (1927a). On the animal remains obtained from the Mugharet el-Emireh in 1925. In: Turville-Petrie, F. (ed). *Researches in Prehistoric Galilee 1925–26*. London: British School of Archaeology in Jerusalem, pp. 9–13.

Bate, D. M. A. (1927b). On the animal remains obtained from the Mugharet el-Zuttiyeh in 1925. In: Turville-Petrie, F. (*op. cit.*), pp. 27–34.

Bate, D. M. A. (1937). Part II—Palaeontology: The fossil fauna of the Wady el-Mughara Caves. In: Garrod, D. A. E. and Bate, D. M. A. (eds). *The Stone Age of Mount Carmel. Excavations at the Wadi el-Mughara. Vol. 1*. Oxford: Clarendon, pp. 136–233.

Bodenheimer, F. S. (1935). *Animal life in Palestine. An Introduction to the Problems of Animal Ecology and Zoogeography*. Jerusalem: Mayer.

Bodenheimer, F. S. (1960). *Animal and Man in Bible lands*. Leiden: Brill.

Bourlière, F. (1963). Specific feeding habits of African carnivores. *Afr. Wildlife* **17**: 21–27.
Butler, B. H., Tchernov, E., Hietala, H. and Davis, S. (1977). Faunal exploitation during the late Epipalaeolithic in the Har Harif. In: Marks, A. E. (ed). *Prehistory and Palaeoenvironments in the Central Negev, Israel, Vol. 2*. Dallas: Southern Methodist University, pp. 327–345.
Bytinski-Salz, H. (1965). Recent findings of hippopotamus in Israel. *Israel J. Zool.* **14**: 38–48.
Chaani, G. (1972). *Studies on the Ecology and the Behaviour of the Wild Boar (Sus scrofa lybicus, Gray 1968) in the Mt. Meiron region.* Tel Aviv University unpublished Masters Thesis.
Clason, A. T. and Buitenhuis, H. (1978). A preliminary report on the faunal remains of Nahr el-Homr, Hadidi and Ta'as in the Tabqa Dam region of Syria. *J. Archaeol. Sci.* **5**: 75–83.
Dorst, J. and Dandelot, P. (1970). *A Field Guide to the Larger Mammals of Africa*. London: Collins.
Eaton, R. L. (1969). Cooperative hunting by cheetahs and jackals and a theory of domestication of the dog. *Mammalia* **33**: 87–92.
Eaton, R. L. (1974). *The Cheetah. The Biology, Ecology and Behaviour of an Endangered Species.* New York: van Nostrand Reinhold.
Ewer, R. F. (1973). *The Carnivores.* London: Weidenfeld and Nicolson.
Fradrich, H. (1974). A comparison of behaviour in the Suidae. In: Geist, V. and Walther, F. (eds). *The Behaviour of Ungulates and its Relation to Management.* Morges: IUCNNR pp. 133–143.
Garrard, A. N. (1980). *Man-Animal-Plant Relationships During the Upper Pleistocene and Early Holocene of the Levant.* Cambridge University, unpublished Ph.D. thesis.
Garrard, A. N., Harvey, C. P. D. and Switsur, V. R. (1981). Environment and settlement during the Upper Pleistocene and Holocene at Jubba in the Great Nefud, Northern Arabia. *Atlal. J. Saudi Arabian Archaeol.* **5**: 137–148.
Gautier, A. (1970). The fauna of Masloukh. *Bulletin du Musee de Beyrouth* **23**: 135.
Goldberg, P. (1976). Upper Pleistocene geology of the Avdat/Aqev area. In: Marks, A. E. (ed). *Prehistory and Palaeoenvironments in the Central Negev, Israel, Vol. 1.*. Dallas: Southern Methodist University, pp. 25–55.
Grayson, C. P. (1974). Pleistocene avifaunas and the overkill hypothesis. *Science* **195**: 691–693.
Groves, C. P. (1974). *Horses, Asses and Zebras in the Wild.* Newton Abbot: David and Charles.
Guggisberg, C. A. W. (1975). *Wild Cats of the World.* Newton Abbot: David and Charles.
Guilday, J. E. (1967). Differential extinction during late Pleistocene and Recent times. In: Martin, P. S. and Wright, H. E. Jr. (eds). *Pleistocene Extinctions: The Search for a Cause.* New Haven: Yale University Press, pp. 121–140.
Haas, G. (1953). On the occurrence of hippopotamus in the Iron Age of the coastal area of Israel (Tell Qasîleh). *Bull. Am. Sch. Orient. Res.* **132**: 30–34.
Haas, G. (1959). Some remarks on *Philistomys roachi* Bate. *Ann. Mag. Nat. Hist.* Series 13, **2**: 688–690.
Haas, G. (1966). *On the Vertebrate Fauna of the Lower Pleistocene Site of Ubeidiya.* Jerusalem: Israel Academy of Sciences and Humanities.
Haas, G. (1967). Bemerkungen veber die fauna der Geula-Hoehle, Carmel. *Quaternaria* **9**: 97–104.

Harrison, D. L. (1964–72). *The Mammals of Arabia*. 3 vols. London: Benn.
Heizer, R. F. (1955). Primitive man as an ecological factor. *Kroeber Anthropol. Soc. Pap.* **13**: 1–31.
Heller, J. (1970). The small mammals of the Geula Cave. *Israel J. Zool.* **19**: 1–49.
Hester, J. J. (1967) The agency of man in animal extinctions. In: Martin, P. S. and Wright, H. E. Jr. (eds). *Pleistocene Extinctions: the search for a cause*. New Haven: Yale University Press, pp. 169–192.
Hooijer, D. A. (1959). Fossil mammals from Jisr Banât Yaqub, south of Lake Huleh, Israel. *Bull. Res. Coun. Israel* **8G**: 177–199.
Hooijer, D. A. (1961a). Middle Pleistocene mammals from Latamne, Orontes Valley, Syria. *Annales Archeologiques Arabe Syrienne* **11**: 117–132.
Hooijer, D. A. (1961b). The Fossil vertebrates of Ksar-Akil, a Palaeolithic rock-shelter in the Lebanon. *Zoologische Verhandelingen* **49**: 4–67.
Horowitz, A. (1971). Climatic and vegetational developments in north-eastern Israel during Upper Pleistocene-Holocene times. *Pollen et Spores* **13**: 255–278.
Huckriede, R. and Wiesemann, G. (1968). Der jungpleistozäne pluvial-see von El Jafr und weitere daten zum Quartär Jordaniens. *Geologica et Palaeontologica* **2**: 73–90.
Hufnagl, E. (1972). *Libyan Mammals*. Harrow: Oleander.
Jelinek, A. J. (1967). Man's role in the extinction of Pleistocene fauna. In: Martin, P. S. and Wright, H. E. Jr. (eds). *Pleistocene Extinctions: the search for a cause*. New Haven: Yale University Press, pp. 193–200.
Kaiser, K., Kempf, E. K., Leroi-Gourhan, A. and Schütt, H. (1973). Quartärstratigraphische untersuchungen aus dem Damaskus-Becken und seiner umgebung. *Zeitschrift für Geomorphologie* New Series **17**: 263–353.
Kistchinski, A. A. (1972). Life history of the Brown Bear (*Ursus arctos* L.) in north-east Siberia. In: Herrero, S. (ed). *Bears—Their Biology and Management*. Morges, IUCNNR, pp. 67–73.
Kowalski, K. (1967). The Pleistocene extinction of mammals in Europe. In: Martin, P. S. and Wright, H. E. Jr. (eds). Pleistocene Extinctions: the search for a cause. New Haven: Yale University Press, pp. 349–64.
Kruuk. H. (1972). *The Spotted Hyaena. A Study of Predation and Social Behaviour*. Chicago: Chicago University Press.
Kruuk, H. (1976). Feeding and social behaviour of the Striped Hyaena (*Hyaena vulgaris* Desmarest). *East Afr. Wildlife J.* **14**: 91–111.
Kruuk, H. and Turner, M. (1967). Comparative notes on predation by lion, leopard, cheetah and wild dog in the Serengeti area, E. Africa. *Mammalia* **31**: 1–27.
Kurtén, B. (1965). The Carnivora of the Palestine caves. *Acta Zoologica Fennica* **107**.
Kurtén B. (1968). *Pleistocene Mammals of Europe*. London: Weidenfeld and Nicolson.
Lernau, H. (1978). Faunal Remains, Strata III-I. In: Amiran, R. (ed). *Early Arad. The Chalcolithic Settlement and Early Bronze City*. Jerusalem: The Israel Exploration Society, pp. 83–113.
Martin, P. S. (1967). Prehistoric overkill. In: Martin, P. S. and Wright. H. E. Jr. (eds). *Pleistocene Extinctions: the search for a cause*. New Haven: Yale University Press, pp. 75–120.
Martin, P. S. and Wright, H. E. Jr. (eds) (1967). *Pleistocene Extinctions. The Search for a Cause*. New Haven: Yale University.
Mech, L. D. (1970). *The Wolf: The Ecology and Behaviour of an Endangered Species*. New York. Natural History Press.
Neal, E. (1977). *Badgers*. Poole: Blandford.
Niklewski, J. and van Zeist, W. (1970). A late Quaternary pollen diagram from north-

western Syria. *Acta Botanica Neerlandica* **19**: 737–754.
Noy, T., Legge, A. J., Higgs, E. S. and Dennell, R. W. (1973). Excavations at Nahel Oren, Israel. *Proc. Prehist. Soc.* **39**: 75–99.
Owen-Smith, R. N. (1974). The social system of the White Rhinoceros. In: Geist, V. and Walther, F. (eds). *The Behaviour of Ungulates and its Relationship to Management.* 2 vol. Morges, IUCNNR, pp. 341–351.
Perkins, D. Jr. (1968). The Pleistocene fauna from the Yabrud Rockshelters. *Annales Archeologiques Arabe Syrienne* **18**: 123–130.
Roberts, T. J. (1977). *The Mammals of Pakistan.* London: Benn.
Saxon, E. C. (1974). The mobile herding economy of Kebarah Cave, Mt. Carmel: an economic analysis of the faunal remains. *J. Archaeol. Sci.* **1**: 27–45.
Schaller, G. B. (1968). Hunting behaviour of the Cheetah in the Serengeti National Park—Tanzania. *East Afri. Wildlife J.* **6**: 95–100.
Schaller, G. B. (1972). *The Serengeti Lion. A study of predator-prey relations.* Chicago: University of Chicago Press.
Schaller, G. B. and Lowther, G. R. (1969). The relevance of carnivore behaviour to the study of early hominids. *Southwestern J. Anthropol.* **25**: 307–341.
Schenkel, R. and Schenkel-Huluger, L. (1969). *Ecology and Behaviour of the Black Rhinoceros (Diceros bicornis L.). A Field Study.* Hamberg: Paul Parey.
Simon, N. (ed). (1966). *Red Data Book. Vol. 1. Mammalia.* Morges: IUCNNR.
Slaughter, B. H. (1967). Animal ranges as a clue to late Pleistocene extinction. In: Martin, P. S. and Wright, H. E. Jr. (eds). Pleistocene Extinctions: the search for a cause. New Haven: Yale University Press, pp. 155–167.
Stewart, D. R. M. (1963–64). The Arabian Oryx (*Oryx leucoryx* Pallas). *East Afr. Wildlife J.* **1**: 103–118, **2**: 168–169.
Talbot, L. M. (1960). A look at threatened species: a report on some animals of the Middle East and Southern Asia which are threatened with extermination. *Oryx* **5**: 153–293.
Tristram, H. B. (1884). *The Survey of Western Palestine. The Fauna and Flora of Palestine.* London: Palestine Exploration Fund.
van den Brink, F. H. (1967). *A Field Guide to the Mammals of Britain and Europe.* London: Collins.
Vaufrey, R. (1951). Mammifères. In: Neuville, R. (ed). Le Paléolithique et le Mésolithique du désert de Judée. *Archives Institute Paleontologie Humaine* **24**, 198–217.
Wilkinson, P. F. (1972). Ecosystem models and demographic hypotheses: predation and prehistory in North America. In: D. L. Clarke (ed), *Models in Archaeology.* London: Methuen, pp. 543–576.
Wyman, J. (1967). The Jackals of the Serengeti. *Animals* **10**: 79–83.

# Author Index

## J



## K



## L



## M

# *Subject Index*

## B



## C

N



O



P

W



Y



Z

# STUDIES IN ARCHAEOLOGY

*Consulting Editor: Stuart Struever*

Department of Anthropology
Northwestern University
Evanston, Illinois

*Charles R. McGimsey III.* **Public Archeology**

*Lewis R. Binford.* **An Archaeological Perspective**

*Muriel Porter Weaver.* **The Aztecs, Maya, and Their Predecessors: Archaeology of Mesoamerica**

*Joseph W. Michels.* **Dating Methods in Archaeology**

*C. Garth Sampson.* **The Stone Age Archaeology of Southern Africa**

*Fred T. Plog.* **The Study of Prehistoric Change**

*Patty Jo Watson (Ed.).* **Archeology of the Mammoth Cave Area**

*George C. Frison (Ed.).* **The Casper Site: A Hell Gap Bison Kill on the High Plains**

*W. Raymond Wood and R. Bruce McMillan (Eds.).* **Prehistoric Man and His Environments: A Case Study in the Ozark Highland**

*Kent V. Flannery (Ed.).* **The Early Mesoamerican Village**

*Charles E. Cleland (Ed.).* **Cultural Change and Continuity: Essays in Honor of James Bennett Griffin**

*Michael B. Schiffer.* **Behavioral Archeology**

*Fred Wendorf and Romuald Schild.* **Prehistory of the Nile Valley**

*Michael A. Jochim.* **Hunter-Gatherer Subsistence and Settlement: A Predictive Model**

*Stanley South.* **Method and Theory in Historical Archeology**

*Timothy K. Earle and Jonathon E. Ericson (Eds.).* **Exchange Systems in Prehistory**

*Stanley South (Ed.).* **Research Strategies in Historical Archeology**

*John E. Yellen.* **Archaeological Approaches to the Present: Models for Reconstructing the Past**

*Lewis R. Binford (Ed.).* **For Theory Building in Archaeology: Essays on Faunal Remains, Aquatic Resources, Spatial Analysis, and Systemic Modeling**

*James N. Hill and Joel Gunn (Eds.).* **The Individual in Prehistory: Studies of Variability in Style in Prehistoric Technologies**

*Michael B. Schiffer and George J. Gumerman (Eds.).* **Conservation Archaeology: A Guide for Cultural Resource Management Studies**

*Thomas F. King, Patricia Parker Hickman, and Gary Berg.* **Anthropology in Historic Preservation: Caring for Culture's Clutter**

*Richard E. Blanton.* **Monte Albán: Settlement Patterns at the Ancient Zapotec Capital**

*R. E. Taylor and Clement W. Meighan.* **Chronologies in New World Archaeology**

*Bruce D. Smith.* **Prehistoric Patterns of Human Behavior: A Case Study in the Mississippi Valley**

*Barbara L. Stark and Barbara Voorhies (Eds.).* **Prehistoric Coastal Adaptations: The Economy and Ecology of Maritime Middle America**

*Charles L. Redman, Mary Jane Berman, Edward V. Curtin, William T. Langhorne, Nina M. Versaggi, and Jeffery C. Wanser (Eds.).* **Social Archeology: Beyond Subsistence and Dating**

*Bruce D. Smith (Ed.)* **Mississippian Settlement Patterns**

*Lewis R. Binford.* **Nunamiut Ethnoarchaeology**

*J. Barto Arnold III and Robert Weddle.* **The Nautical Archeology of Padre Island: The Spanish Shipwrecks of 1554**

*Sarunas Milisauskas.* **European Prehistory**

*Brian Hayden (Ed.).* **Lithic Use-Wear Analysis**

*William T. Sanders, Jeffrey R. Parsons, and Robert S. Santley.* **The Basin of Mexico: Ecological Processes in the Evolution of a Civilization**

*David L. Clarke.* **Analytical Archaeologist: Collected Papers of David L. Clarke. Edited and Introduced by His Colleagues**

*Arthur E. Spiess.* **Reindeer and Caribou Hunters: An Archaeological Study**

*Elizabeth S. Wing and Antoinette B. Brown.* **Paleonutrition: Method and Theory in Prehistoric Foodways**

*John W. Rick.* **Prehistoric Hunters of the High Andes**

*Timothy K. Earle and Andrew L. Christenson (Eds.).* **Modelling Change in Prehistoric Economics**

*Thomas F. Lynch (Ed.).* **Guitarrero Cave: Early Man in the Andes**

*Fred Wendorf and Romuald Schild.* **Prehistory of the Eastern Sahara**

*Henri Laville, Jean-Philippe Rigaud, and James Sackett.* **Rock Shelters of the Perigord: Stratigraphy and Archaeological Succession**

*Duane C. Anderson and Holmes A. Semken, Jr. (Eds.).* **The Cherokee Excavations: Holocene Ecology and Human Adaptations in Northwestern Iowa**

*Anna Curtenius Roosevelt.* **Parmana: Prehistoric Maize and Manioc Subsistence along the Amazon and Orinoco**

*Fekri A. Hassan.* **Demographic Archaeology**

*G. Barker.* **Landscape and Society: Prehistoric Central Italy**

*Lewis R. Binford.* **Bones: Ancient Men and Modern Myths**

*Richard A. Gould and Michael B. Schiffer (Eds.).* **Modern Material Culture: The Archaeology of Us**

*Muriel Porter Weaver.* **The Aztecs, Maya, and Their Predecessors: Archaeology of Mesoamerica, 2nd edition**

*Arthur S. Keene.* **Prehistoric Foraging in a Temperate Forest: A Linear Programming Model**

*Ross H. Cordy.* **A Study of Prehistoric Social Change: The Development of Complex Societies in the Hawaiian Islands**

*C. Melvin Aikens and Takayasu Higuchi.* **Prehistory of Japan**

*Kent V. Flannery (Ed.).* **Maya Subsistence: Studies in Memory of Dennis E. Puleston**

*Dean R. Snow (Ed.).* **Foundations of Northeast Archaeology**

*Charles S. Spencer.* **The Cuicatlán Cañada and Monte Albán: A Study of Primary State Formation**

*Steadman Upham.* **Polities and Power: An Economic and Political History of the Western Pueblo**

*Carol Kramer.* **Village Ethnoarchaeology: Rural Iran in Archaeological Perspective**

*Michael J. O'Brien, Robert E. Warren, & Dennis E. Lewarch (Eds.).* **The Cannon Reservoir Human Ecology Project: An Archaeological Study of Cultural Adaptations in the Southern Prairie Peninsula**

*in preparation*

*Vincas P. Steponaitis.* **Ceramics, Chronology, and Community Patterns: An Archaeological Study at Moundville**

*Merrilee H. Salmon.* **Philosophy and Archaeology**

*Jonathon E. Ericson and Timothy K. Earle (Eds.).* **Contexts for Prehistoric Exchange**

*William J. Folan, Ellen R. Kintz, and Laraine A. Fletcher.* **Coba: A Classic Maya Metropolis**